Data Assimilation for the

G000153164

NATO Science Series

A Series presenting the results of scientific meetings supported under the NATO Science Programme.

The Series is published by IOS Press, Amsterdam, and Kluwer Academic Publishers in conjunction with the NATO Scientific Affairs Division

Sub-Series

I. Life and Behavioural Sciences	IOS Press
II. Mathematics, Physics and Chemistry	Kluwer Academic Publishers
III. Computer and Systems Science	IOS Press
IV. Earth and Environmental Sciences	Kluwer Academic Publishers
V. Science and Technology Policy	IOS Press

The NATO Science Series continues the series of books published formerly as the NATO ASI Series.

The NATO Science Programme offers support for collaboration in civil science between scientists of countries of the Euro-Atlantic Partnership Council. The types of scientific meeting generally supported are "Advanced Study Institutes" and "Advanced Research Workshops", although other types of meeting are supported from time to time. The NATO Science Series collects together the results of these meetings. The meetings are co-organized bij scientists from NATO countries and scientists from NATO's Partner countries – countries of the CIS and Central and Eastern Europe.

Advanced Study Institutes are high-level tutorial courses offering in-depth study of latest advances in a field.
Advanced Research Workshops are expert meetings aimed at critical assessment of a field, and iden-tification of directions for future action.

As a consequence of the restructuring of the NATO Science Programme in 1999, the NATO Science Series has been re-organised and there are currently five sub-series as noted above. Please consult the following web sites for information on previous volumes published in the Series, as well as details of ear-lier sub-series.

http://www.nato.int/science
http://www.wkap.nl
http://www.iospress.nl
http://www.wtv-books.de/nato-pco.htm

Series IV: Earth and Environmental Sciences – Vol. 26

Data Assimilation for the Earth System

edited by

Richard Swinbank
Met Office,
Bracknell, U.K.

Victor Shutyaev
Institute of Numerical Mathematics,
Russian Academy of Sciences, Moscow, Russia

and

William Albert Lahoz
Data Assimilation Research Centre,
University of Reading, Reading, U.K.

Kluwer Academic Publishers

Dordrecht / Boston / London

Published in cooperation with NATO Scientific Affairs Division

Proceedings of the NATO Advanced Study Institute on
Data Assimilation for the Earth System
Acquafredda, Maratea, Italy
19 May–1 June 2002

A C.I.P. Catalogue record for this book is available from the Library of Congress.

ISBN 1-4020-1592-5 (HB)
ISBN 1-4020-1593-3 (PB)

Published by Kluwer Academic Publishers,
P.O. Box 17, 3300 AA Dordrecht, The Netherlands.

Sold and distributed in North, Central and South America
by Kluwer Academic Publishers,
101 Philip Drive, Norwell, MA 02061, U.S.A.

In all other countries, sold and distributed
by Kluwer Academic Publishers,
P.O. Box 322, 3300 AH Dordrecht, The Netherlands.

Printed on acid-free paper

Printed in the Netherlands.

Table of Contents

PREFACE

This book is based on a set of lectures presented at the NATO (North Atlantic Treaty Organisation) Advanced Study Institute (ASI) on *Data Assimilation for the Earth System*, which was held during late May 2002. The ASI grew out of a long-held concern that there was little teaching available in data assimilation, even though it has become central to modern weather forecasting, and is becoming increasingly important in a range of other earth science disciplines.

Over recent years a few teaching initiatives have been started, for example a data assimilation module in the European Centre for Medium-range Weather Forecasts (ECMWF) meteorological training course. However, we still felt that there was a continuing unfulfilled demand for training in data assimilation that focused on the application of data assimilation techniques to a range of earth sciences.

An additional consideration was the wealth of earth observation satellite data that is now coming on stream, as outlined elsewhere in this book. Data assimilation is the best, or, arguably, the only feasible technique available to make use of the enormous variety and volume of data that are being relayed from these satellite instruments.

To meet the need for advanced training in data assimilation, we submitted a proposal to NATO for an Advanced Study Institute on the subject. We are very grateful to NATO that they accepted our proposal and provided generous financial support. The number of applications to attend the ASI substantially exceeded the number of student places available, confirming the continuing and expanding demand for training in this field. A particular feature of the NATO sponsorship agreement was that 40% of the students at the summer school came from NATO partner countries and Mediterranean Dialogue countries. This meant that a high proportion of the students came from countries where data assimilation is still in its infancy. It is our hope that the ASI has helped the technique to become more established in earth science applications in those countries.

Like the ASI, this book is aimed at graduate students and post-doctoral and other researchers who are working on data assimilation across a range of earth sciences. The opening chapters survey the basic theory of data assimilation, forming the theoretical framework for the remainder of the book. The next set of chapters explores some of the more advanced theory of data assimilation, taking the advanced reader to some of the frontiers of current research work. In the second half of the book, the applications of data assimilation are outlined. For those not familiar with meteorology, atmospheric chemistry, oceanography and land surface processes, some of the basic concepts in those subject areas are described before moving on to detailed discussions of the applications.

The ASI could not have taken place without strong support from the organising committee and lecturers. As well as the editors of this volume, the organising committee comprised: Alan O'Neill, who is the director of the Data Assimilation Research Centre (DARC) in Reading; Olivier Talagrand, one of the founding fathers of meteorological data assimilation; and Boris Khattatov, a leading expert on assimilation of chemical constituents. We were fortunate to enrol an international team of lecturers who are each experts in their own particular areas of data assimilation. We would like to thank all the lecturers for presenting their material both in lectures at the ASI, and in written form in this book.

In addition to lectures and tutorials presented by the lecturing staff, a key component of the ASI was a set of computer-based practicals. We would like to thank the three teaching assistants, Matt Huddleston, Amos Lawless and Stefano Migliorini, who assisted students with the computer practicals. Amos played a particularly important role in coordinating the practicals, and presenting some of the practical exercises. We also thank Andy Heaps, who organised the computers for the ASI and served as computer manager. As editors, we felt that we could not easily present material from the computer practicals in this book. However, we encourage interested readers to visit the web-site at http://darc.nerc.ac.uk/asi_progs, where the relevant computer software, and associated documentation has been made available.

We are particularly grateful to all the various organisations which provided financial support that made the Advanced Study Institute possible. We have already mentioned generous backing by NATO, the main sponsor of the summer school. In addition, NASA supported students travel costs; we would particularly like to thank Phil DeCola for securing this generous support. We would like to thank Boris Khattatov and Marilena Stone of the Atmospheric Chemistry Division at the National Center for Atmospheric Research (NCAR), who administered the NASA grant. Centre National de la Recherche Scientifique (CNRS) and Centre National d'Etudes Spatiales (CNES) both provided grants to support students and teaching assistants at the summer school. We would also like to thank the various organisations where the lecturers work, for allowing the lecturers time to prepare the lectures, attend the ASI and write their contributions to this book. Jan Fillingham, of DARC at the University of Reading, provided essential administrative support, particularly in the months leading up to the ASI, and we are especially grateful to her.

We also thank Maria Armiento and the staff of the Hotel Villa del Mare in Acquafredda di Maratea. Their hospitality made a major contribution to the success of the ASI. We are particularly grateful for their assistance with two major problems: arranging bus transport when the Italian railways called a strike to coincide with the start of the ASI, and for arranging rental of replacement computer monitors when the equipment ordered from the UK did not arrive. We also appreciated the social events that they arranged, including the evening soccer match against the hotel staff – which the ASI team looked like winning at one point.

Finally, we would like to thank the students for their enthusiastic participation in the summer school. The awards ceremony, in which the students highlighted special talents of some of the lecturers, was particularly memorable. Although the main aim of the ASI was to provide advanced teaching to the students, the lecturers also found it a very stimulating experience.

Richard Swinbank
Victor Shutyaev
William Lahoz

LIST OF CONTRIBUTORS

Dr Pierre Gauthier
Atmospheric Environment Service,
ARMA/SEA (5e Etage),
2121 Route Transcanadienne,
Dorval P.Q.H9P 1J3,
Canada
email: pierre.gauthier@ec.gc.ca

Dr Paul R. Houser
Hydrological Services Branch,
Code 974,
NASA Goddard Space Flight Center,
Greenbelt MD 20771 ,
USA
email: houser@hsb.gsfc.nasa.gov

Dr William A. Lahoz
DARC, Department of Meteorology,
PO Box 243,
Earley Gate,
Reading RG6 6BB,
UK
email: wal@met.reading.ac.uk

Prof Nancy K. Nichols
Department of Mathematics,
University of Reading,
PO Box 220,
Reading, Berks RG6 6AX,
UK
email: nkn@po.cwru.edu

Dr Oleg M. Pokrovsky
Principal Scientist & Head of Laboratory ,
Main Geophysical Observatory,
Karbyshev str.7,
St.Petersburg, 194021,
Russia
email: pokrovsk@cnrm.meteo.fr

Dr Keith Haines
ESSC, University of Reading,
PO Box 238,
Earley Gate,
Reading RG6 6AL ,
UK
email: kh@mail.nerc-essc.ac.uk

Dr Boris Khattatov
NCAR,
P.O. Box 3000,
Boulder, CO 80307-3000,
USA
email: boris@ucar.edu

Dr Peter Lynch
Assistant Director,
Met Eireann,
Glasnevin Hill ,
Dublin 9 ,
Ireland
email: peter.lynch@met.ie

Prof Alan O'Neill
DARC, Department of Meteorology,
PO Box 243,
Earley Gate,
Reading RG6 6BB,
UK
email: alan@met.rdg.ac.uk

Dr Richard B. Rood
Acting Chief, ESDCD,
Code 930,
NASA Goddard Space Flight Center,
Greenbelt MD 20771 ,
USA
email: Richard.B.Rood.1@gsfc.nasa.gov

Prof Victor Shutyaev
INM, Russian Academy of Sciences ,
Gubkina 8,
117951 GSP-1,
Moscow,
Russia
email: shutyaev@inm.ras.ru

Dr Richard Swinbank
NWP, Met Office,
London Road,
Bracknell RG12 2SZ,
UK
email: richard.swinbank@metoffice.com

Dr Olivier Talagrand
LMD, École Normale Supérieure,
24 Rue Lhomond,
72531 PARIS Cedex 05,
France
email: talagran@lmd.ens.fr

Dr Jean-Noel Thépaut
ECMWF,
Shinfield Park,
Reading RG29AX ,
UK
email: jean-noel.thepaut@ecmwf.int

INTRODUCTION

R. SWINBANK
NWP Division, Met Office
Bracknell, Berkshire, UK

1. Introduction

Over recent decades there have been an extraordinary advances across many earth science disciplines, thanks to better observations and understanding of the earth system. This is nowhere better illustrated than in meteorology, where the skill of the three-day forecast now is similar to the skill of a one-day prediction about twenty years ago. Despite these advances there are continuing requirements for improvements; for example, a one-day forecast still cannot give us good enough quantitative rainfall predictions. While major developments have taken place, there are still many advances which need to be made over the coming decades.

One of the fundamental requirements for skilful weather forecasts is the accurate specification of initial conditions for the forecast. In modern times, weather forecasts are produced using a numerical model that embodies the laws of physics that govern the evolution of the atmosphere. Before running the forecast itself, it is necessary to ingest information from a range of weather observations into the numerical model, in order to produce the best possible estimate of the initial state of the atmosphere. This process is referred to as *data assimilation*. More generally, the aim of data assimilation is to ascertain the best estimate of the state of a system, by combining information from observations of that system with an appropriate model of the system.

While data assimilation has attained a degree of maturity in its application to meteorology, it can also be applied to a range of other earth sciences. Much of this book is focused on the use of data assimilation in meteorology, but we also describe its application to oceanography, atmospheric chemistry and land surface processes. However, those disciplines are by no means the limit of the potential uses for data assimilation in earth sciences; instead we hope that this book might serve as an inspiration for the application of data assimilation to other fields.

Before moving on to modern data assimilation techniques and applications, it is helpful to give a perspective on historical developments in data analysis, and how data assimilation came to be so important. Since data assimilation techniques have been relatively well developed in meteorology, I will focus on how atmospheric data analysis techniques have developed over the years, and how they have grown up into the modern data assimilation methods. Lynch (2001) gives a good overview of the historical development of weather forecasting. In his seminal book on atmospheric data analysis, Daley (1991) gives an interesting account of early techniques.

R. Swinbank et al. (eds.), Data Assimilation for the Earth System, 1–8.

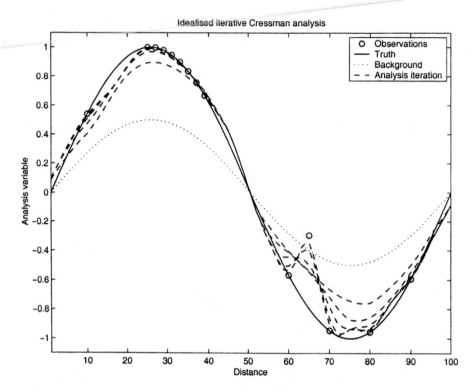

Figure 1. Idealised Cressman analysis, calculated from the observations and background values shown. The length scale L is reduced from 50 to 3, by a factor of about 2/3 on each iteration.

2. Historical perspective

2.1. BEGINNINGS

Instruments for measuring basic weather parameters, such as temperature, pressure, wind and rainfall were developed by the 18th century. While particular individuals were making regular weather observations, it was only after the development of the telegraph that it became possible to exchange weather data in anything approaching real time. Once it became possible to obtain a synoptic view of the weather (i.e. the weather occurring over a region at a particular time), interest then turned to the question of whether it might be possible to forecast the weather. Techniques for subjective analysis of the weather at a particular time were developed during the mid 19th century (by, for example, Admiral Fitzroy, founder of the British Meteorological Office). The initial approaches were entirely subjective, and the resulting attempts at forecasting the weather were not entirely successful.

Modern numerical weather prediction (NWP) can be said to date from Richardson's famous experiment (Richardson, 1922). This was first attempt to carry out a numerical weather forecast, using the equations of motion of the atmosphere. In order to provide initial conditions for the forecast, he started by gridding a subjective analysis, using some of the very earliest measurements of the free troposphere over continental Europe. Unfortunately, the forecast was not a success; P. Lynch discusses reasons for the failure later in this volume, in the chapter "Introduction to Initialization".

It was almost 30 more years until a similar experiment was carried out by Charney et al (1950) using ENIAC, the first multipurpose digital computer. The results were more successful, although setting up the initial conditions (once again from a subjective analysis) was rather time-consuming.

2.2. OBJECTIVE ANALYSIS

Panofsky (1949) carried out the first attempt at objective analysis of meteorological data. The basis of the technique was fitting polynomials to the observation values in a fairly ad hoc manner. This was developed further by Gilchrist and Cressman (1954), who introduced a region of influence for each observation, and suggested the use of a background field (from a previous forecast).

In the approach of Bergthorsson and Döös (1955) the background field plays a more central role – their technique was based on an analysis of observation minus background differences, rather than the observation values themselves. They attempted to optimise the weights given to each observation based on the accuracy of different observation types, as compiled on a database. Later variations on the technique involved multiple iterations of the analysis - the Successive Correction Method, SCM (e.g. Cressman, 1959). In Cressman's scheme, each observation was given a weight that varied with the distance r between each observation and grid point, within a radius of influence L:

$$w(r) = \frac{L^2 - r^2}{L^2 + r^2} \quad (r \leq L).$$ (2.1)

It was found helpful to reduce L with successive iterations. The first iterations fit the observations on a large scale, while later iterations fit small-scale variations in the observations. With appropriate tuning the scheme can fit a set of observations in a realistic manner, as illustrated in Fig. 1. However, since the end result of the iterations is to fit the observations as closely as possible, it can equally well fit erroneous data.

2.3. STATISTICAL METHODS

Perhaps the most important breakthrough was the adoption of statistical interpolation techniques. While this approach can be traced back to Kolmogorov (1941), it came to prominence in meteorology with the book by Gandin (1963). This put the hitherto pragmatic approach to data analysis onto a proper statistical basis, usually referred to as *Optimal Interpolation* (or OI). The weights given to the observations were properly related to the observation errors. At the same time, the background field was no longer

4

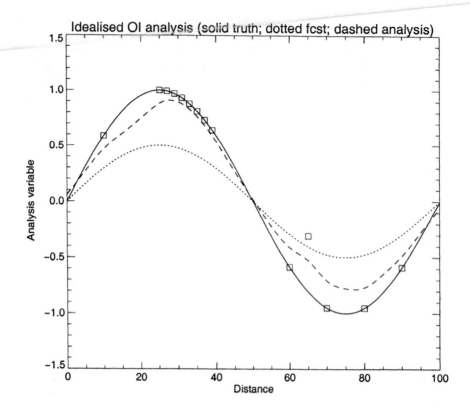

Figure 2. Idealised OI solution to the same data as shown in Fig.1, assuming equal observation and background errors, uncorrelated observation errors, and a background error correlation length of 5.

just the starting point for the calculation of the analysis, but instead was recognised as being another useful source of information, with its own error characteristics. Arguably, this is the key idea in the development of data assimilation

For a simple illustration of the principles underlying OI, consider two unbiased, independent estimates x_1, x_2 of the same quantity x. Assuming that we know the errors (standard deviations) of each of those estimates, which we denote σ_1 and σ_2 respectively, we can combine them to get an improved estimate, x^a:

$$x^a = \frac{\sigma_2^2 x_1 + \sigma_1^2 x_2}{\sigma_1^2 + \sigma_2^2}.$$
(2.2)

The optimal estimate of x is obtained by weighting x_1, x_2 in inverse proportion to their respective error variances (so the less accurate estimate is given less weight). An alternative way of writing this expression is

$$x^a = x_1 + \frac{\sigma_1^2}{\sigma_1^2 + \sigma_2^2}\left(x_2 - x_1\right). \tag{2.3}$$

We can generalise this approach for the case where we have a set of observations, and background data, each with specified error characteristics. In the recommended standard notation (Ide et al, 1997), optimal interpolation theory gives the following expression for the analysis x^a, in terms of the background field x^b and its covariance $\mathbf{B}$ and the observations y^o and their error covariance $\mathbf{R}$:

$$\mathbf{x}^a = \mathbf{x}^b + \mathbf{BH}^T \left(\mathbf{HBH}^T + \mathbf{R}\right)^{-1} \left(\mathbf{y}^o - \mathbf{Hx}^b\right). \tag{2.4}$$

$\mathbf{H}$ is usually referred to as the observation operator; it is an operator which derives synthetic observation values from the model state $\mathbf{x}$ (we have assumed $\mathbf{H}$ is linear). In the simplest case, $\mathbf{H}$ is an interpolation of the appropriate model fields to the observation location. The form of Eq. 2.4 corresponds exactly to the simple case in Eq. 2.3; the difference is that this equation is written in terms of vectors, rather than scalars. The analysis $\mathbf{x}^a$ is calculated from the background $\mathbf{x}^b$ plus a correction term, which is a weighting times the *innovation vector* (the difference between the true observations $\mathbf{y}^o$ and observations simulated from the background field $\mathbf{Hx}^b$).

Equation 2.4 shows that the analysis increment $(\mathbf{x}^a\text{-}\mathbf{x}^b)$ can be found by first calculating $(\mathbf{HBH}^T\text{+}\mathbf{R})^{-1}(y^o\text{-}\mathbf{Hx}^b)$ in observation space. This solution is then transformed back to physical space by multiplying by $\mathbf{BH}^T$. So this equation demonstrates that the region influenced by each observation is not just an arbitrary tuning parameter, but is instead intimately related to the structure of $\mathbf{B}$, the background error covariance. Figure 2 shows an example of the OI solution to the same idealised problem as shown in Fig 1. The OI solution takes better account of the background field. Where the observations are dense they are fitted well, and less attention is paid to the erroneous observation.

Once computer power made them feasible, data assimilation schemes based on OI were implemented in several operational centres in the late 1970s, e.g. ECMWF (Lorenc, 1981) and the Met Office (Lyne et al, 1982). However, the first implementations had to make major approximations to make the calculations feasible. The ECMWF scheme was based on small analysis volumes, while in the Met Office scheme a maximum of 8 observations influenced each grid-point. These drastic simplifications meant that initial implementations of OI were, in fact, rather sub-optimal. In later years this gave the term "Optimal Interpolation" a rather bad name within the meteorological community.

Over about the next 20 years, data assimilation schemes continued to develop, based on various approximations to solving the basic OI equations (see Lorenc, 1986). Later analysis schemes provided improved approximations. For example, the Analysis Correction scheme of Lorenc et al (1991) is, in a sense, a hybrid between OI and SCM. However the key breakthrough has probably been the adoption of variational techniques to determine the solution to the OI equations (e.g. Le Dimet and Talagrand, 1986). The variational approach is based the minimization of a cost function that quantifies the mismatch between a model state and both observations and a background state. In our simple example, we could take x_1 to be an observation of a particular quantity and x_2 an independent (background) estimate of the same quantity. One can then write the expression for a cost function J, quantifying the mismatch between an arbitrary value x and our two estimates:

$$J(x) = \frac{1}{2} \left\{ \frac{(x-x_1)^2}{\sigma_1^2} + \frac{(x-x_2)^2}{\sigma_2^2} \right\}. \tag{2.5}$$

It is straightforward to show that the minimum of J is given by Eq 2.2 (or 2.3).

The first operational implementation of a three-dimensional variational data assimilation (3D-VAR) system was by the National Centers of Environmental Prediction (NCEP) (Parrish and Derber, 1992), followed by the European Centre for Medium-range Weather Forecasts (ECMWF) a few years later (Courtier et al, 1998). Since then they have been implemented by several other centres; of particular note is the implementation of the "dual" PSAS (Physical-space Statistical Analysis System)at the NASA Data Assimilation Office (Cohn et al, 1998). Variational methods are explored in much more detail in later chapters of this book. Further information about atmospheric data assimilation is given by Daley (1991), and the recent textbook by Kalnay (2003).

2.4. NUMERICAL MODELS

In parallel to the development of data analysis methods, the availability of modern computers has also allowed numerical weather prediction models to develop in leaps and bounds. Particularly over the past 30 years, numerical models of the atmosphere have increased dramatically in both resolution and sophistication. W. Lahoz gives an overview of modelling techniques in the chapter *Atmospheric modelling*.

The increasing accuracy of numerical models means that forecast fields are now a crucial component of the data assimilation system. In many cases, the background field from an NWP model will be more accurate than observations being assimilated, so it is increasingly important that the data assimilation systems correctly account for the background error covariances as well as the observation error covariances. At present, data assimilation systems use rather crude representation of the error covariances, but current research work is aimed at improving the specification of the backrgound error covariances.

Numerical models are a crucial part of four-dimensional variational (4D-VAR) data assimilation systems. Linearised models, and their adjoints, are used to fit observations over a time window, rather than at a specific analysis time. The first operational 4D-VAR system was implemented by ECMWF (Courtier et al, 1994), and several other centres are developing similar systems.

3. Modern Context

As outlined above, much of development of data assimilation systems has historically been driven by the requirement for improved weather forecasts. Following the development of modern, sophisticated, meteorological data assimilation systems, they are increasingly becoming important for a range of other applications, described more fully in subsequent chapters in this book.

Because of the continual improvement to data assimilation systems and changes to the observation network, it has not proved practical to use the assimilated data produced from operational systems for studies of long-term variability of the atmospheric circulation. Instead, in the early 1990s, several centres embarked on "Reanalysis Projects", in which several decades of observations are processed by the same data assimilation system.

While most of the effort in meteorological data assimilation has concentrated on the lower atmosphere, or troposphere, several initiatives have extended the method to stratospheric data. Rood et al (1990) developed the pioneering STRATAN data assimilation system. Swinbank and O'Neill (1994) extended the then operational Met Office data assimilation system into the stratosphere, primarily to support validation and interpretation of measurements from the Upper Atmosphere Research Satellite. The assimilation of stratospheric data has been extended to a range of constituents. Several systems have been built to assimilate ozone, while other systems have treated a range of constituents using models with sophisticated representations of atmospheric chemistry.

In parallel to the development of atmospheric data assimilation systems, there have been major developments in the application of assimilation techniques to the ocean. In many ways this is a more challenging problem - not least because of the relative sparsity of observations.

Historically, the description of land surface processes in weather and climate models has been very crude. Because of the need to improve land surface models for climate simulation, and for applications like flood forecasting, the assimilation of land-surface data is now a fast-developing field; recent developments are described in this book.

Alongside these developments in data assimilation, there is a wealth of remote sensing observations from research satellites that have just been launched, or will be launched over next few years. NASA launched the EOS (Earth Observing System) Terra research satellite in late 1999, followed by Aqua in spring 2002, and the follow-up Aura mission will be launched soon. The European Envisat (Environment Satellite) was also launched early in 2002, and the Japanese ADEOS-II (Advanced Earth Observation Satellite 2) in late 2002. In addition there will be a wide range of new types of operational meteorological satellites from both NOAA and EUMETSAT.

Over the coming years, there will be an enormous range of measurements of diverse aspects of the earth system. Data Assimilation is perhaps the best technique, or arguably the only practical technique, to make use of the enormous variety and volume of data that are being relayed from these satellite instruments.

Acknowledgements. I would like to thank Alan O'Neill and Peter Lynch for their constructive comments on this manuscript, and Clive Wilson for information on the improvements in forecast skill.

References

Bergthorsson, P. and B. Döös, 1955: Numerical weather map analysis, *Tellus*, **7**, 329-340.

Charney, J., R. Fjortoft and J. von Neumann, 1950: Numerical integration of the barotropic vorticity equation, *Tellus*, **2**, 237-254.

Cohn, S.E., A. da Silva, J. Guo, M. Sienkiewicz and D. Lamich, 1998: Assessing the effects of data selection with the DAO Physical-space Statistical Analysis System, *Mon. Weather Rev.* , **126**, 2912-2926.

Courtier P, E. Andersson, W. Heckley, J. Pailleux, D. Vasiljevic, M. Hamrud, A. Hollingsworth, F. Rabier and M. Fisher, 1998: The ECMWF implementation of three-dimensional variational assimilation (3D-Var). I: Formulation. *Quart. J. R. Meteorol. Soc.*, **124**, 1783-1807.

Courtier, P., J.N. Thepaut and A. Hollingsworth, 1994: A strategy for operational implementation of 4D-Var, using an incremental approach, *Quart. J. R. Meteorol. Soc.*, **120**, 1389-1408.

Cressman, G, 1959: An operational objective analysis system, *Mon Wea. Rev.*, **87**, 367-374.

Daley, R., 1991: *Atmospheric Data Analysis*, Cambridge University Press, Cambridge, 457pp.

Gandin, L., 1963: *Objective analysis of meteorological fields*, Gridromet, Leningrad. (English translation Israel Program for Scientific Translation, Jerusalem, 1965).

Gilchrist, B. and G. Cressman, 1954: An experiment in objective analysis, *Tellus*, **6**, 309-318.

Ide, K., Courtier, P., Ghil, M., and Lorenc, A.C., 1997: Unified notation for data assimilation: Operational, sequential and variational. *J. Met. Soc. Japan*, **75**, 181-189.

Kalnay, E., 2003: *Atmospheric Modeling, Data Assimilation and Predictibility*, Cambridge University Press, Cambridge, 341pp.

Le Dimet, F-X. and O. Talagrand, 1986: Variational algorithms for analysis and assimilation of meteorological observations: Theoretical Aspects. *Tellus*, **38A**, 97-110.

Lorenc, A.C., 1981: A global three-dimensional multivariate statistical analysis scheme, *Mon. Wea. Rev.*, **109**, 701-721.

Lorenc, A.C., 1986: Analysis methods for numerical weather prediction, *Quart. J. R. Meteorol. Soc.*, **112**, 1177-1194.

Lorenc, A.C., R.S. Bell and B. Macpherson, 1991: The Meteorological Office Analysis Correction Data Assimilation scheme, *Quart. J. R. Meteorol. Soc.*, **117**, 59-89.

Lynch, P., 2001: Weather Forecasting from woolly art to solid science, *Meteorology at the Millenium*, 106-119, Academic Press.

Lyne, W.H., R. Swinbank and N.T. Birch, 1982: A data assimilation experiment, with results showing the atmospheric circulation during the FGGE special observing periods, *Quart. J. R. Met. Soc.*, **108**, 575-594.

Panofsky, H., 1949: Objective weather-map analysis, *J. Appl. Meteor*, **6**, 386-392.

Parrish, D.F. and J.C. Derber, 1992: The National Meteorological Center's spectral statistical interpolation analysis scheme. *Mon. Wea. Rev.*, **120**, 1747-1763.

Richardson, L., 1922: *Weather prediction by numerical process*, Cambridge University Press, Cambridge.

Rood, R.B., P.A. Newman, L.R. Lait, D.J. Lamich, K.R. Chan, 1990: Stratospheric temperatures during AASE: results from STRATAN. *Geophys. Res. Lett.*, **17**, 337-340.

Swinbank, R. and A. O'Neill, 1994: A stratosphere-troposphere data assimilation system, *Monthly Weather Review* , **122**, 686-702.

DATA ASSIMILATION: AIMS AND BASIC CONCEPTS

N.K. NICHOLS
*Department of Mathematics, The University of Reading,
Box 220, Reading, RG6 6AX United Kingdom*

1. Introduction

Atmosphere and ocean systems can be simulated effectively by discrete numerical models and, provided that the initial states of the system are known, accurate forecasts of future dynamical behaviour can be determined. Complete information defining all of the states of the system at a specified time are, however, rarely available. Moreover, both the models and the measured data contain inaccuracies and random noise. In this case, observations of the system measured over an interval of time can be used in combination with the model equations to derive estimates of the expected values of the states. The problem of constructing a 'state-estimator,' or 'observer,' for these systems can be treated by using feedback design techniques from control theory. For the very large nonlinear systems arising in climate, weather and ocean prediction, however, traditional control techniques are not practicable and 'data assimilation' schemes are used instead to generate accurate state-estimates (see, for example, Daley, 1994; Bennett, 1992).

The aim of data assimilation is to incorporate measured observations into a dynamical system model in order to produce accurate estimates of all the current (and future) state variables of the system. The problem can be stated as follows.

Problem 1 *Given a (noisy) discrete model of the dynamics of a system, find estimates of the system states from (noisy) observations.*

The most significant properties of the data assimilation problem are that the models are very large and nonlinear, with order $O(10^7)$ state variables. The number of observations is also large, of order $O(10^5 - 10^6)$.

There are two approaches to this problem. The first uses a 'dynamic observer,' which gives a *sequential data assimilation scheme*, and the second uses a 'direct observer', which gives a *four-dimensional data assimilation*

9

R. Swinbank et al. (eds.), Data Assimilation for the Earth System, 9–20.
© 2003 *Kluwer Academic Publishers. Printed in the Netherlands.*

scheme. In the next section these two approaches are illustrated for a simple linear time-invariant system. In the following sections data assimilation schemes for a full nonlinear system are developed and examples are presented.

2. Data Assimilation in Linear Systems

We first examine a discrete linear time-invariant system model. The dynamical equations describing the evolution of the states from time t_k to time t_{k+1} are given by

$$\mathbf{x}_{k+1} = F\mathbf{x}_k + G\mathbf{u}_k, \quad k = 0, \ldots, N-1, \tag{1}$$

where $\mathbf{x}_k \in \mathbb{R}^n$ denotes the model states and $\mathbf{u}_k \in \mathbb{R}^m$ denotes the model inputs at time t_k, and $F \in \mathbb{R}^{n \times n}$, $G \in \mathbb{R}^{n \times m}$ are the system matrices. The input vector contains the known forcing functions that drive the system, and it is assumed that for given inputs and for any initial state $\mathbf{x}_0$ given at time t_0, the system equations uniquely determine the states of the system for all future times.

The observations are linearly related to the system states by the equations

$$\mathbf{y}_k = H\mathbf{x}_k, \quad k = 0, \ldots, N-1, \tag{2}$$

where $\mathbf{y}_k \in \mathbb{R}^p$ is a vector of p observations at time t_k and $H \in \mathbb{R}^{p \times n}$ is the observation matrix, which includes transformations and grid interpolations.

In the case where $\mathbf{x}_0$ is not known (accurately), the objective of the assimilation schemes is to use measured data over time to estimate correctly the states of the system at times t_k, $k > 0$. Two classes of methods can be applied.

2.1. SEQUENTIAL DATA ASSIMILATION

With sequential assimilation, *a priori* estimates for the initial states $\mathbf{x}_0$ are chosen and the model is evolved forward to the time t_k where the first observations are available. The predicted states of the system at this time are known as the background states and are denoted by $\mathbf{x}_k^b$. The difference between the predicted observation vector given by the background states and the vector of measured observations at this time ($H\mathbf{x}_{k+1}^b - \mathbf{y}_{k+1}$), known as the 'innovation vector,' is then used to make a correction to the background state vector in order to obtain improved state estimates $\mathbf{x}_k^a$, known as the analysis states. The model is then evolved forward again from the analysis states to the next time where an observation is available and the process is repeated.

Mathematically this procedure may be written

$$\mathbf{x}_{k+1}^b = F\mathbf{x}_k^a + G\mathbf{u}_k, \tag{3}$$

$$\mathbf{x}_{k+1}^a = \mathbf{x}_{k+1}^b + K(H\mathbf{x}_{k+1}^b - \mathbf{y}_{k+1}). \tag{4}$$

The matrix $K \in \mathbb{R}^{n \times p}$, known as the 'gain matrix,' must be chosen to ensure that the analysis states converge to the true states of the system over time. This is always possible if the system is 'observable.' Conditions for this property to hold are well-known. (See, for example, Barnett and Cameron, 1985.)

The system (3)-(4) forms a modified dynamical system for the analysis states which can be written

$$\mathbf{x}_{k+1}^a = (F + KHF)\mathbf{x}_k^a + (G + KHG)\mathbf{u}_k - K\mathbf{y}_{k+1}. \tag{5}$$

This system has different properties from the original discrete system model (1) and the stability of its response depends upon the spectrum of the modified system matrix $F + KHF$. The choice of the gain matrix K therefore determines the behaviour of the analysed states over time and this choice characterizes the data assimilation scheme.

2.2. FOUR-DIMENSIONAL DATA ASSIMILATION

In contrast to sequential data assimilation, which evolves the model one step at a time and updates the estimated states each time an observation is available, the four-dimensional assimilation schemes use all the observations available over a given time window to provide improved estimates for all the states in that window. Thus 'time' is the fourth dimension in the problem.

Mathematically we may rearrange the system equations (1)-(2) in order to express the observations over time in terms of the initial (unknown) states $\mathbf{x}_0$. Using the relation

$$\mathbf{y}_k = H\mathbf{x}_k = HF\mathbf{x}_{k-1} + HG\mathbf{u}_{k-1}, \tag{6}$$

we may rewrite the system equations at points t_k, $k = 0, 1, \ldots, N-1$, in the form

$$H\mathbf{x}_0 = \mathbf{y}_0, \tag{7}$$

$$HF\mathbf{x}_0 = \mathbf{y}_1 - HG\mathbf{u}_0, \tag{8}$$

$$HF^2\mathbf{x}_0 = \mathbf{y}_2 - HG\mathbf{u}_1 - HFG\mathbf{u}_0, \tag{9}$$

$$\vdots$$

$$HF^{N-1}\mathbf{x}_0 = \mathbf{y}_{N-1} - \sum_{j=0}^{N-2} HF^j G\mathbf{u}_{N-2-j}. \tag{10}$$

These equations form an $Np \times N$ over-determined system of linear equations for the unknown initial states that may be solved by the method of least squares. The problem may be written

$$\min_{\mathbf{x}_0} \|V\mathbf{x}_0 - \mathbf{r}\|_2^2, \qquad (11)$$

where $V^T = [H^T, (HF)^T, (HF^2)^T, \ldots, (HF^{N-1})^T]$, $\mathbf{r}$ denotes the right-hand-side of the equations (7)–(10) and $\|\mathbf{z}\|_2$ denotes the L_2-vector norm of $\mathbf{z}$, given by $(\mathbf{z}^T\mathbf{z})^{\frac{1}{2}}$. The matrix V is of full rank if the system is 'observable' and $N \geq n$. The least square problem then has a unique solution.

Using the system equations, we may write the linear least-square problem (11) as a constrained least square problem

$$\min_{\mathbf{x}_0} \frac{1}{2} \sum_{k=0}^{N-1} (H\mathbf{x}_k - \mathbf{y}_k)^T (H\mathbf{x}_k - \mathbf{y}_k), \qquad (12)$$

subject to the conditions that $\mathbf{x}_k$, $k = 1, \ldots, N-1$, satisfy the model equations (1) starting with initial data $\mathbf{x}_0$. The problem is very large and may be solved using an iterative procedure.

In this approach the initial states are treated as parameters that must be selected to minimise the mean square errors between the observations predicted by the model and the measured observations over the entire time window. The initial data is adjusted to different positions in order to achieve the best fit, using an efficient iterative minimisation algorithm.

2.3. SUMMARY

Two approaches to the problem of data assimilation are illustrated here in the case of a discrete linear deterministic dynamical model. These same approaches apply to models of the atmosphere and oceans, but the models are nonlinear and contain uncertainties. In practice the model equations are *not accurate*, the input data are *not accurate*, the observational data are *not accurate*, and the computation is *not accurate*! In the next section we develop sequential and four-dimensional assimilation schemes for treating nonlinear systems with observations that contain random noise. In the chapter *Treating Model Error in 3-D and 4-D Data Assimilation*, we consider nonlinear stochastic systems containing both random and systematic model errors.

3. Data Assimilation in Nonlinear Dynamical Systems

A variety of models are used to describe systems arising in atmosphere and ocean applications, as well as in other physical, biological and economic fields. These range from simple linear, deterministic, continuous ordinary differential equation models to sophisticated nonlinear stochastic partial differential or discrete models. The data assimilation schemes, with minor modifications, can be applied to any general model.

Data assimilation schemes are described here for a system modelled by the discrete nonlinear equations

$$\mathbf{x}_{k+1} = \mathbf{f}_k(\mathbf{x}_k, \mathbf{u}_k), \quad k = 0, \ldots, N - 1, \tag{13}$$

where $\mathbf{x}_k \in \mathbb{R}^n$ denotes the model states and $\mathbf{u}_k \in \mathbb{R}^{m_k}$ denotes the m_k known inputs to the system at time t_k, and $\mathbf{f}_k : \mathbb{R}^n \times \mathbb{R}^{m_k} \to \mathbb{R}^n$ is a nonlinear function describing the evolution of the states from time t_k to time t_{k+1}.

The observations are assumed to be related to the system states by the equations

$$\mathbf{y}_k = \mathbf{h}_k(\mathbf{x}_k) + \boldsymbol{\delta}_k, \quad k = 0, \ldots, N - 1, \tag{14}$$

where $\mathbf{y}_k \in \mathbb{R}^{p_k}$ is a vector of p_k observations at time t_k and $\mathbf{h}_k : \mathbb{R}^n \to \mathbb{R}^{p_k}$ is a nonlinear function that includes transformations and grid interpolations. The observational errors $\boldsymbol{\delta}_k \in \mathbb{R}^{p_k}$ are assumed to be unbiased, serially uncorrelated, Gaussian random vectors with covariance matrices $R_k \in \mathbb{R}^{p_k \times p_k}$.

Prior estimates, or 'background estimates,' $\mathbf{x}_0^b$ of the initial states $\mathbf{x}_0$ are assumed to be known and the initial random errors $(\mathbf{x}_0 - \mathbf{x}_0^b)$ are assumed to be Gaussian with covariance matrix $B_0 \in \mathbb{R}^{n \times n}$. The observational errors and the errors in the prior estimates are assumed to be uncorrelated.

The data assimilation scheme, or 'state-estimator,' for this model is defined by the system

$$\mathbf{x}_{k+1}^b = \mathbf{f}_k(\mathbf{x}_k^a, \mathbf{u}_k) \tag{15}$$
$$\mathbf{x}_{k+1}^a = \mathbf{x}_{k+1}^b + K_{k+1}(\mathbf{h}_k(\mathbf{x}_{k+1}^b) - \mathbf{y}_{k+1}), \tag{16}$$

for $k = 0, \ldots, N - 1$. As for the linear model, these equations represent a prediction for the background states from the model equations, followed by an update for the analysed states using the innovations $(\mathbf{h}_k(\mathbf{x}_{k+1}^b) - \mathbf{y}_{k+1})$. The gain matrices K_k must be chosen to ensure that the analysed states $\mathbf{x}_k^a$ converge to the expected values $E(\mathbf{x}_k)$ of the true states as $k \to \infty$ for *any* initial background state $\mathbf{x}_0^b$. The choice of the gain matrices characterizes the data assimilation scheme. The primary question in designing the scheme therefore concerns how K_k should be selected.

For the 'optimal' analysis, we aim to find the best estimates $\mathbf{x}_k^a$ for the expected values of the true states $\mathbf{x}_k$, $k = 0, \ldots, N - 1$, given observations $\mathbf{y}_k$, $k = 0, \ldots, N - 1$, subject to the model equations (13) and prior estimates $\mathbf{x}_0^b$. Under the statistical assumptions made here, the optimal analysis is given by the maximum likelihood *a priori* Bayesian estimate of the system states. The problem reduces to minimising the square error between the observations predicted by the model and the measured observations, weighted by the inverse of the covariance matrices, over the assimilation window (Lorenc, 1986). The data assimilation problem is then defined explicitly as follows.

Problem 2 *Minimise, with respect to* $\mathbf{x}_0$, *the objective function*

$$\mathcal{J} = \frac{1}{2}(\mathbf{x}_0 - \mathbf{x}_0^b)^T B_0^{-1}(\mathbf{x}_0 - \mathbf{x}_0^b) +$$

$$+ \frac{1}{2}\sum_{k=0}^{N-1}(\mathbf{h}_k(\mathbf{x}_k) - \mathbf{y}_k)^T R_k^{-1}(\mathbf{h}_k(\mathbf{x}_k) - \mathbf{y}_k), \qquad (17)$$

subject to $\mathbf{x}_k, \ k = 1, \ldots, N - 1$, *satisfying the system equations (13) with initial states* $\mathbf{x}_0$.

The model is assumed to be 'perfect' and the system equations are treated as strong constraints on the minimisation problem. The states $\mathbf{x}_k$ that satisfy the model equations (13) are uniquely determined by the initial states of the system and therefore can be written explicitly in terms of $\mathbf{x}_0$, as in the linear case. Substituting into the objective function (17) then allows the optimisation problem to be expressed in terms of the initial states alone. The assimilation problem, Problem 2, thus becomes an unconstrained weighted least squares problem where the initial states are the required control variables in the optimisation.

The aim is to determine the analysis of the initial states that produces the best fit, in a statistical sense, between the observations predicted by the analysis and the measured observations over the assimilation window. The first term in the objective function constrains the choice of the initial states and acts as a 'regularising' term. Since the covariance matrices B_0, R_k, $k = 0, 1, \ldots, N-1$, are of full rank, the resulting least square problem necessarily has a unique solution.

The data assimilation problem, Problem 2, can be solved directly to give a sequential assimilation scheme, or it can be solved indirectly to give a four-dimensional 'variational' assimilation scheme. These two techniques are described in the next subsections.

3.1. SEQUENTIAL ASSIMILATION SCHEMES

To derive the 'optimal' sequential assimilation scheme for the nonlinear system, we assume that we have prior estimates for the background states

$\mathbf{x}_k^b$ at time step t_k with error covariance matrix

$$B_k = E((\mathbf{x} - \mathbf{x}_k^b)(\mathbf{x} - \mathbf{x}_k^b)^T). \tag{18}$$

The optimal analysis $\mathbf{x}_k^a$ then solves the problem

$$\min_{\mathbf{x}}[\frac{1}{2}(\mathbf{x} - \mathbf{x}_k^b)^T B_k^{-1}(\mathbf{x} - \mathbf{x}_k^b) + \frac{1}{2}(\mathbf{h}_k(\mathbf{x}) - \mathbf{y}_k)^T R_k^{-1}(\mathbf{h}_k(\mathbf{x}) - \mathbf{y}_k)]. \tag{19}$$

The optimal solution is given by

$$\mathbf{x}_k^a = \mathbf{x}_k^b + K_k(\mathbf{h}_k(\mathbf{x}_k^b) - \mathbf{y}_k), \tag{20}$$

where

$$K_k = B_k H_k^T (H_k B_k H_k^T + R_k)^{-1} \tag{21}$$

and $H_k = \frac{\partial \mathbf{h}_k}{\partial \mathbf{x}}|_{\mathbf{x}_k^b}$. (The matrix H_k is the 'Jacobian' of the function $\mathbf{h}_k$ with respect to $\mathbf{x}$ evaluated at $\mathbf{x}_k^b$, and thus $H_k(\mathbf{x} - \mathbf{x}_k^b)$ approximates $(\mathbf{h}_k(\mathbf{x}) - \mathbf{h}_k(\mathbf{x}_k^b))$ for small $\|\mathbf{x} - \mathbf{x}_k^b\|_2$).

For linear systems the solution (20)-(21) gives the exact optimal analysis, but for nonlinear systems this solution gives only a first order approximation to the optimal due to the linearisation H_k of the nonlinear observation operator that is used (Lorenc, 1986, 1988; Thacker, 1996).

In evolving the 'optimal' analysis sequentially, two computational difficulties arise. The first is that the background covariance matrices B_k are required at each time step. These matrices can be propagated forward in time from the initial background error covariance matrix B_0 using an extended Kalman filter (EKF) technique (Kalman, 1961). For full-scale weather and ocean systems, however, the EKF requires the computation of matrices of order $O(10^7 \times 10^7)$ at every time step, making it very expensive to implement. Moreover, for nonlinear models the EKF is highly sensitive to computational round-off errors and may become unstable as a dynamical system (Bierman, 1977).

The second difficulty in implementing the optimal assimilation scheme sequentially is that in order to compute the analysis $\mathbf{x}_k^a$ at each time step, we must find $B_k H_k^T \mathbf{w}_k^a$, where $\mathbf{w}_k^a$ solves the linear equations

$$(H_k B_k H_k^T + R_k)\mathbf{w}_k^a = (\mathbf{h}_k(\mathbf{x}_k^b) - \mathbf{y}_k). \tag{22}$$

This is a very large inverse problem with $O(10^5 - 10^6)$ variables to find.

In practice most operational sequential assimilation schemes avoid these two difficulties by using approximations that can be implemented efficiently. A variety of schemes have been developed that differ mainly in the detailed steps of the procedures.

Sequential data assimilation schemes used operationally include:

– *Successive Correction.* In these schemes, the feedback gain K_k is not chosen optimally, but is designed to smooth observations into the states at all spatial grid points within some radius of influence of each observation (Bergthorsen and Doos, 1955). An iterative process is used to determine the analysis. The Cressman scheme is an example (Cressman, 1959). The iterations converge to a result that is consistent with observational error but may not be consistent with the dynamical system equations. Over time the analysis states may not converge to the expected values of the true states. These schemes are generally not effective in data sparse regions.

– *Optimal Interpolation or Statistical Interpolation.* These schemes approximate the optimal solution by replacing the optimal gain matrix K_k in (20) by the approximation

$$\tilde{K}_k = \tilde{B} H_k^T (H_k \tilde{B} H_k^T + R_k)^{-1},$$

where $\tilde{B}$ has a 'fixed' structure for all k. (See Ghil and Malanotte-Rizzoli, 1991.) The matrix $\tilde{B}$ is generally defined by an isotropic correlation function (dependent only on the distance between spatial grid points and observational points), with the correlation lengths adjusted empirically. To simplify the inversion step, the gain is further modified to have a block structure by using innovations only in small regions around grid points to obtain the analysis states. The inversion problem then reduces to solving a number of much smaller systems of equations.

– *Analysis Correction.* In these schemes, approximations to the optimal analysis states are computed iteratively, as in the Successive Correction method. The procedure is designed, however, to ensure that the iterates converge to the approximate 'optimal' analysis that is obtained by replacing the optimal gain K_k in (20) by the gain $\tilde{K}_k$, as defined for the optimal interpolation scheme (Bratseth, 1986; Lorenc, Bell and MacPherson, 1991). This scheme is effective across data sparse regions and the analysis produced remains consistent with the dynamical equations.

– *3DVAR.* These schemes apply iterative minimisation methods directly to the variational problem (19) (Rabier *et al.*, 1993). The covariance matrix B_k is replaced by the approximation $\tilde{B}$, as defined for optimal interpolation. The solution converges to the analysis obtained by replacing the optimal gain K_k by $\tilde{K}_k$ in (20). Minimisation techniques used commonly are Preconditioned Conjugate Gradient methods and Quasi-Newton methods. The properties of the analysis are similar to those obtained by the Analysis Correction method, but the iteration procedure is more efficient.

– *3DPSAS.* In these schemes iterative minimisation methods are applied to the dual variational problem

$$\min_{\mathbf{w}}[\frac{1}{2}(\mathbf{w}^T H_k \tilde{B} H_k + R_k)\mathbf{w} - \mathbf{w}^T(\mathbf{h}_k(\mathbf{x}) - \mathbf{y}_k)].$$

The iterates converge to the solution $\mathbf{w}_k^a$ of the system (22) and the resulting analysis states converge to $\mathbf{x}_k^a = \tilde{B} H_k^T \mathbf{w}_k^a$, which approximates the 'optimal' solution as in the 3DVAR scheme (Cohn et al., 1998). The advantage is that this scheme operates in the 'observation space,' which is of lower dimension than the state space. Additional work is needed, however, in order to reconstruct the analysis states.

In summary, most operational sequential data assimilation schemes aim to approximate the optimal analysis by replacing the background error covariance matrix by an approximation that is fixed over time and by simplifying the inversion problem and/or solving the inversion iteratively. Simple examples illustrating the use of these schemes can be found in Martin et al. (1999).

3.2. FOUR-DIMENSIONAL ASSIMILATION SCHEMES

The full four-dimensional data assimilation problem is currently treated in operational centres using four-dimensional variational schemes (4DVAR) (Sasaki, 1970; Talagrand, 1981; Rabier et al., 2000). In these schemes the constrained minimisation problem, Problem 2, is solved iteratively by a gradient optimisation method. The problem is first reduced to an unconstrained problem using the method of Lagrange. Necessary conditions for the solution to the unconstrained problem then require that a set of adjoint equations together with the system equations (13) must be satisfied. The adjoint equations are given by

$$\lambda_N = 0, \tag{23}$$

$$\lambda_k = F_k^T(\mathbf{x}_k)\lambda_{k+1} - H_k^T R_k^{-1}(\mathbf{h}_k(\mathbf{x}_k) - \mathbf{y}_k), \quad k = N - 1, \ldots, 0, \tag{24}$$

where $\lambda_k \in \mathbb{R}^n$, $j = 0, \ldots, N$, are the adjoint variables and $F_k \in \mathbb{R}^{n \times n}$ and $H_k \in \mathbb{R}^{n \times p_k}$ are the Jacobians of $\mathbf{f_k}$ and $\mathbf{h_k}$ with respect to $\mathbf{x}_k$. The adjoint variables λ_k measure the sensitivity of the objective function (17) to changes in the solutions $\mathbf{x}_k$ of the state equations for each k.

The gradient of the objective function (17) with respect to the initial data $\mathbf{x}_0$ is then given by

$$\nabla_{\mathbf{x}_0} \mathcal{J} \equiv B_0^{-1}(\mathbf{x}_0 - \mathbf{x}_0^b) - \lambda_0. \tag{25}$$

At the optimal, the gradient (25) is required to be equal to zero. Otherwise this gradient provides the local descent direction needed in the iteration procedure to find an improved estimate for the optimal initial states.

Each step of the gradient iteration process requires one forward solution of the model equations, starting from the current best estimate of the initial states, and one backward solution of the adjoint equations. The estimated initial conditions are then updated using the computed gradient direction. This process is expensive, but it is operationally feasible, even for very large systems.

A dual approach (4DPSAS) in which the minimisation is performed in observation space is also possible (Courtier, 1987). In this scheme, as in the three dimensional 3DPSAS method, a dual four-dimensional variational problem is solved using a gradient iteration method, and the analysis states are then reconstructed from the dual variables.

To make these variational methods more efficient, an 'incremental' approach may be used in which the forward solution of the nonlinear model equations in the iterative procedure is replaced by the forward solution of an approximate 'incremental' linear system (Courtier, Thepaut and Hollingsworth, 1994). In the variational problem the unknown states are replaced by $\mathbf{x}_k = \mathbf{x}_k^b + \delta\mathbf{x}_k$ and the innovations are denoted $\mathbf{d}_k = -(\mathbf{h}_k(\mathbf{x}_k) - \mathbf{y}_k)$. The variational problem then becomes (to first order)

$$\min_{\delta\mathbf{x}_0}[\frac{1}{2}\delta\mathbf{x}_k^T B_0^{-1}\delta\mathbf{x}_k + \frac{1}{2}\sum_{k=0}^{N-1}(H_k\delta\mathbf{x}_k - \mathbf{d}_k)^T R_k^{-1}(H_k\delta\mathbf{x}_k - \mathbf{d}_k)], \qquad (26)$$

subject to the tangent linear model (TLM) equations

$$\delta\mathbf{x}_{k+1} = F_k\delta\mathbf{x}_k, \qquad (27)$$

where $F_k \in \mathbb{R}^{n \times n}$ and $H_k \in \mathbb{R}^{n \times p_k}$ are the Jacobians of $\mathbf{f_k}$ and $\mathbf{h_k}$.

The incremental procedure requires first a forward solution to the nonlinear model (13), starting from the initial background states $\mathbf{x}_0^b$, in order to determine the background states $\mathbf{x}_k^b$ and the innovations $\mathbf{d}_k$ for each k. Then an initial choice for $\delta\mathbf{x}_0$ is made and a forward solution to the tangent linear model (27) is found. The backward solution of the corresponding adjoint system and the gradient of the linearised objective function (26) are then determined, and the minimisation algorithm is used to find an improved estimate for $\delta\mathbf{x}_0$. The linear process is repeated to obtain further updates to $\delta\mathbf{x}_0$. The success of this procedure depends on how well the tangent linear model approximates the nonlinear model. After a number of iterations of the linearised model, the background solution to the nonlinear model is usually updated by a forward solve of the nonlinear equations, starting from the incremented initial states. The full incremental variational procedure thus consists of an inner and outer iteration process.

The primary difficulty in implementing variational assimilation schemes is the need to develop a full adjoint model for the system. The adjoint equations are related theoretically to the linearised state equations, and the system matrix of the adjoint model is given directly by F_k^T, where F_k is the system matrix of the linearised model. The adjoint equations can thus be generated directly from the linearised system equations. Automatic differentiation techniques can be applied to the forward solution code to generate the adjoint code (Griewank and Corliss, 1991). Alternatively an approximate adjoint system can be obtained by discretising a continuous linear or adjoint model of the nonlinear dynamics (Lawless, 2001). This approach has the advantage that additional approximations can be incorporated into the linearisation of the system equations.

Four-dimensional variational data assimilation schemes are currently in operational use and are under further development. Examples illustrating the use of these schemes can be found in Griffith (1997) and Lawless (2001).

4. Conclusions

The aims and basic concepts of data assimilation for atmosphere and ocean systems are described here. Two approaches to the problem of data assimilation, sequential and four-dimensional assimilation, are introduced. A variety of assimilation schemes for discrete nonlinear system models are derived and practical implementation issues are discussed. For all of these schemes, the model equations are assumed to be 'perfect' representations of the true dynamical system. In practice the models contain both systematic errors and random noise. In the chapter *Treating Model Error in 3-D and 4-D Data Assimilation* in this volume, we discuss assimilation techniques for treating stochastic model error and systematic bias errors.

References

Barnett, S. and Cameron, R.G. (1985) *Introduction to the Mathematical Theory of Control*, 2nd edition, Clarendon Press, Oxford, UK.

Bennett, A.F. (1992) *Inverse Methods in Physical Oceanography*, Cambridge University Press, Cambridge, UK.

Bergthorsson, P. and Doos, B.R. (1955) Numerical weather map analysis, *Tellus*, **7**, 329–340.

Bierman, G.L. (1977) *Factorization Methods for Discrete Sequential Estimation*, Mathematics in Science and Engineering, V. 128, Academic Press, New York.

Bratseth, A.M. (1986) Statistical interpolation by means of successive corrections, *Tellus*, **38A**, 439–447.

Cohn, S. E., da Silva, A., Guo, J., Sienkiewicz, M. and Lamich, D. (1998) Assessing the effects of data selection with the DAO physical-space statistical analysis system, *Monthly Weather Review*, **126**, 2913–2926.

Courtier, P. (1997) Dual formulation of four-dimensional variational assimilation, *Quart. J. Roy. Met. Soc.*, **123**, 2449–2461.

Courtier, P., Thepaut, J-N. and Hollingsworth, A. (1994) A strategy for operational implementation of 4D-Var, using an incremental approach, *Quart. J. Roy. Met. Soc.*, **120**, 1367–1387.

Cressman, G. (1959) An optimal objective analysis system, *Monthly Weather Review*, **87**, 367–374.

Daley, R. (1994) *Atmospheric Data Analysis*, Cambridge University Press, Cambridge, UK.

Ghil, M. and Malanotte-Rizzoli, P. (1991) Data assimilation in meteorology and oceanography, *Adv. Geophys.*, **33**, 141–266.

Griewank, A. and Corliss, G.F. (1991) *Automatic Differentiation of Algorithms*, SIAM, Philadelphia.

Griffith, A.K. (1997) *Data Assimilation for Numerical Weather Prediction Using Control Theory*, The University of Reading, Department of Mathematics, PhD Thesis.

Kalman, R.E. (1961) A new approach to linear filtering and prediction problems, *Transactions of the ASME, Series D*, **83**, 35–44.

Lawless, A.S. (2001) *Development of Linear Models for Data Assimilation in Numerical Weather Prediction*, Department of Mathematics, The University of Reading, Ph.D. Thesis.

Lorenc, A.C. (1986) Analysis methods for numerical weather prediction, *Quart. J. Roy. Met. Soc.*, **112**, 1177-1194.

Lorenc, A.C. (1988) Optimal nonlinear objective analysis, *Quart. J. Roy. Met. Soc.*, **114**, 205–240.

Lorenc, A.C., Bell, R.S. and Macpherson, B. (1991) The Met. Office analysis correction data assimilation scheme, *Quart. J. Roy. Met. Soc.*, **117**, 59–89.

Martin, M.J., Nichols, N.K. and Bell, M.J. (1999) *Treatment of Systematic Errors in Sequential Data Assimilation*, Meteorological Office, Ocean Applications Division, Tech. Note, No. 21.

Rabier, F., Courtier, P., Pailleux, J., Talalgrand, O. and Vasiljevic, D. (1993) A comparison between four-dimensional variational assimilation and simplified sequential assimilation relying on three-dimensionsal variational analysis, *Quart. J. Roy. Met. Soc.*, **119**, 845–880.

Rabier, F., Järvinen, H., Klinker, E., Mahfouf, J.-F. and Simmons, A. (2000) The ECMWF operational implementation of four-dimensional variational assimilation. I: Experimental results with simplified physics, *Quart. J. Roy. Met. Soc.*, **126**, 1143-1170.

Sasaki, Y. (1970) Some basic formulisms on numerical variational analysis, *Monthly Weather Review*, **98**, 875–883.

Talagrand, O. (1981) A study on the dynamics of four-dimensional data assimilation, *Tellus*, **33**, 43–60.

Thacker, W.C. (1996) Relationships between statistical and deterministic methods of data assimilation, in Y.K. Sasaki (ed.), *Variational Methods in the Geosciences*, Elsevier, New York.

BAYESIAN ESTIMATION. OPTIMAL INTERPOLATION. STATISTICAL LINEAR ESTIMATION

O. TALAGRAND

Laboratoire de Météorologie Dynamique / CNRS
École Normale Supérieure, Paris, France

1. Introduction

The purpose of data assimilation can be described as to evaluate as accurately as possible the state of the atmospheric (or oceanic) flow, using all available relevant information. Depending on the particular application that is being considered, one may want to evaluate the state of the flow at a given time, or alternatively the evolution of the flow over a given period of time. As for the available information, it essentially consists of two components:

(a) The observations proper, which may be distributed more or less regularly in both space and time, and may vary in nature, accuracy, as well as in spatial and temporal resolution (the system of meteorological observations is described in this book in the chapter *Observing the Atmosphere* by R. Swinbank). Some observations may be 'direct' in that they bear on the physical quantities, such as temperature, pressure or velocity components, which are to be used for describing the state of the flow. Others may on the contrary be 'indirect', in that they bear on complex combinations (often, line integrals in space) of those physical quantities of interest. This is for instance the case of radiances measured by satellites, which are weighted integrals of temperature and/or humidity profiles. The use of such 'indirect' observations will necessarily require some form of 'inversion', either explicit or implicit. And all observations, whether direct or indirect, will be affected by some form of errors. These errors may be ordinary instrumental errors. They may also be sampling, or *representativeness* errors, due to the fact that the observations are not performed at the spatial and temporal resolution at which one wants to describe the flow.

(b) The physical laws governing the evolution of the flow. These are particularly important, since in many situations, one is led to use observations that are distributed over periods of time over which the evolution of the flow cannot be neglected. It is actually for referring to such situations that the word assimilation is particularly used, the word *analysis* being preferably used when dealing with observations that are synchronous, or distributed over a period of time short enough so that the evolution of the flow can be neglected. The physical laws governing the flow are in practice available in the form of a discretised, and necessarily

R. Swinbank et al. (eds.), Data Assimilation for the Earth System, 21–35.

approximate, numerical model (see, for example, the chapter *Atmospheric modelling* by W. Lahoz). Assimilation can therefore be described as a process through which observations are merged together with a numerical model of the dynamical evolution of the flow.

If observations on the one hand, and relevant physical laws on the other, constitute the basic information to be processed in assimilation, additional sources of information must also often be used. The most common example has to do with geostrophic balance, which dominates the midlatitude atmospheric and oceanic flow. If appropriate measures are not taken to ensure that the estimated fields are also in geostrophic balance, the consequence may be the presence of unrealistic high frequency oscillations in those fields. Such oscillations may be particularly damageable, for instance in the early phases of a numerical weather prediction. To be true, geostrophic balance is a consequence of the physical laws governing the flow, and even moderately realistic models of the atmospheric or oceanic flow do asymptotically tend to geostrophic balance. But assimilation is most often performed over periods of time that are short enough so that the model cannot by itself filter out observational noise, making it necessary to impose as an additional explicit constraint that the estimated flow must be in appropriate balance. This particular aspect of assimilation, known as *initialisation*, is discussed in the two chapters by P. Lynch. Similar questions, which have so far not been studied in as much detail as the question of geostrophic balance, arise in other aspects of the atmospheric or oceanic flow, for instance concerning the balance between the atmospheric temperature and humidity fields.

Data assimilation is one of many *inverse problems* that are encountered in many fields of science and technology. Two features are specific to assimilation of meteorological and oceanographical observations, and make it a particularly difficult problem. Firstly, the dynamics of the atmosphere and the ocean which, as has been said, must explicitly be taken into account in the assimilation process, is highly nonlinear, and even chaotic. Secondly, the numerical dimension of the problems to be solved is extremely large. As an example, the state dimension of present numerical weather prediction models (*i.e.*, the number of independent parameters which define the model flow at a given time, and must be explicitly specified in order to start a numerical prediction) lies now in the range 10^6-10^7, while the number of individual meteorological observations that are introduced into the assimilation over a 24-hour period lies in the range 10^5-10^6. Both numbers, especially the latter, can be expected to increase significantly in the coming years. If one takes into account the additional unavoidable constraint that the forecast must every day be ready in time ('no right for a second try'), those numbers give an idea of the demands for efficiency and robustness that are imposed on algorithms for operational assimilation of meteorological observations.

2. Bayesian Estimation

All data used in assimilation will always be affected with some uncertainty, as will be the fields produced by the assimilation. It is of course highly desirable to quantify that uncertainty, in order first to be able to weight appropriately the data in the estimation process, and second to know the degree of reliability that can be granted to the estimated fields. A convenient way to quantify uncertainty is through probabilities. It is thus natural to describe the data on the one hand, the estimated fields on the other, in terms of *probability distributions*. This then leads to state the problem of assimilation in terms of *conditional*, or *bayesian*, estimation. In that perspective, and to put it briefly, the purpose of assimilation is to determine the conditional probability distribution of the state of the flow, given the data.

As a simple example, consider the following situation. A scalar quantity x is to be determined from two observations of the form

$$z_1 = x + \zeta_1 \tag{2.1a}$$

$$z_2 = x + \zeta_2 \tag{2.1b}$$

In these expressions, ζ_1 and ζ_2 are observational errors, whose exact values are of course unknown, but which are assumed to be gaussian, mutually independent, each with expectation 0, and with respective variances s_1^2 and s_2^2, *i.e.* with respective probability density functions

$$p_1(\zeta) \propto \exp[- (\zeta^2)/2s_1^2] \tag{2.2a}$$

$$p_2(\zeta) \propto \exp[- (\zeta^2)/2s_2^2] \tag{2.2b}$$

where $\propto$ denotes proportionality. The unknown x will be equal to a given value ξ if, and only if, $\zeta_1 = z_1 - \xi$ and $\zeta_2 = z_2 - \xi$. The corresponding probability is equal to the product $p_1(z_1 - \xi) \, p_2(z_2 - \xi)$, which can be written, after standard algebraic manipulations, as

$$p_1(z_1 - \xi) \, p_2(z_2 - \xi) \propto \exp[- (\xi - x^a)^2/2s^2] \tag{2.3}$$

where

$$x^a = s^2 (z_1/s_1^2 + z_2/s_2^2] \tag{2.4a}$$

$$1/s^2 = 1/s_1^2 + 1/s_2^2 \tag{2.4b}$$

These equations define the conditional, or *a posteriori* probability distribution for x, given z_1 and z_2. The corresponding density will be denoted $p(x=\xi \,|\, z_1, z_2)$. It is equal, up to a multiplicative normalisation constant, to the expression in the right-hand side of eq. (2.3), which shows that the *a posteriori* distribution is gaussian, with expectation x^a and variance s^2. If we call *accuracy* the inverse of a variance, eqs (2.4) show that the conditional expectation x^a is the average of the two observations z_1 and z_2, weighted in proportion of their accuracies, and that the accuracy of the conditional distribution is the sum of the accuracies of the individual observations. This implies that the *a posteriori* variance s^2 is less than either one of the variances s_1^2 or s_2^2. A noteworthy feature is also that the *a posteriori* variance is independent of the observations z_1 and z_2. When the quantity x, or the errors ζ_1 and ζ_2, vary, the conditional probability distribution is translated along the x-axis, but does not change shape.

For another example, assume now that the errors ζ_1 and ζ_2 in (2.1), while still being independent, are both distributed according to the same exponential law

$$q(\zeta) = (1/2a) \exp(-|\zeta|/a) \tag{2.5}$$

where a is a constant. This distribution has expectation 0 and variance $2a^2$. The same computation as before shows that the *a posteriori* probability density is now constant over the interval between the observations z_1 and z_2, and varies like $\exp(-2|\xi|/a)$ outside that interval. More precisely, assuming without loss of generality that $z_1 \leq z_2$:

$$p(x=\xi | z_1, z_2) = A \exp[-2(z_1-\xi)/a] \qquad \text{if } \xi < z_1 \tag{2.6a}$$

$$p(x=\xi | z_1, z_2) = A \qquad \text{if } z_1 \leq \xi \leq z_2 \tag{2.6b}$$

$$p(x=\xi | z_1, z_2) = A \exp[-2(\xi-z_2)/a] \qquad \text{if } z_2 < \xi \tag{2.6c}$$

where the normalisation constant A is equal to $1/(a+z_2-z_1)$. The corresponding expectation is equal to $(z_1+z_2)/2$, while the variance is equal to

$$\text{Var}(x|z_1, z_2) = a^2 (2\delta^3/3 + \delta^2 + \delta + 1/2) / (1 + 2\delta) \tag{2.7}$$

where $\delta = (z_2-z_1)/(2a)$. The *a posteriori* variance now depends on the values of the observations. It is minimum, equal to $a^2/2$ (*i.e.*, to half the value that would obtain for gaussian errors with the same variance), when $\delta=0$. It increases with increasing δ, to become infinite for infinite δ. This means that the *a posteriori* variance can be *larger* than the *a priori* variance. This result, which may be deemed paradoxical, only means that variance alone may not be a good measure of uncertainty, and that estimation of uncertainty requires care and discernment.

A general definition of a conditional probability distribution is as follows. We assume from now on that we want to determine a vector x which describes the state of the flow. The vector x belongs to *state space*, denoted by S, with dimension n. The available data make up a vector z, with dimension m, belonging to *data space* D. The data and state vectors are linked through a relation of the form

$$F(z, x, \zeta) = 0 \tag{2.8}$$

where ζ is a random vector representing uncertainty. Eqs (2.1), with $x = x$, $z = (z_1, z_2)^T$, $\zeta = (\zeta_1, \zeta_2)^T$, are of form (2.8). The conditional probability that x be equal to some given value ξ is clearly proportional to the probability that ζ be such that $F(z, \xi, \zeta) = 0$. The proportionality coefficient must be such that the conditional probabilities, integrated over all possible values of ξ, sum up to 1. Thus

$$p(x=\xi | z) = p[F(z, \xi, \zeta) = 0] / \int p[F(z, \xi', \zeta) = 0] \, d\xi' \tag{2.9}$$

The term on the right-hand side is unambiguous only if, for any ζ, there is at most one value of ξ such that $F(z, \xi, \zeta) = 0$. If there were several such values, it would not be known to which one the probability of ζ must be assigned. The condition that, for given ζ, $F(z, \xi, \zeta) = 0$ for at most one value of ξ means there are enough data to determine all components of x. It will be called the *determinacy* condition.

Eq. (2.1a), provided a probability distribution is known for the error ζ_1, defines a probability distribution for x in which z_1 is a constant. If a joint probability distribution is known for the vector $(\zeta_1, \zeta_2)^T$, a probability distribution is then defined for the variable $y = x + \zeta_2$, for which eq. (2.2a) defines in addition a numerical value. More generally, if the determinacy condition is verified, a data vector z of the form (2.8) will define a joint probability distribution, which we denote $p(x, y)$, for some vector $(x^T, y^T)^T$, where x is the state vector, and y is a complementary set of data. It will also define a numerical value for the vector y. In these conditions, the conditional probability distribution for x, given y, is equal to (using simplified, but unambiguous notations)

$$p(x \mid y) = \frac{p(x, y)}{p(y)} \qquad (2.10)$$

where $p(y) \equiv \int p(x, y) \, dx$. Inverting the roles of x and y, this can be rewritten as

$$p(x \mid y) = \frac{p(y \mid x) \, p(x)}{p(y)} \qquad (2.11)$$

where $p(x) \equiv \int p(x, y) \, dy$. Eq. (2.11) expresses what is known as *Bayes' theorem*. It often provides the most convenient way for defining the conditional probability distribution for the state vector x. Eq. (2.10) is actually often taken as the definition of a conditional probability distribution. It however requires an *a priori* explicit probability distribution $p(x, y)$ which, as shown by the example of eqs (2.8-9), can remain implicit.

As a final example, we consider the case of a data vector z of the form

$$z = \Gamma x + \zeta \qquad (2.12)$$

where Γ is a known $(m \times n)$-matrix. The determinacy condition is then that the null space of Γ is void or, equivalently, that $\mathrm{rank}\,\Gamma = n$. We assume the 'error' ζ to be gaussian, with expectation 0 and covariance matrix S. It can be shown that the conditional probability distribution for x, given z, is then also gaussian, with expectation

$$x^a = (\Gamma^T S^{-1} \Gamma)^{-1} \Gamma^T S^{-1} z \qquad (2.13a)$$

and covariance matrix

$$P^a = (\Gamma^T S^{-1} \Gamma)^{-1} \qquad (2.13b)$$

These formulæ generalise eqs (2.3-4). Two remarkable properties can be mentioned.

Firstly, the expectation x^a depends linearly on the data vector z. Secondly, and analogously to what was already noted on eq. (2.4b), the covariance matrix P^a is independent of z.

Bayesian estimation defines a systematic and rigorous approach to assimilation of observations. Its full-scale implementation, at least in applications such as are encountered in meteorology and oceanography, is however simply impossible. As already said, numerical models for large-scale meteorological prediction now have state dimension n in the range 10^6-10^7. The simplest way to numerically define a probability distribution over a state with dimension n is to divide each of the n coordinate axes into a number of intervals, and to specify the integrated probability on each of the volume elements built on those intervals. Assuming that each coordinate axis is divided into 10 intervals (which may arguably be said to be very crude), this defines 10^n volume elements, which, for the values of n that have just been mentioned, goes well beyond the capabilities of any present computer. And even if the necessary computer power became available, it would not be possible, at least for a long time, to put it to an efficient use. The uncertainty affecting many of the data to be used in the assimilation is itself very poorly known. Little is quantitatively known for instance on the errors affecting numerical weather prediction models, whose lifetime is much shorter than the time necessary to accumulate reliable statistics on their performance. So, even if enough computing power was available, it would not be possible, at least in the present state of knowledge of the errors affecting the data, to estimate the conditional probability distribution $p(x|z)$ with any degree of accuracy.

These considerations do not mean that the bayesian approach is useless. But it is useful only to the extent that it defines general guidelines along which to develop an assimilation system, and along which to evaluate its results. In that respect, it is absolutely necessary. But it is also necessary, in view of the stringent limitations that are encountered in any practical application, to make drastic simplifying assumptions in the definition of an assimilation algorithm. Two main lines have so far been followed in that respect.

In the simplest approach, which began with assimilation itself, what is being sought is simply a reasonable approximation of the state of the flow, together with some crude estimate of the associated error. Most algorithms developed along that approach are based on *statistical linear estimation*, which uses formulæ of form (2.13), with innumerable variants and simplifying assumptions as to the definition of the matrices Γ and S, and as to the numerical algorithms used for computing the 'analysis' x^a and (a small number of components of) the associated error covariance matrix P^a. Most standard assimilation algorithms, such as 'optimal interpolation', Kalman filter and smoother, variational analysis and assimilation (both in their three-dimensional, '3D-Var', and four-dimensional, '4D-Var', forms, and in their 'primal' and 'dual' formulations), are built on statistical linear estimation. The various (and fundamentally equivalent) forms of statistical linear estimation will be described in the following sections.

It is only much more recently that a significantly different form of assimilation, namely *ensemble assimilation*, has been developed. The output of ensemble assimilation is not an 'analysed' state together with some estimate of the associated error, but an ensemble of points in state space, whose dispersion is meant to represent the conditional probability distribution $p(x|z)$. Ensemble assimilation is thus a form of Monte-Carlo approximation. In present applications, the size of the analysed ensembles lies typically in the range from a few tens to a few hundreds. The great advantage of ensemble assimilation is that it fundamentally eliminates any need for a linear approximation of form (2.12). Ensemble assimilation will not be described in these notes. Basic references are the papers by Evensen and van Leeuwen (2000) and Houtekamer and Mitchell (2001).

3. Statistical Linear Estimation. Optimal Interpolation.

3.1 A REMINDER OF A FEW BASIC FACTS ABOUT RANDOM VARIABLES, VECTORS AND FUNCTIONS

This sub-section presents a number of basic facts concerning random variables, vectors and functions. Mathematical expectation will be denoted by $E(.)$. A random variable is said to be *centred* if it has expectation equal to 0. Given a random variable x, we will denote $x' = x - E(x)$. Given two random scalar variables x and y, the *covariance* of x and y is the expectation $E(x'y')$. Covariance, being linear with respect to each of its two arguments, defines an inner product on the space of random variables with finite variance. A Euclidean structure, and the associated geometry, are thus defined on that space. *Orthogonality* is then equivalent to decorrelation $[E(x'y') = 0]$. All the notions of standard geometry, in particular the notion of an orthogonal projection onto a subspace, are valid in that geometric description.

Given two random vectors x and y, with respective dimensions n and p, and respective coordinates $(x_i)_{i=1,...,n}$ and $(y_l)_{l=1,...,p}$, the *covariance matrix* of x and y is the $(n{\times}p)$-matrix with entries $E(x_i'y_l')$. It will be denoted by $E(x'y'^T)$. The covariance matrix $E(x'x'^T)$ of a vector x with itself is symmetric non-negative. It is singular if, and only if, a linear relationship holds with probability 1 between the components of x.

Consider now a *random function* defined over some spatial domain $\mathcal{R}$, *i.e.* a collection of random variables, denoted $\phi(\xi)$, where ξ belongs to $\mathcal{R}$ (in meteorological or oceanographical applications, $\mathcal{R}$ is a two- or three-dimensional Euclidean space, or the surface of the terrestrial sphere). We assume the variables $\phi(\xi)$ to be centred. The *covariance function* of the random function ϕ is defined for any two points ξ_1 and ξ_2 belonging to $\mathcal{R}$ by

$$C_\phi(\xi_1, \xi_2) \equiv E[\phi(\xi_1)\,\phi(\xi_2)] \tag{3.1}$$

The covariance function $C_\phi(\xi_1, \xi_2)$ is said to be *homogeneous* if it is a function of the vector difference $\xi_2 - \xi_1$ only, *i.e.* if $C_\phi(\xi_1, \xi_2) = H(\xi_2 - \xi_1)$. On the surface of the sphere, along which the notion of a vector difference is not defined, a covariance function $C_\phi(\xi_1, \xi_2)$ is said to be homogeneous if it depends only on the angular distance between ξ_1 and ξ_2.

The notion of a covariance function generalises the notion of a covariance matrix. A covariance function $C_\phi(\xi_1, \xi_2)$ defined on a space $\mathcal{R}$ has the property that, given any N

points ξ_1, ..., ξ_N in $\mathcal{R}$, the ($N \times N$)-matrix with entries $C_\phi(\xi_i, \xi_j)$ ($i, j = 1, ..., N$) is symmetric non-negative. That property characterises covariance functions, and functions that possess it are said to be of *positive type*. Homogeneous functions of positive type, *i.e.* functions of positive type of the form $C(\xi_1, \xi_2) = H(\xi_2 - \xi_1)$ are in turn characterised by the fact that the Fourier transform of H is real and non-negative. This important result, which identifies a large class of covariance functions, is known as the *Bochner-Khintchin theorem*. Fourier transforms are not defined along the surface of a sphere, but an analogous result holds in terms of expansion in Legendre polynomials. More precisely, let $C(\xi_1, \xi_2) = H(\delta)$, where δ is the angular distance between ξ_1 and ξ_2, be a function defined along the surface of a sphere. It is of positive type if and only $H(\delta) = \Sigma_{n=0, ...} a_n P_n(\cos\delta)$, where P_n is the Legendre polynomial of degree n, and a_n's ≥ 0.

Given two random functions $\phi(\xi)$ and $\psi(\xi)$, the *cross-covariance function* of ϕ and ψ is defined as

$$C_{\phi\psi}(\xi_1, \xi_2) \equiv E[\phi(\xi_1)\ \psi(\xi_2)] \tag{3.2}$$

3.2 OPTIMAL INTERPOLATION

Consider the following problem. A meteorological field, denoted by $\phi(\xi)$, has been observed at p locations ξ_j ($j = 1, ..., p$), providing p measurements of the form

$$y_j = \phi(\xi_j) + \varepsilon_j \tag{3.3}$$

where ε_j is an observational error. We will denote by $\boldsymbol{y} = (y_1, ..., y_p)^\mathsf{T}$ the vector consisting of the observations. We want to use those observations to determine the value $x = \phi(\xi)$ at some point ξ. We look for x in the form

$$x^a = \alpha + \Sigma_j \beta_j y_j = \alpha + \beta^\mathsf{T} \boldsymbol{y} \tag{3.4}$$

where β is a p-vector with components $\beta_1, ..., \beta_p$. The constant quantities α and β are to be determined so as to minimise in some sense the estimation error $x^a - x$. More precisely, we look for the values of α and β which minimise the statistical quadratic error $E[(x^a - x)^2]$. In terms of the geometrical interpretation of covariance, this amounts to looking for the point of form (3.4) which lies closest to x. The points of form (3.4) span the space $\mathcal{P}$ defined by the variables y_j and the constant random variable. The value that is being looked for is therefore the orthogonal projection, in the sense of covariance, of x onto the space $\mathcal{P}$. It is characterised by the fact that the error $x^a - x$ must be uncorrelated with the y_j's and the constant random variable. The condition that the error is uncorrelated with the constant random variable reads

$$E(x^a - x) = 0 \tag{3.5}$$

This equation expresses that the estimate x^a is *unbiased*. Carried into eq. (3.4), it leads successively to

$$\alpha = E(x) - \Sigma_j \, \beta_j \, E(y_j) \qquad (3.6)$$

and

$$x^a = E(x) + \Sigma_j \, \beta_j \, y_j' \qquad (3.7)$$

where, as already said, a prime (') denotes a centred variable. The condition that x^a-x is uncorrelated with $y_l = y_l' + E(y_l)$ reads

$$\Sigma_j \, \beta_j \, E(y_j'y_l') = E(x'y_l') \qquad (3.8)$$

or, in matrix form

$$E(y'y'^T)\beta = E(x'y') \qquad (3.9)$$

Solving this equation for β, and carrying the result into eq. (3.7) lead to

$$x^a = E(x) + E(x'y'^T) \, [E(y'y'^T)]^{-1} \, y' \qquad (3.10)$$

As for the error $E[(x^a$-$x)^2]$, which is minimised by the choice of β, the Pythagorean theorem shows that it is equal to

$$E[(x^a$-$x)^2] = E(x^2) - E[(x^a)^2] = E(x'^2) - E(x'y'^T) \, [E(y'y'^T)]^{-1} \, E(y'x') \qquad (3.11)$$

The meaning of those two equations is clear. In the absence of any knowledge on y, the variance minimizing estimate of x is simply the expectation $E(x)$. The second term on the right-hand side of eq. (3.10) thus expresses the change in our knowledge of x resulting from the knowledge of y. That term is 0 when x and y are uncorrelated $[E(x'y'^T)=0]$. As for the error variance (3.11), it is equal, in the absence on any knowledge on y, to the variance $E(x'^2)$. The second term on the right-hand side of eq. (3.11) therefore represents the gain in the knowledge of x that results from the knowledge of y. That gain is also 0 when x and y are uncorrelated.

Coming back to the original description (3.3), assume the observational errors ε_j's are centred, have the same variance r, are mutually uncorrelated, and are uncorrelated with the observed field ϕ. The covariance $E(y_j'y_l')$ is then equal to

$$E(y_j'y_l') = E\{[\phi(\xi_j) + \varepsilon_j] \, [\phi(\xi_l) + \varepsilon_l]\} = C_\phi(\xi_j, \xi_l) + r\delta_{jl} \qquad (3.12a)$$

where δ_{jl} is the Kronecker symbol. The last expression on the right-hand side is obtained by developing the product in the expectation, and then taking the expectation of each term according to the hypotheses that have been made on the errors ε_j's. Similarly the covariance $E(x'y_l')$ is equal to

$$E(x'y_l') = C_\phi(\xi, \xi_l) \qquad (3.12b)$$

It is thus seen that the determination of the analysed value (3.10) requires the prior knowledge of the expectation and covariance function of the field ϕ, and of the observational errors ε_i's (possible covariances between the field ϕ and the observational errors must also be taken into account).

The estimation procedure that has just been described, summed up in eqs (3.10-11), is known in meteorological and oceanographical applications as *optimal interpolation*. Originally introduced in meteorology by Eliassen (1954), and further developed by Gandin (1965), it has played a major role in the development of numerical weather prediction. An important aspect of optimal interpolation is of course the need for *a priori* specification of the covariance function of the field to be estimated.

Optimal interpolation is said to be *univariate* if, as in the description given above, the observations bear on the field to be estimated. It is said to be *multivariate* if observations of one physical field are used for estimating one or several other fields (for instance, if observations of the wind field are used to determine the geopotential, or *vice-versa*). Multivariate optimal interpolation requires the specification of the cross-covariance functions between the observed fields and the fields to be estimated.

Optimal interpolation possesses many interesting properties, all of which will not be discussed here. We refer to Lorenc (1981) for a detailed discussion of many aspects of optimal interpolation. Let us denote by $\gamma = (\gamma_j)_{j=1, \ldots, p}$ the vector $[E(y'y'^T)]^{-1} y'$ which appears on the right-hand side of eq. (3.10). It is independent of the variable x to be estimated, which appears only in the matrix $E(x'y'^T)$. Denoting now by $\phi^a(\xi)$ the field value estimated at point ξ, eq. (3.10) becomes, taking eq. (3.12b) into account

$$\phi^a(\xi) = E[\phi(\xi)] + \Sigma_{j=1, \ldots, p} \, \gamma_j \, C_\phi(\xi, \, \xi_j) \tag{3.13}$$

The analysed field $\phi^a(\xi)$ (or more precisely its deviation from its expectation) is therefore a linear combination of the p functions $C_\phi(\xi, \, \xi_j)$. Each observation y_l influences the analysed field through a spatial pattern that coincides, up to a multiplicative constant, with the covariance $C_\phi(\xi, \, \xi_l)$. The latter is called the *representer* associated with the observation y_l.

Optimal interpolation thus appears as a form of *function fitting*, in which the estimated field is a linear combination of a set of basic functions. But unlike many approaches in which the basic functions are chosen *a priori*, the condition of optimality, in the sense of least-variance, imposes here that the basic functions must be 'cross-sections' of the covariance function of the field under consideration. This property, and the notion of representers, extends easily to the case of multivariate optimal interpolation. The representers are then cross-covariance functions.

Eqs (3.10) and (3.11) have been written for the estimation of one scalar variable. They easily extend to the case when it is an entire state vector x, with dimension n (for instance the values of a given physical field at the grid-points of a numerical weather prediction model), that has to be estimated. With obvious notations, eq. (3.10) becomes

$$x^a = E(x) + E(x'y'^T) \, [E(y'y'^T)]^{-1} \, y' \tag{3.14}$$

while the covariance matrix of the associated estimation error reads

$$\boldsymbol{P}^a \equiv E[(\boldsymbol{x}^a\text{-}\boldsymbol{x})\,(\boldsymbol{x}^a\text{-}\boldsymbol{x})^\mathsf{T}] = E(\boldsymbol{x}\boldsymbol{x}^\mathsf{T}) - E[(\boldsymbol{x}^a)\,(\boldsymbol{x}^a)^\mathsf{T}] = E(\boldsymbol{x}'\boldsymbol{x}'^\mathsf{T}) - E(\boldsymbol{x}'\boldsymbol{y}'^\mathsf{T})\,[E(\boldsymbol{y}'\boldsymbol{y}'^\mathsf{T})]^{-1}\,E(\boldsymbol{y}'\boldsymbol{x}'^\mathsf{T})$$

$$(3.15)$$

Eq. (3.14) contains no more than eq. (3.10), taken over all components of $\boldsymbol{x}$. But eq. (3.15), which is an ($n\times n$)-matrix equation, contains more than eq. (3.11). In addition to the variances of the estimation errors on the components of $\boldsymbol{x}$, which make up its diagonal, its off-diagonal entries are the covariances between the estimation errors on the different components of $\boldsymbol{x}$.

4. The *Best Linear Unbiased Estimate* and its Various Forms

Consider again the problem of determining the state vector $\boldsymbol{x}$ from a data vector $\boldsymbol{z}$ of the form given by eq. (2.12), which we rewrite here

$$\boldsymbol{z} = \Gamma\boldsymbol{x} + \zeta \tag{4.1}$$

We assume that the determinacy condition rank $\Gamma = n$ is verified. As concerns the 'error' ζ, we only assume it is centred and has finite covariance matrix $\boldsymbol{S}$, but with no other hypothesis as to its probability distribution. We look for an estimate of x which depends linearly on ζ, subject to the following two conditions

(a) The estimate is independent of the choice of the origin in state space
(b) The variance of the corresponding estimation error is minimum.

The solution to that problem is the estimate

$$\boldsymbol{x}^a = (\Gamma^\mathsf{T}\boldsymbol{S}^{-1}\,\Gamma)^{-1}\,\Gamma^\mathsf{T}\boldsymbol{S}^{-1}\,\boldsymbol{z} \tag{4.2a}$$

The covariance matrix of the corresponding error is

$$\boldsymbol{P}^a = (\Gamma^\mathsf{T}\boldsymbol{S}^{-1}\,\Gamma)^{-1} \tag{4.2b}$$

These equations are identical with eqs (2.13) obtained in the case when the error ζ is gaussian, but *their meaning is different*. While eqs (2.13) define a complete conditional probability distribution for given z, eq. (4.2a) defines only one particular estimate for x. That estimate, depending on the probability distribution of ζ, may, or may not, coincide with the conditional expectation of x. As for eq. (4.2b), it defines the covariance of the estimation error $x\text{-}x^a$, taken over all possible realisations of the error ζ. The difference, in case the error ζ is not gaussian, between conditional estimation and linear estimation as given by eqs (4.2), is visible on data of form (2.1), with exponential probability distribution (2.5) for ζ. The conditional expectation, which is then equal to $(z_1+z_2)/2$, depends linearly on the data, and is equal to the estimate (4.2a). But the conditional variance (2.7), which depends on the difference $z_2\text{-}z_1$, is not equal to the value given by eq. (4.2b) (that value is equal to a^2). It is only the expectation of the variance (2.7) over all values of the difference z_2-z_1 which is equal to the value (4.2b).

The estimate (4.2a) is called the *Best Linear Unbiased Estimate*, or *BLUE*, of the state vector x from the data vector z. It is defined independently of whether or not the error ζ is gaussian. When ζ is gaussian, the *BLUE* coincides with the conditional expectation of x. Because of its mathematical simplicity and its relatively low computational cost, the *BLUE* is very often used in estimation problems of many kinds. In fact, unless it is quantitatively known by how much the probability distribution for the error ζ differs from gaussianity, there is no reason to resort to anything more complicated or costly than the *BLUE*.

It is easily verified that the *BLUE* (4.2a) minimizes the following scalar function, defined on state space

$$\mathcal{J}(\xi) \equiv (1/2) \, [\Gamma\xi - z]^{\mathrm{T}} \, S^{-1} \, [\Gamma\xi - z] \tag{4.3}$$

The significance of this expression is clear. For given ξ belonging to state space, $\Gamma\xi$ is what an exact data acquisition system, if applied on ξ, would produce. The difference $\Gamma\xi - z$ is therefore the misfit between the data z and the data that would correspond to ξ. $\mathcal{J}(\xi)$ measures the magnitude of that misfit, normalized by the inverse covariance matrix S^{-1}. This choice of normalization makes the quantity $\mathcal{J}(\xi)$ physically non-dimensional, thus allowing to combine together data of different physical nature. It also weights the data in inverse proportion to the errors that affect them, thus giving a large weight to accurate data, and a small weight to inaccurate ones. The function $\mathcal{J}(\xi)$ will be called the *objective function* associated with the estimation of the state vector x from the data vector z.

An additional very important property of the *BLUE*, which can be verified, *e.g.*, from expression (4.2a), is that it is invariant in a linear change of coordinates in either data space or state space. This also means that the *BLUE* is independent of the choice of a norm in either space.

In many applications, especially in meteorological or oceanographical applications, the data vector z fundamentally consists of two parts

(a) A *prior*, or *background* estimate for the state vector x, of the form

$$x^b = x + \zeta^b \tag{4.4a}$$

where ζ^b is an error. The background estimate is usually a forecast coming from the past. But it can be for instance a climatological estimate.

(b) An additional set of data, making up a vector y with dimension $p = m$-n, of the form

$$y = Hx + \varepsilon \tag{4.4b}$$

where H is a known matrix, and ε is an error. The vector y is often a set of observations, but may contain other data, for instance numerical values assigned to 'balance' constraints. For convenience, we will call it the *observation vector*, belonging to *observation space*.

It is often legitimate to assume that the background and observation errors are uncorrelated

$$E(\zeta^b \varepsilon^{\mathrm{T}}) = 0 \qquad (4.4c)$$

Assuming the data to be of the form (4.4a-c) may seem restrictive. It is not. If the determinacy condition rank $\Gamma = n$ is verified, the data vector z in eq. (4.1) can always be transformed to the form (4.4a-c) (this is a particular case of a general remark made in Section 2 before introducing eq. 2.10).

Setting

$$\boldsymbol{B} \equiv E(\zeta^b \zeta^{b\mathrm{T}}), \qquad \boldsymbol{R} \equiv E(\varepsilon \varepsilon^{\mathrm{T}}) \qquad (4.5)$$

and using (4.4c), eqs (4.2) yield

$$x^a = [\boldsymbol{B}^{-1} + \boldsymbol{H}^{\mathrm{T}} \boldsymbol{R}^{-1} \boldsymbol{H}]^{-1} [\boldsymbol{B}^{-1} x^b + \boldsymbol{H}^{\mathrm{T}} \boldsymbol{R}^{-1} y] \qquad (4.6a)$$

and

$$\boldsymbol{P}^a = [\boldsymbol{B}^{-1} + \boldsymbol{H}^{\mathrm{T}} \boldsymbol{R}^{-1} \boldsymbol{H}]^{-1} \qquad (4.6b)$$

Eq. (4.6a) is in turn transformed into

$$x^a = x^b + \boldsymbol{P}^a \boldsymbol{H}^{\mathrm{T}} \boldsymbol{R}^{-1} [y - \boldsymbol{H} x^b] \qquad (4.6c)$$

In this equation, the analysed field x^a is expressed as the sum of the background x^b and of a correction which is itself proportional to the vector

$$d \equiv y - \boldsymbol{H} x^b \qquad (4.7)$$

That vector is the difference between the observation vector and what the observation operator $\boldsymbol{H}$ would produce if it were applied on the background. It is therefore the discrepancy between the observation and the background. If the vector d was equal to 0, *i.e.* if the observation was exactly compatible with the background, there would no reason to modify the latter. It is therefore natural, in the linear approach followed here, that the correction applied on the background be a linear function of the vector d. That is exactly what is expressed by eq. (4.6c). The vector d is called the *innovation vector*. As for the matrix $\boldsymbol{P}^a \boldsymbol{H}^{\mathrm{T}} \boldsymbol{R}^{-1}$ that multiplies the innovation vector, it is called the *gain matrix*.

The objective function (4.3) is transformed through (4.4-5) into

$$\mathscr{J}(\xi) = (1/2) [\xi - x^b]^{\mathrm{T}} \boldsymbol{B}^{-1} [\xi - x^b] + (1/2) [\boldsymbol{H}\xi - y]^{\mathrm{T}} \boldsymbol{R}^{-1} [\boldsymbol{H}\xi - y] \qquad (4.8)$$

If probability distributions are known for ζ^b and ε in eqs (4.4), probability distributions are defined for x and y, and it is possible to estimate x from y through orthogonal projection of form (3.14-15). The result is

$$x^a = x^b + BH^T[HBH^T + R]^{-1}[y - Hx^b] \qquad (4.9a)$$

$$P^a = B - BH^T[HBH^T + R]^{-1}HB \qquad (4.9b)$$

These equations are exactly equivalent to eqs (4.6). Eq. (4.9a) is of the same form as eq. (4.6c), in that it defines the analysed state as the sum of the background and of a correction that is proportional to the innovation vector. But the gain matrix has a different expression. Similarly, eq. (4.9b) gives an expression for the analysis error covariance matrix which is different from, but exactly equivalent to expression (4.6b).

The analysed state x^a can be determined by explicit minimisation of the objective function (4.8), which is defined on state space. Alternatively, the vector $[HBH^T + R]^{-1}$ $[y-Hx^b]$ that appears on the right-hand side of eq. (4.9a) can be obtained by maximisation of the following objective function

$$K(\mu) \equiv -(1/2)\,\mu^T[HBH^T + R]\,\mu + d^T\mu \qquad (4.10)$$

which is defined on the dual of the observation space (the minus sign in front of the quadratic term in 4.10, and the corresponding fact that the objective function has to be maximised, are conventional).

We have obtained four sets of different linear equations for the analysed state x^a and the associated error covariance matrix P^a, namely, eqs (3.14-15), (4.2), (4.6) and (4.9), respectively. In spite of apparent differences (for instance, data operators Γ and H are explicitly present in eqs 4.2 to 4.9, but not in 3.14-15), these four sets of equations are exactly equivalent, and define the *BLUE* of the state vector x from the available data. As already said, and with innumerable variants as to the choice of the data, the *a priori* specification of the first- and second-order statistical moments of the errors affecting those data, and the numerical algorithms that are used for actually carrying out the necessary computations, these equations are at the basis of most analysis and assimilation algorithms used in meteorology and oceanography. The only exceptions so far are first the already mentioned ensemble assimilation, in which some components at least cannot be described as particular applications of the general theory of the *BLUE*. And second some algorithms used for *quality control* of observations, *i.e.* for detection of erroneous observations (see Lorenc, 1997, and also the chapter *Observation Monitoring and Quality Control* by P. Gauthier in this volume). Apart from these exceptions, all presently existing analysis and assimilation algorithms can be described as particular applications of the *BLUE*. Some algorithms perform an explicit minimisation or maximisation of an objective function of form (4.3), (4.8) or (4.10). They are called *variational algorithms*.

References

Eliassen, A. (1954) *Provisional report on calculation of spatial covariance and autocorrelation of the pressure field*, Report no 5, Videnskaps-Akademiets Institutt for Vaer - Og Klimaforskning, Oslo, Norway, 12 pp., Reprinted in L. Bengtsson, M. Ghil, and E. Källén (eds.), 1981 *Dynamic Meteorology. Data Assimilation Methods*, Springer-Verlag, New York, USA, pp. 319-328.

Evensen, G. and van Leeuwen, P. J. (2000) An ensemble Kalman smoother for nonlinear dynamics, *Mon. Wea. Rev.* **128**, 1852-1867.

Gandin, L. S. (1965) *Objective Analysis for Meteorological Fields* (translated from the Russian), Israel Program for Scientific Translations, Jerusalem, Israel.

Houtekamer, P. L. and Mitchell, H. L. (2001) A sequential ensemble Kalman filter for atmospheric data assimilation, *Mon. Wea. Rev.* **129**, 123-137.

Lorenc, A. C. (1981) A global three-dimensional multivariate statistical interpolation scheme, *Mon. Wea. Rev.* **109**, 701-721.

Lorenc, A. C. (1997) Quality control, in Data Assimilation, Proceedings of Seminar, September 1996, ECMWF, Reading, pp. 251-274.

VARIATIONAL ASSIMILATION. ADJOINT EQUATIONS

O. TALAGRAND

Laboratoire de Météorologie Dynamique / CNRS
École Normale Supérieure, Paris, France

1. Variational assimilation

The basic theory of statistical linear estimation has been described in the chapter *Bayesian Estimation. Optimal Interpolation. Statistical Linear Estimation* (which will hereafter be referred to as Part I). A number of formulæ, which are mutually equivalent, have been established for the *BLUE* x^a of the state vector, given a vector of data z. Some of these formulæ express x^a as the minimiser of a scalar objective function defined on state space. If the data vector is of the form

$$z = \Gamma x + \zeta \qquad (1.1)$$

then the *BLUE* minimises the objective function

$$\mathcal{J}(\xi) \equiv (1/2) \, [\Gamma \xi - z]^{\mathrm{T}} \, S^{-1} \, [\Gamma \xi - z] \qquad (1.2)$$

where $S = E(\zeta \zeta^{\mathrm{T}})$ (see eqs 4.1 and 4.3 of Part I). If the data vector (1.1) consists of a background x^b and of an additional observation vector y, *viz.*,

$$x^b = x + \zeta^b \qquad (1.3a)$$

$$y = Hx + \varepsilon \qquad (1.3b)$$

with uncorrelated errors ζ^b and ε, then the objective function (1.2) takes the form

$$\mathcal{J}(\xi) = (1/2) \, [\xi - x^b]^{\mathrm{T}} \, B^{-1} \, [\xi - x^b] + (1/2) \, [H\xi - y]^{\mathrm{T}} \, R^{-1} \, [H\xi - y] \qquad (1.4)$$

where

$$B \equiv E(\zeta^b \zeta^{b\mathrm{T}}), \quad R \equiv E(\varepsilon \varepsilon^{\mathrm{T}}) \qquad (1.5)$$

(see eqs 4.4 and 4.8 of Part I). Alternatively, the *BLUE* corresponding to data (1.3) can be expressed as

$$x^a = x^b + BH^{\mathrm{T}} w \qquad (1.6)$$

R. Swinbank et al. (eds.), Data Assimilation for the Earth System, 37–53.
© 2003 *Kluwer Academic Publishers. Printed in the Netherlands.*

where the vector w maximises the objective function

$$\mathcal{K}(\mu) \equiv -(1/2)\,\mu^{\mathrm{T}}\,[HBH^{\mathrm{T}} + R]\,\mu + d^{\mathrm{T}}\mu \tag{1.7}$$

which is defined on the dual of the observation space, and where d is the innovation vector

$$d \equiv y - Hx^{b} \tag{1.8}$$

Variational assimilation consists in evaluating the analyzed state x^{a} by explicit minimisation (or maximisation) of an objective function of form (1.2) or (1.4) (or 1.7). A typical situation is the following: the background x^{b} in (1.3a) is a recent forecast valid at some time t, while the vector y in (1.3b) consists of observations performed at that same time. Minimisation of the objective function (1.4) will produce an analysis for that time. Variational assimilation, or rather variational analysis, when implemented in such a situation, is very often referred to as *3D-Var*, in order to stress that it solves a three-dimensional problem in space.

A different situation is encountered when one wants to assimilate observations that are distributed in time. Let us assume observations are available at successive times $k = 0, 1, \ldots, M$, of the form

$$y_{k} = H_{k}x_{k} + \varepsilon_{k} \tag{1.9}$$

where x_{k} is the exact state of the flow at time k, H_{k} is a linear observation operator and ε_{k} is an observational error, with covariance matrix R_{k}. The observational errors are assumed to be uncorrelated in time. Assume in addition that the temporal evolution of the flow is described by the equation

$$x_{k+1} = M_{k}x_{k} + \eta_{k} \tag{1.10}$$

where M_{k} is a known linear operator and η_{k} is a random variable. It is worth commenting briefly on the meaning of that equation. The operator M_{k} contains all what is known of the dynamics of the flow. In particular, it is all what is available in case one wants to make a forecast. It will be called the *model* of the flow. As for the variable η_{k}, it represents that part of the reality that the model cannot represent, and will be called the *model error*.

We finally assume that a background x_{0}^{b}, with error covariance matrix B, and error uncorrelated with the observational errors in eq. (1.9), is available at time 0.

The purpose of assimilation can then be stated as follows. Determine the history of the flow between times 0 and M, using all the available data, which consist of the observations (1.9), the dynamical equation (1.10) and the background x_{0}^{b}. A first approach is to ignore model error. Any initial condition ξ_{0} at time 0 defines a model solution

$$\xi_{k+1} = M_{k}\xi_{k}, \quad k = 0, \ldots, M\text{-}1 \tag{1.11}$$

The objective function

$$\mathcal{J}(\xi_0) = (1/2) \, [\xi_0 - x_0^b]^T \, B^{-1} \, [\xi_0 - x_0^b] + (1/2) \, \Sigma_{k=0,\ldots,M} \, [H_k\xi_k - y_k]^T \, R_k^{-1} \, [H_k\xi_k - y_k] \qquad (1.12)$$

which is of the general form (1.4), measures the distance between the model solution (1.11) and the data. Minimisation of $\mathcal{J}(\xi_0)$ will define the initial condition of the model solution that fits the data most closely. This is called *strong constraint four-dimensional variational assimilation*, often abbreviated in *strong constraint 4D-Var*. The words 'strong constraint' stress the fact that the model equation (1.11) must be exactly verified by the sequence of estimated state vectors.

Ignoring the model error is of course an approximation. If it is taken into account, eq. (1.10) defines an additional set of 'noisy' data. We assume the model error η_k in (1.10) to have covariance matrix Q_k, to be uncorrelated in time and to be uncorrelated with observation and background errors. Eq. (1.2) then gives the following expression for the objective function defining the *BLUE* of the sequence of states (x_k), $k=0,\ldots,M$

$$\begin{aligned}
\mathcal{J}(\xi_0, \ldots, \xi_M) = &\; (1/2) \, [\xi_0 - x_0^b]^T \, B^{-1} \, [\xi_0 - x_0^b] \\
&+ (1/2) \, \Sigma_{k=0,\ldots,M} \, [H_k\xi_k - y_k]^T \, R_k^{-1} \, [H_k\xi_k - y_k] \\
&+ (1/2) \, \Sigma_{k=0,\ldots,M-1} \, [\xi_{k+1} - M_k\xi_k]^T \, Q_k^{-1} \, [\xi_{k+1} - M_k\xi_k]
\end{aligned} \qquad (1.13)$$

The objective function is now a function of the whole sequence of states (ξ_k), $k=0,\ldots,M$. Minimisation of an objective function of the form (1.13), where the model equations are present as noisy data to be fitted by the analysed fields like any other data, is called *weak constraint variational assimilation*.

All objective functions (1.4), (1.12) and (1.13) are particular cases of the general expression (1.2). Function (1.7) is of a different form, but, as already said, its maximisation is equivalent to the minimisation of (1.4).

We have assumed, in agreement with the theory of linear estimation, the operator Γ in eq. (1.1) (or, equivalently, the operators H, H_k or M_k in eqs 1.3b, 1.9 or 1.10) to be linear. But expressions (1.4), (1.12) and (1.13) remain meaningful if those operators are nonlinear, even if they do not define exactly quadratic functions. And if the nonlinearity of the operators is in a sense small enough, minimisation of the corresponding objective functions is likely to produce a physically useful estimate of the state of the flow. One can note however that the dual objective function (1.7), whose expression explicitly contains the transpose H^T, is defined only for a linear observation operator H (which implies in particular, in case of four-dimensional assimilation, a linear model 1.11).

2. The principle of adjoint equations

2.1. GRADIENT METHODS

Variational assimilation aims at minimising an objective function of one of the forms defined in the previous section (from now on, we will speak of minimisation, even though the function 1.7 has been defined in such a way that it is to be maximised). The functions we will consider can be linear or nonlinear. We will make a slight change of notation, and will systematically denote by ξ, and will call *control variable*, the argument of the function to be minimised (in expression 1.13, the control variable is the whole sequence $\xi_0, \ldots, \xi_M$, while it is the vector μ in expression 1.7). The control variable belongs to *control space*, whose dimension will be denoted by n. We will denote by $\partial\mathcal{J}/\partial\xi$ the gradient of $\mathcal{J}$ with respect to ξ, *i.e.* the n-vector whose components are the partial derivatives of $\mathcal{J}$ with respect to the components ξ_i of ξ, *viz.*,

$$\frac{\partial\mathcal{J}}{\partial\xi} = \left(\frac{\partial\mathcal{J}}{\partial\xi_i}\right)_{i=1,\ldots,n} \tag{2.1}$$

The gradient is equal to 0 at the minimum of the objective function. One way to determine the minimum could conceivably be (as is actually often done in simple small dimension problems) to determine analytical expressions for the components of the gradient, and then to solve a system of n scalar equations for the minimising components of ξ. In meteorological and oceanographical applications, the complexity of the computations defining the objective function (in 4D-Var, these calculations include the temporal integration of a numerical dynamical model of the flow over the assimilation window) makes it totally inconceivable even to obtain analytical expressions for the gradient. Another way to proceed is to implement an iterative minimisation algorithm, which determines a sequence of successive approximations $\xi^{(p)}$ of the minimising value of ξ, *viz.*,

$$\xi^{(p+1)} = \xi^{(p)} - \boldsymbol{D}^{(p)} \tag{2.2}$$

where $\boldsymbol{D}^{(p)}$ is at every iteration an appropriately chosen vector in control space. One possibility is to choose $\boldsymbol{D}^{(p)}$ along the direction of the local gradient $\partial\mathcal{J}/\partial\xi$. Algorithms which are based on that choice, called *steepest descent* algorithms, turn out however not to be numerically very efficient. In other algorithms, the vector $\boldsymbol{D}^{(p)}$ is determined as a combination of the local gradient and of a number of gradients computed at previous steps of the iteration (2.2) (see, *e.g.*, Bonnans *et al.*, 2002). All minimisation methods that are efficient in large dimension are of the form (2.2), and require the explicit determination, at each iteration step, of the local gradient $\partial\mathcal{J}/\partial\xi$. They are called *gradient methods*. Since one cannot hope to obtain an analytical expression for the gradient, it must be determined numerically. One possibility could be to determine it by finite differences, by imposing in turn a perturbation $\Delta\xi_i$ on all components ξ_i of the control vector, and approximating the partial derivative $\partial\mathcal{J}/\partial\xi_i$ by the difference quotient

$$\frac{\partial\mathcal{J}}{\partial\xi_i} \approx \frac{\mathcal{J}(\xi + \Delta\xi_i) - \mathcal{J}(\xi)}{\Delta\xi_i} \tag{2.3}$$

This however would require n explicit computations of the objective function, *i.e.*, in the case of four-dimensional assimilation, n integrations of the assimilating model. Although that has actually been done for variational assimilation of meteorological observations, in an experimental setting, with a relatively small dimension model (Hoffman, 1986), it would clearly be impossible in any practical application.

The method of *adjoint equations* allows numerical computation of the gradient of a scalar function at a cost that is at most a few times the cost of the direct computation of that function. Adjoint equations are an extremely powerful mathematical and numerical tool. They are central to the theory of *optimal control*, *i.e.* the theory of how the behaviour of a physical system can be controlled by acting on some of its components (see for instance the book by Lions, 1971). Adjoint equations can also be used for solving mathematical problems as such (see the lecture notes by V. P. Shutyaev in this book). The use of adjoint equations in meteorological and oceanographical applications was advocated at an early stage of development of numerical modelling of the atmosphere and ocean (see, *e.g.*, Marchuk, 1974). We will describe here how the adjoint method can be used for determining gradients of compound functions. We will first demonstrate the adjoint method in the finite dimensional, 'algebraic' case, which is in principle sufficient for numerical applications, and where it fundamentally reduces to a systematic use of the chain rule for differentiation of compound functions. We will then demonstrate the adjoint method on functions defined through integration of differential equations. That second part will highlight a number of interesting properties of the adjoint method.

2.2. THE ADJOINT METHOD. THE ALGEBRAIC CASE

Let us consider a numerical process of the form

$$v = G(u) \qquad (2.4)$$

where u is the *input* vector, with dimension n, and components u_i ($i = 1, ..., n$), and v is the *output* vector, with dimension m and components v_j ($j = 1, ..., m$). The developments that follow are very general, but are of particular interest when the computations represented by eq. (2.4) are coded on a computer, but are too complex for explicit analytical manipulations. Assuming the process G to be differentiable, a perturbation $\delta u = (\delta u_i)$ results to first order in a perturbation $\delta v = (\delta v_j)$ such that

$$\delta v_j = \sum_{i=1}^{n} \frac{\partial v_j}{\partial u_i} \delta u_i \qquad (2.5a)$$

or, in matrix form

$$\delta v = G' \delta u \qquad (2.5b)$$

where G' is the matrix of partial derivatives $\partial v_j / \partial u_i$, or *Jacobian matrix* of the process G.

Consider now a scalar function $\mathcal{J}(v)$ of the output vector v. It is a compound function of the input vector u. The partial derivatives of $\mathcal{J}$ with respect to the components of u read, through the chain rule

$$\frac{\partial \mathcal{J}}{\partial u_i} = \sum_{j=1}^{m} \frac{\partial v_j}{\partial u_i} \frac{\partial \mathcal{J}}{\partial v_j} \tag{2.6a}$$

The partial derivatives making up the Jacobian G' are again present in this equation, but the summation is now performed on the components of v, and not on the components of u as in eq. (2.5a). Eq. (2.6a) thus expresses a multiplication by the transpose Jacobian G'^{T}. In terms of gradient vectors, eq. (2.6a) reads

$$\frac{\partial \mathcal{J}}{\partial u} = G'^{T} \frac{\partial \mathcal{J}}{\partial v} \tag{2.6b}$$

It is interesting to compare eqs (2.5) and (2.6). The former state that the direct Jacobian G', applied on any perturbation of the input, produces the corresponding first-order perturbation on the output. The latter state that the transpose Jacobian G'^{T}, applied on the gradient of any function with respect to the output, produces the gradient of the same function with respect to the input.

The adjoint method consists in computing the gradient $\partial \mathcal{J}/\partial u$ through eq. (2.6). More precisely, computation (2.4) is first performed from the input value u under consideration. If $\mathcal{J}$ is a 'simple' function of the output v, the gradient $\partial \mathcal{J}/\partial v$ is then easily determined. Computation (2.6) is then performed. The numerical cost of computation (2.6) is the same as the cost of computation (2.5), *i.e.* the cost of a first-order differentiation of the basic equation (2.4). That cost is at most the cost of a few explicit computations of the form (2.4). More precisely, it can be shown that, if eq. (2.4) represents a finite sequence of arithmetic operations (which will always be the case in a real numerical application), the operation count of the adjoint computation (2.6) is at most four times the operation count of computation (2.4), that ratio being reduced to two if one considers only multiplications and divisions. This is true independently of the dimensions of the input and the output. For large values of the input dimension n, a tremendous gain is therefore achieved, through use of eq. (2.6), over direct perturbation of the input.

Let us assume the process (2.4) is the composition of a number of more elementary processes, *i.e.*,

$$G = G_N \, o \, \dots \, o \, G_2 \, o \, G_1 \tag{2.7}$$

where the symbol o denotes composition. Each of the components on the right-hand side of this equation can be thought as representing for instance a one-timestep integration of a numerical dynamical model. Each is of these steps can in turn be decomposed into more elementary components, down to the individual arithmetic operations. According to the chain rule, the Jacobian G' is the numerical product of the Jacobians of the individual components

$$G' = G'_N \dots G'_2 \, G'_1 \tag{2.8a}$$

The transpose Jacobian G'^{T} is then the product of the transpose elementary Jacobians, taken in reverse order

$$G'^{T} = G'_1{}^{T} G'_2{}^{T} \dots G'_N{}^{T} \tag{2.8b}$$

This equation shows that the adjoint computation (2.6b) must be performed in reversed order of the direct computation (2.4). In particular, if the latter includes the temporal integration of a dynamical model into the future, the corresponding adjoint computation will include a form of backward integration into the past.

If the basic process (2.4) is nonlinear, the Jacobian G', and its transpose G'^{T}, depend on the local value of the input vector u. More precisely, it is easily seen that all quantities that appear as operands in nonlinear operations in the direct computation (2.4) appear as coefficients in the adjoint computation (2.6b). This means that those quantities must either be stored in memory in the course of the direct integration performed before the adjoint integration, or be recomputed in the course of the latter. In the first option, the storing requirements may be very high, and are the price to be paid for the tremendous gain in computing time permitted by the adjoint approach. In the second option, the gain in computing time is reduced.

What has just been described is what can be called the *algebraic* form of the adjoint approach, which, as already said, is in principle sufficient for numerical applications. In particular, the foregoing developments are sufficient for developing the adjoint of a given computer code. A code being available for performing the basic computation (2.4) for given input u, the adjoint code will perform computation (2.6b) for given u and $\partial \mathcal{J}/\partial v$. The adjoint code can be developed from the direct code, using systematic rules that directly result from the above developments (see, *e.g.*, Talagrand, 1991, Giering and Kaminski, 1998).

There is however much more in the adjoint approach than what has just been described, as we are going to see on a didactic example built on a simple partial differential equation.

2.3. THE ADJOINT METHOD. THE CASE OF DIFFERENTIAL EQUATIONS

Consider the one-dimensional nonlinear advection-diffusion equation

$$\frac{\partial v}{\partial t} + v \frac{\partial v}{\partial x} - \varkappa \frac{\partial^2 v}{\partial x^2} = 0 \tag{2.9}$$

In this equation, $v(x, t)$ is a scalar velocity field, which is a function of one spatial scalar coordinate x and of time t. The parameter $\varkappa$ is a positive diffusion coefficient, and integration of eq. (2.9) is well-posed only for increasing t. We consider the solutions of eq. (2.9) in the rectangular domain $\mathcal{D} = [x_1, x_2] \times [t_1, t_2]$ in (x, t)-space. Eq. (2.9), being first-order with respect to t, and second-order with respect to x, requires one boundary condition along the t-direction, and two conditions along the x-direction. A unique solution v is therefore defined by the specification of v at initial time $t = t_1$, and along the two lateral boundaries $x = x_1$ and $x = x_2$. The part of the boundary of $\mathcal{D}$ consisting of the three segments $t = t_1$, $x = x_1$ and $x = x_2$ will be called the *input boundary*, and denoted by Γ_i. Given a scalar function u defined along Γ_i, there exists a unique solution of eq. (2.9) such that $v = u$ along Γ_i.

Consider now a field $v_0(x,t)$ of 'observations' over the domain $\mathcal{D}$. For each solution v of eq. (2.9), we measure the misfit between that solution and the observations by the scalar function

$$\mathcal{G}(v) = \frac{1}{2} \iint_{\mathcal{D}} (v - v_0)^2 \, dx dt \tag{2.10}$$

$\mathcal{G}$ is a compound function of the input u, and our goal is to explicitly determine the gradient of $\mathcal{G}$ with respect to u, *i.e.* to explicitly state the relationship between a perturbation δu of u, and the corresponding first-order variation $\delta\mathcal{G}$ of $\mathcal{G}$. More precisely, the perturbation δu induces a perturbation δv on v which evolves to first-order according to the *tangent linear equation*, obtained by differentiating eq. (2.9) with respect to v, *viz.*,

$$\frac{\partial \delta v}{\partial t} + \frac{\partial(v \delta v)}{\partial x} - \varkappa \frac{\partial^2 \delta v}{\partial x^2} = 0 \tag{2.11}$$

and integrated from the condition $\delta v = \delta u$ along Γ_i.

The perturbation δv induces in turn a perturbation $\delta\mathcal{G}$ of the objective function (2.10), which is to first-order equal to

$$\delta\mathcal{G} = \iint_{\mathcal{D}} (v - v_0) \delta v \, dx dt \tag{2.12}$$

Our goal is to relate explicitly and directly the variations of u and $\mathcal{G}$, by writing $\delta\mathcal{G}$ in the form

$$\delta\mathcal{G} = \int_{\Gamma_i} \gamma \delta u \, ds \tag{2.13}$$

where ds is the length element along Γ_i. If a function γ can be defined along Γ_i that verifies eq. (2.13), we will say that function is the *gradient* of $\mathcal{G}$ with respect to u. That definition is an obvious generalisation of the definition of a finite-dimensional gradient.

To that end, we multiply the tangent linear equation (2.11) by a scalar field $\lambda(x, t)$ (which is at this stage arbitrary), integrate over the whole space-time domain $\mathcal{D}$, and combine the result with eq. (2.12), thereby obtaining the expression

$$\delta\mathcal{G} = \iint\limits_{\mathcal{D}} [(v - v_0)\delta v - \lambda(\frac{\partial\delta v}{\partial t} + \frac{\partial(v\delta v)}{\partial x} - \varkappa\frac{\partial^2\delta v}{\partial x^2})]dxdt \tag{2.14}$$

where the minus sign in front of λ is arbitrary, but convenient. We then proceed to remove the differentiation operators acting on δv, by performing integrations by parts. Considering for instance the term $\lambda(\partial\delta v/\partial t)$, it becomes, after integration by parts with respect to t and integration over $\mathcal{D}$

$$\iint\limits_{\mathcal{D}} \lambda\frac{\partial\delta v}{\partial t} dxdt = \int\limits_{x_1}^{x_2} \lambda\delta v dx\bigg|_{t_1}^{t_2} - \iint\limits_{\mathcal{D}} \frac{\partial\lambda}{\partial t}\delta v dxdt \tag{2.15}$$

The other two terms in eq. (2.14) that involve derivatives of δv can be transformed in a similar way, with the diffusion term, being second-order, requiring two successive integrations by parts. Performing these transformations and reordering terms lead to

$$\delta\mathcal{G} = \iint\limits_{\mathcal{D}} [\frac{\partial\lambda}{\partial t} + v\frac{\partial\lambda}{\partial x} + \varkappa\frac{\partial^2\lambda}{\partial x^2} + v - v_0]\delta v dxdt$$

$$- \int\limits_{x_1}^{x_2} \lambda\delta v dx\bigg|_{t_1}^{t_2} - \int\limits_{t_1}^{t_2} v\lambda\delta v dt\bigg|_{x_1}^{x_2} + \int\limits_{t_1}^{t_2} \varkappa[\lambda\frac{\partial\delta v}{\partial x} - \delta v\frac{\partial\lambda}{\partial x}]dt\bigg|_{x_1}^{x_2} \tag{2.16}$$

This expression is valid for any function $\lambda(x, t)$. We are going to choose $\lambda(x, t)$ in such a way as to transform expression (2.16) into an expression of form (2.13). The space-time integral can be cancelled by imposing that $\lambda(x, t)$ verify the following partial differential equation

$$\frac{\partial\lambda}{\partial t} + v\frac{\partial\lambda}{\partial x} + \varkappa\frac{\partial^2\lambda}{\partial x^2} + v - v_0 = 0 \tag{2.17}$$

The first boundary term in eq. (2.16) itself consists of two terms, at times t_1 and t_2 respectively. The term at time t_1 is defined along the input boundary, and is of form (2.13). We eliminate the term at time t_2 by imposing that λ be zero at that time, $viz.$,

$$\lambda(x, t_2) = 0, \qquad x_1 \leq x \leq x_2 \tag{2.18a}$$

The other boundary terms are defined along the input boundary, but one term in the last integral depends on δv through the derivative $\partial\delta v/\partial x$, which is incompatible with form

(2.13). We eliminate that term by imposing to λ to be zero along the lateral boundaries, *viz.*,

$$\lambda(x_1, t) = 0, \qquad t_1 < t \le t_2 \tag{2.18b}$$

$$\lambda(x_2, t) = 0 \qquad t_1 < t \le t_2 \tag{2.18c}$$

This happens to also cancel the second boundary term in eq. (2.16). With conditions (2.17-18) on λ, eq. (2.16) reduces to

$$\delta \mathscr{I} = \int_{x_1}^{x_2} \lambda(x,t_1)\delta v(x,t_1)dx + \int_{t_1}^{t_2} \varkappa \frac{\partial \lambda}{\partial x}(x_1,t)\delta v(x_1,t)dt - \int_{t_1}^{t_2} \varkappa \frac{\partial \lambda}{\partial x}(x_2,t)\delta v(x_2,t)dt \tag{2.19}$$

This expression is of form (2.13), and shows that the gradient γ, considered as a function of x and t, is equal to

$$\gamma(x, t_1) = \lambda(x, t_1) \qquad x_1 < x < x_2 \tag{2.20a}$$

$$\gamma(x_1, t) = \varkappa(\partial \lambda/\partial x)(x_1, t) \qquad t_1 < t < t_2 \tag{2.20b}$$

$$\gamma(x_2, t) = -\varkappa(\partial \lambda/\partial x)(x_2, t) \qquad t_1 < t < t_2 \tag{2.20c}$$

Before we proceed any further, it is necessary to check that the function $\lambda(x, t)$ is unambiguously defined by conditions (2.17-18). Eq. (2.17), like the basic and tangent linear equations (2.9) and (2.11), is first-order with respect to t, and second-order with respect to x. But the sign of the diffusion term has now changed with respect to the sign of the time derivative, and eq. (2.17) defines a well-posed problem for backward integration with respect to t. The specification of λ according to conditions (2.18), *i.e.* at the final time t_2, and along the two spatial boundaries x_1 and x_2, therefore unambiguously defines the solution $\lambda(x, t)$ and the corresponding gradient of $\mathscr{I}$ along the input boundary.

Eq. (2.17) is the *adjoint equation* of the tangent linear equation (2.11). For convenience, it is often called the adjoint of the basic equation (2.9), although that denomination is not perfectly correct. The process for obtaining the gradient of the objective function (2.10) with respect to the input u is now well-defined:

a) Starting from u, integrate the basic equation (2.9) forward in time, thereby defining a solution $v(x, t)$.

b) Starting from the boundary conditions (2.18), integrate the adjoint equation (2.17) backward in time. The gradient of $\mathscr{I}$ with respect to u is then defined by eqs (2.20).

There is an obvious analogy between the foregoing development and the finite-dimension development of subsection 2.2. The input u to the advection-diffusion equation (2.9) corresponds to the input $\boldsymbol{u}$ to the process (2.4). The analogue of that process is the integration of (2.9), the corresponding solution $v(x, t)$ being the analogue of the output $\boldsymbol{v}$ of (2.4). The analogues of the first-order perturbation computation (2.5b), and of the gradient $\partial\mathscr{J}/\partial v$ are, respectively, the integration of the tangent linear equation (2.11) from given δu, and the difference $v\text{-}v_0$. Finally, the analogue of the transpose computation (2.6b) is the integration of the adjoint equation (2.17).

The adjoint equation (2.17) consists of a number of terms that are linear with respect to λ, and of an inhomogeneous term $v\text{-}v_0$. The former are in a one-to-one correspondence with the terms in the tangent linear equation (2.11), and are determined by the basic dynamical equation (2.9) (and by the particular solution v of that equation that is being considered). The inhomogeneous term, on the other hand, is the local gradient of the objective function (2.10), by which it is entirely determined. The basic dynamical equation on the one hand, the objective function of which the gradient is sought on the other, thus contribute in distinct ways to the adjoint equation. Assume in particular that the observed field $v_0(x, t)$ is a solution of eq. (2.9), and look at what the adjoint equation produces for $v=v_0$. The inhomogeneous term is identically zero, and so is the adjoint solution $\lambda(x, t)$, which is defined by conditions (2.18). The gradient (2.20) is also identically zero. This is to be expected, since the objective function (2.10) reaches its absolute minimum, if v_0 is a solution of (2.9), when $v=v_0$.

The basic solution $v(x, t)$ appears explicitly in the adjoint equation. It appears in the adjoint of the advection term. This is so because that term is nonlinear in v. For a linear basic equation, the tangent linear equation, and the corresponding (homogeneous) adjoint equation are independent of the basic solution that is considered. The basic solution $v(x, t)$ also appears in the forcing term of eq. (2.17). This is so because the objective function (2.10) is nonlinear in v. What we say here is the same fact that has already been noted about the algebraic adjoint, namely that quantities that appear nonlinearly in the direct computations appear explicitly in the adjoint computations.

We have considered gradients with respect to initial and lateral boundary conditions. The adjoint approach can be used for computing the gradient of any function of the output of the direct computations with respect to any input parameter. This is perfectly clear in the algebraic approach described in the previous subsection. In the case of eq. (2.9), one might be interested in determining the derivative of the objective function with respect to the diffusion coefficient $\varkappa$. The way to proceed is as follows. A possible variation $\delta\varkappa$ of $\varkappa$ results in an additional term $-\delta\varkappa(\partial^2 v/\partial x^2)$ in the tangent linear equation (2.11). In the computations that follow, that term, which is directly proportional to $\delta\varkappa$, does not have to be modified, and simply produces the following additional term in the expression (2.19) for the first-order variation $\delta\mathscr{J}$

$$\delta \mathcal{G} = ... + \delta \varkappa \iint_{\mathcal{D}} \lambda \frac{\partial^2 v}{\partial x^2} dx dt \qquad (2.21)$$

where the dots stand for the right-hand side of expression (2.19), and λ is the same solution of the adjoint equation (2.17-18) as before. Eq. (2.21) shows that the partial derivative of the objective function $\mathcal{G}$ with respect to the diffusion coefficient $\varkappa$ is equal to

$$\frac{\partial \mathcal{G}}{\partial \varkappa} = \iint_{\mathcal{D}} \lambda \frac{\partial^2 v}{\partial x^2} dx dt \qquad (2.22)$$

This example is particularly simple in that the diffusion coefficient is present as a simple multiplicative coefficient in the basic equation (2.9). A possibly more interesting situation is the one in which the diffusion coefficient $\varkappa$ is a function of the spatial coordinate x, and the diffusion term in (2.9) reads $-\partial/\partial x[\varkappa(\partial v/\partial x)]$. It is left to the reader to determine how the adjoint equation (2.17) is then modified, and to establish the expression for the partial derivative of the objective function with respect to the diffusion coefficient at point x.

It is seen that the same adjoint solution defines the gradient of the objective function with respect to all input parameters. Three components are present in the problem we have considered here: the basic dynamical equation (2.9), which defines the link between the input u and the output v, the objective function (2.10) of which the gradient is sought, and the input parameters with respect to which that gradient is sought. As already mentioned, the basic equation defines the homogeneous part of the adjoint equation, while the objective function defines the inhomogeneous forcing term in that equation. And the partial derivative of the objective function with respect to any parameter is a linear function of the adjoint solution, as shown by eqs (2.20) and (2.22). That structure is absolutely general.

The adjoint of the diffusion term in the basic equation (2.9) is diffusive for the backward integration of the adjoint equation (2.17). Technically, that is due to the fact that the diffusion term, being a second-order derivative, requires two integrations by parts, and therefore two changes of sign, for the definition of its adjoint. The time derivative in (2.9), which is of first-order, requires on the other hand one change of sign only. But there is a deeper significance to the adjoint of diffusion. Diffusion in (2.9) tends to smooth out spatial variations of the field v, and to make it tend asymptotically to a uniform field, independent of the initial state. Therefore, if one considers the evolution of the field from some initial time τ_1 to some posterior time τ_2, the sensitivity of the state at time τ_2 with respect to the state at time τ_1 will decrease when τ_2 increases, τ_1 being kept fixed. But the sensitivity will also decrease if τ_1 decreases, τ_2 being kept fixed. What the integration of the adjoint equation produces is essentially the sensitivity of a fixed final state with respect to an initial state at an ever decreasing prior initial time. It is therefore necessary that diffusion remain diffusive in the backward adjoint integration.

That last remark emphasises, if need be, that the adjoint integration is by no means a backward integration of the basic equation (2.9), nor of the tangent linear equation (2.11). Where the direct basic integration produces values of pressure, temperature or wind velocity components, the corresponding adjoint integration produces values of partial derivatives with respect to pressure, temperature or wind velocity components. It is left as an exercise to the reader to identify in which conditions the homogeneous part of the adjoint equation is identical with the basic dynamical equation (or with its tangent linear equation).

The basic problem that has been considered here, before the problem of determining the gradient of the objective function $\mathcal{J}$ with respect to the input u, is minimisation of $\mathcal{J}$ subject to the constraint (2.9). That problem can be solved by the classical method of *Lagrange multipliers*. One introduces Lagrange multipliers associated with the constraint to be satisfied, namely eq. (2.9). Since the constraint is a scalar condition to be verified over the domain $\mathcal{D}$, the Lagrange multipliers consist of a scalar field defined over $\mathcal{D}$, which we also denote $\lambda(x, t)$. One then forms the *Lagrangian*

$$\mathcal{L}(v, \lambda) = \frac{1}{2} \iint_{\mathcal{D}} (v - v_0)^2 dxdt - \iint_{\mathcal{D}} \lambda(\frac{\partial v}{\partial t} + v \frac{\partial v}{\partial x} - \varkappa \frac{\partial^2 v}{\partial x^2}) dxdt \qquad (2.23)$$

which is a function of v and λ. The theory of Lagrange multipliers says that making the objective function (2.10) stationary, subject to constraint (2.9), is equivalent to looking for couples (v, λ) that make the Lagrangian stationary with respect to both v and λ. The first-order variation of $\mathcal{L}$ resulting from a variation δv of v is formally identical with the right-hand side of eq. (2.14). The theory of adjoint equations is in essence the same as the theory of Lagrange multipliers, from which it can be derived (see, *e.g.*, Thacker and Long, 1988). In particular, the couple (v, λ) that makes $\mathcal{L}$ stationary consists of the solution v of (2.9) that minimises the objective function (2.10) and of the associated adjoint solution. The emphasis has been put above on the determination of the gradient of $\mathcal{J}$ with respect to the input u, while the standard presentation of the method of Lagrange multipliers concentrates on characterising the constrained minimum of $\mathcal{J}$. But these are only two aspects of the same basic theory. One can note that the adjoint solution at the minimum, in addition to verifying the boundary conditions (2.18), must be such that the gradient of $\mathcal{J}$ with respect to the input is equal to 0, *i.e.* must make the quantities (2.20) equal to 0. The minimising solution v is thus characterised by the property that the corresponding adjoint solution λ verifies what can be called 'generalised zero boundary conditions', *i.e.*, λ is equal to 0 along the whole boundary of $\mathcal{D}$, and has zero normal derivatives $\partial \lambda / \partial x$ along the two lateral boundaries $x = x_1$ and $x = x_2$.

3. Practical implementation

The first attempt at using the adjoint approach for variational assimilation of meteorological observations was made by Penenko and Obraztsov (1976), on a simple one-level linear atmospheric model, and with synthetic data. Other attempts were made by Lewis and Derber (1985) and Le Dimet and Talagrand (1986). Courtier and Talagrand (1987) first used real data, while Thacker and Long (1988) made the first attempt at using adjoint equations for variational assimilation of oceanographical observations. Thépaut and Courtier (1991) first used a full primitive equation meteorological model. These early works showed that variational assimilation of observations was numerically feasible at an acceptable cost, and produced physically realistic results. Variational assimilation was progressively applied to more and more complex numerical models. It was introduced in 1997 in operational prediction, in the strong-constraint formulation, at the European Centre for Medium-range Weather Forecasts (Klinker et al., 2000), and in 2000 at the French Meteorological Service (Météo-France). In both places, operational implementation of variational assimilation has resulted in significant improvements of the ensuing forecasts. Some of these improvements were due to side effects not directly linked to the variational character of the assimilation, but others, especially in a number of specific meteorological situations, were due to better consistency between the assimilated states and the dynamics of the atmospheric flow. Other meteorological services are developing variational assimilation in view of operational implementation.

Similar developments have taken place in oceanography, and variational assimilation using the adjoint of oceanographic circulation models is now commonly used for many diverse applications (although not so far for operational oceanographic prediction). In addition to the already mentioned early paper by Thacker and Long (1988), see, e.g., Morrow and De Mey (1995), Greiner and Périgaud (1996), Weaver and Anderson (1997). One can mention in particular the paper by Egbert et al. (1994), who used the dual approach through maximisation in dual observation space of an objective function of form (1.7). In that approach, each iteration requires first a backward integration of the adjoint model, followed by a forward integration of the tangent linear model (see also Louvel, 2001). Variational assimilation has also extended to other fields of geophysics and environmental sciences, such as for instance atmospheric chemistry (Fisher and Lary, 1995, Elbern et al., 1997, Errera and Fonteyn, 2001).

For all its advantages, the adjoint approach nevertheless possesses a number of distinct disadvantages. The major one of these is probably the need for developing the adjoint code which performs computation (2.6) (or analogously, integrates the adjoint equation 2.17). Not only must the adjoint code be developed, but it must be carefully validated, since experience shows that even minor errors in the computed gradient can significantly degrade the efficiency of the minimisation (if not totally inhibit it). Writing the adjoint of a code at the same time as the direct code involves only a rather small amount of additional work (10% or 20%). But developing the adjoint of an already existing code can require a substantial amount of work, and can be a very tedious and time-consuming task. On the other hand, as already said, the systematic approach that has been described in subsection 2.2 leads to perfectly defined 'adjoint' coding rules, which make the development of an adjoint code, if lengthy and tedious, at least totally straightforward. These rules are at the basis of 'adjoint compilers', i.e. software pieces that are intended at automatically developing the adjoint of a given code (see, e.g., Rostaing et al., 1993, Giering and Kaminski, 1998). The adjoint of a particular piece of code is independent of the other pieces, and automation of writing the adjoint instructions of a sequence of coding instructions, which is a purely local operation, is relatively easy. Other aspects, like the choice and management of 'nonlinear' variables to be kept in memory from the direct integration, or to be recomputed in the course of the adjoint integration, require a global view of the code, and are more difficult to automate. For that reason, the use of those software pieces still require experience of adjoint coding as well as some preparatory work, but they are nevertheless extremely useful, and very substantially reduce the amount of time and work necessary for developing the adjoint of an atmospheric or oceanic circulation model.

The adjoint approach is used in assimilation of meteorological and oceanographical observations for numerically solving, through an iterative minimisation process, an optimisation problem. Now, as explained above, what the adjoint equations really do is simply computing the gradient of one scalar output of a numerical process with respect to (potentially all) the input parameters of that process. As such, the adjoint approach can be used for sensitivity studies of outputs with respect to inputs, independently of any optimisation or minimisation. It will be useful to use the adjoint approach when the number of output parameters whose sensitivity is sought is smaller than the number of input parameters with respect to which the sensitivity is sought (in the inverse case, direct perturbation of the input parameters will be more economical). Actually, the first proponents of the use of the adjoint approach in meteorology and oceanography had primarily sensitivity studies in mind (Marchuk, 1974, Hall et al., 1982). Adjoint models have been used to perform sensitivity studies of many different kinds: sensitivity of the atmospheric flow with respect to initial or lateral boundary conditions (Errico and Vukisevic, 1992, Rabier et al., 1992, Gustafsson et al., 1998), sensitivity of the global oceanic circulation to parameters (Marotzke et al., 1999), sensitivity of biogeochemical processes (Waelbroeck and Louis, 1995), sensitivity of atmospheric chemical processes (Zhang et al., 1998). Two specific applications seem particularly promising. The first one is the determination of the dominant singular perturbation modes of the flow, i.e. of the modes that amplify most rapidly over a period of time (Lacarra and Talagrand, 1988, Farrell, 1989, Urban, 1993, Mu, 2000). This allows accurate analysis of the instability of a particular meteorological situation. The second application is the determination of the components of the flow to which a particular feature of the future evolution of the flow (such as for instance the deepening of a depression) is most sensitive. This allows one to 'target' observations in order to optimise the prediction of the feature under consideration. This has been implemented successfully, in particular on the occasion of the Fronts and Atlantic Storm-Track Experiment (FASTEX), performed in 1997 (see, e.g., Langland et al., 1999, or Bergot and Doerenbecher, 2002).

References

Bergot, T. and Doerenbecher, A. (2002) A study on the optimization of the deployment of targeted observations using adjoint-based methods, Q. J. R. Meteorol. Soc., **128**, 1689-1712.

Bonnans, J.-F., Gilbert, J.-C., Lemaréchal, C. and Sagatizabal, C. (2002) Numerical Optimization - Theoretical and Practical Aspects, Springer Verlag, Berlin.

Courtier, P. and Talagrand, O. (1987) Variational assimilation of meteorological observations with the adjoint vorticity equation. II: Numerical results, Q. J. R. Meteorol. Soc. 113, 1329-1347.

52

Egbert, G. D., Bennett, A. F. and Foreman, M. G. C. (1994) Topex/Poseidon tides estimated using a global inverse model, *J. Geophys. Res.* **99**, 24,821-24,852.

Elbern, H., Schmidt, H. and Ebel, A. (1997) Variational Data Assimilation for Tropospheric Chemistry Modeling, *J. Geophys. Res.* **102** (D13), 15,967-15,985.

Errera, Q. and Fonteyn, D. (2001) Four-dimensional variational chemical assimilation of CRISTA stratospheric measurements, *J. Geophys. Res.* **106** (D11), 12,253-12,265.

Errico, R. M. and Vukisevic, T. (1992) Sensitivity Analysis Using an Adjoint of the PSU-NCAR Mesoscale Model, *Mon. Wea. Rev.* **120**, 1644-1660.

Farrell, B. F. (1989) Optimal Excitation of Baroclinic Waves, *J. Atmos. Sci.* **46**, 1193-1206.

Fisher, M. and Lary, D. J. (1995) Lagrangian four-dimensional variational data assimilation of chemical species, *Q. J. R. Meteorol. Soc.* **121**, 1681-1704.

Giering, R. and Kaminski, T. (1998) Recipes for Adjoint Code Construction, *Trans. Math. Software* **24**, 437-474.

Greiner, É. and Périgaud, C. (1996) Assimilation of Geosat Altimetric Data in a Nonlinear Shallow-Water Model of the Indian Ocean by Adjoint Approach. Part II: Some Validation and Interpretation of the Assimilated Results, *J. Phys. Oceanogr.* **26**, 1735-1746.

Gustafsson, N., Källen, E. and Thorsteinsson, S. (1998) Sensitivity of forecast errors to initial and lateral boundary conditions, *Tellus* **50A**, 167-185.

Hall, M. C. G., Cacuci, D. G. and Schlesinger, M. E. (1982) Sensitivity analysis of a radiative-convective model by the adjoint method, *J. Atmos. Sci.* **39**, 2038-2050.

Hoffman, R. N. (1986) A four-dimensional analysis exactly satisfying equations of motion, *Mon. Wea. Rev.* **114**, 388-397.

Klinker, E., Rabier, F., Kelly, G. and Mahfouf, J.-F. (2000) The ECMWF operational implementation of four-dimensional variational assimilation. III: Experimental results and diagnostics with operational configuration, *Q. J. R. Meteorol. Soc.* **126**, 1191-1215.

Lacarra, J.-F. and Talagrand, O. (1988) Short-range evolution of small perturbations in a barotropic model, *Tellus* **40A**, 81-95.

Langland, R. H., Gelaro, R., Rohaly, G. D. and Shapiro, M. A. (1999) Targeted observations in FASTEX: Adjoint-based targeting procedures and data impact experiments in IOP17 and IOP18, *Q. J. R. Meteorol. Soc.* **125**, 3241-3270.

Le Dimet, F.-X. and Talagrand, O. (1986) Variational algorithms for analysis and assimilation of meteorological observations: theoretical aspects, *Tellus* **38A**, 97-110.

Lewis, J. M. and Derber, J. C. (1985) The use of adjoint equations to solve a variational adjustment problem with advective constraints, *Tellus* **37A**, 309-322.

Lions, J.-L. (1971) *Optimal control of systems governed by partial differential equations* (translated from the French), Springer-Verlag, Berlin.

Louvel, S. (2001) Implementation of a dual variational algorithm for assimilation of synthetic altimeter data in the oceanic primitive equation model MICOM, *J. Geophys. Res.* **106** (C5), 9199-9212.

Marchuk, G. I. (1974) *Numerical solution of the problems of dynamics of the atmosphere and the ocean* (in Russian), Gidrometeoizdat, Leningrad.

Marotzke, J., Giering, R., Zhang, K. Q., Stammer, D., Hill, C. and Lee, T. (1999) Construction of the adjoint MIT ocean general circulation model and application to Atlantic heat transport sensitivity, *J. Geophys. Res.* **104** (C12), 29,529-29,547.

Morrow, R. and De Mey, P. (1995) Adjoint assimilation of altimetric, surface drifter and hydrographic data in a QG model of the Azores current, *J. Geophys. Res.* **100** (C12), 25,007-25,025.

Mu M. (2000) Nonlinear singular vectors and nonlinear singular values, *Science in China* (ser. D) **43**, 375-385.

Penenko, V. V. and Obraztsov, N. N. (1976) A variational initialization method for the fields of the meteorological elements (English translation), *Soviet Meteorol. Hydrol.* no 11, 1-11.

Rabier, F., Courtier, P. and Talagrand, O. (1992) An Application of Adjoint Models to Sensitivity Analysis, *Beitr. Phys. Atmosph.* **65**, 177-192.

Rostaing, N., Dalmas, S. and Galligo, A. (1993) Automatic Differentiation in Odyssée, *Tellus* **45A**, 558-568.

Talagrand, O. (1991) The Use of Adjoint Equations in Numerical Modeling of the Atmospheric Circulation, in A. Griewank and G. F. Corliss (eds), *Automatic Differentiation of Algorithms: Theory, Implementation, and Application*, Society for Industrial and Applied Mathematics, Philadelphia, United States, pp. 169-180.

Thacker, W. C. and Long, R. B. (1988) Fitting dynamics to data, *J. Geophys. Res.* **93**, 1227-1240.

Thépaut, J.-N. and Courtier, P. (1991) Four-dimensional variational data assimilation using the adjoint of a multilevel primitive-equation model, *Q. J. R. Meteorol. Soc.* **117**, 1225-1254.

Urban, B. (1993) A method to determine the theoretical maximum error growth in atmospheric models, *Tellus* **45A**, 270-280.

Waelbroeck, C. and Louis, J.-F. (1995) Sensitivity analysis of a model of CO_2 exchange in tundra ecosystems by the adjoint method, *J. Geophys. Res.* **100** (D2), 2801-2816.

Weaver, A. T. and Anderson, D. L. T. (1997) Variational Assimilation of Altimeter Data in a Multilayer Model of the Tropical Pacific Ocean, *J. Phys. Oceanogr.* **27**, 664-682.

Zhang, Y., Bischof, C. H., Easter, R. C. and Wu, P.-T. (1998) Sensitivity analysis of a mixed-phase chemical mechanism using automatic differentiation, *J. Geophys. Res.* **103** (D15), 18,953-18,979.

CONTROL OPERATORS AND FUNDAMENTAL CONTROL FUNCTIONS IN DATA ASSIMILATION

V. SHUTYAEV

Institute of Numerical Mathematics
Russian Academy of Sciences
Gubkina 8, 119991, GSP-1, Moscow, Russia

1. Statement of the Problem

Consider mathematical model of a physical process that is described by the evolution problem

$$
\begin{cases}
\dfrac{d\varphi}{dt} + A(t)\varphi = f, \quad t \in (0, T) \\[2mm]
\varphi|_{t=0} = u,
\end{cases}
\tag{1.1}
$$

where $\varphi = \varphi(t)$ is an unknown function, $A(t)$ is an operator (generally, nonlinear) acting for each t in the Hilbert space X, $u \in X$, and $f = f(t)$ is a prescribed function.

Introduce the functional

$$
J(u) = \frac{\alpha}{2}(u - \hat{\varphi}^\circ, u - \hat{\varphi}^\circ) + \frac{1}{2} \int\limits_0^T (C(\varphi - \hat{\varphi}), \varphi - \hat{\varphi}) dt,
\tag{1.2}
$$

where $\alpha = const \geq 0$, C is a linear operator, and $(\cdot, \cdot)$ is an inner product in X. The function $\hat{\varphi} = \hat{\varphi}(t)$, as a rule, is determined by a priori observation data, $\hat{\varphi}^\circ \in X$. Here φ is a solution of (1.1) with $\varphi(0) = u$.

Consider problem (1.1) with an unknown function $u \in X$ in the initial condition. The *variational data assimilation problem* can be formulated as follows: find φ and u such that they satisfy (1.1) and, on the set of solutions to equation (1.1), functional (1.2) takes the minimum value. Write this

R. Swinbank et al. (eds.), Data Assimilation for the Earth System, 55–63.
© 2003 *Kluwer Academic Publishers. Printed in the Netherlands.*

problem as

$$
\begin{cases}
\dfrac{d\varphi}{dt} + A(t)\varphi = f, & t \in (0, T) \\[2mm]
\varphi(0) = u \\[2mm]
J(u) = \inf_{\tilde{u} \in H} J(\tilde{u}).
\end{cases}
\tag{1.3}
$$

Problems in the form (1.3) were analyzed by Pontryagin (1962), Lions (1968) (see also Sasaki (1970), Marchuk and Penenko (1978), Le Dimet and Talagrand (1986), Navon (1986), Glowinski and Lions (1994), Agoshkov and Marchuk (1993), Marchuk and Zalesny (1993), Marchuk and Shutyaev (1994), Marchuk *et al.* (1996, 2001), Shutyaev (1995, 2001) etc.). To solve (1.3) a number of approaches may be used (see, e.g. Marchuk *et al.*, 2001). In the following, we assume that $A(t)$ is a linear operator.

The necessary optimality condition (Lions, 1968) reduces problem (1.1) to the system for finding the functions φ, φ^*, u:

$$
\frac{d\varphi}{dt} + A(t)\varphi = f, \ t \in (0, T); \ \varphi(0) = u
\tag{1.4}
$$

$$
-\frac{d\varphi^*}{dt} + A(t)^*\varphi^* = C(\hat{\varphi} - \varphi), \ t \in (0, T); \ \varphi^*(T) = 0
\tag{1.5}
$$

$$
\alpha(u - \hat{\varphi}^\circ) - \varphi^*(0) = 0,
\tag{1.6}
$$

where $A(t)^*$ is the operator adjoint to $A(t)$.

2. Control Operator

Let us introduce the operator $L : X \to X$ defined through the successive solutions of the following problems:

$$
\begin{cases}
\dfrac{d\psi}{dt} + A(t)\psi &= 0, \ t \in (0, T) \\[2mm]
\psi|_{t=0} &= v,
\end{cases}
\tag{2.1}
$$

$$
\begin{cases}
-\dfrac{d\psi^*}{dt} + A^*(t)\psi^* &= -C\psi, \ t \in (0, T) \\[2mm]
\psi^*|_{t=T} &= 0,
\end{cases}
\tag{2.2}
$$

$$
Lv = \alpha v - \psi^*(0).
\tag{2.3}
$$

We define also $F \in X$ as the result of the successive solutions of the following problems:

$$
\begin{cases}
\dfrac{d\phi}{dt} + A(t)\phi &= f, \ t \in (0, T) \\[2mm]
\phi|_{t=0} &= 0,
\end{cases}
\tag{2.4}
$$

$$\begin{cases} -\dfrac{d\phi^*}{dt} + A^*(t)\phi^* = C(\hat{\varphi} - \phi), & t \in (0, T) \\ \phi^*|_{t=T} = 0, \end{cases} \tag{2.5}$$

$$F = \alpha\hat{\varphi}^\circ + \phi^*(0), \tag{2.6}$$

where $f, \hat{\varphi}, \hat{\varphi}^\circ$ are introduced in (1.4)–(1.6).

Then, the system (1.4)–(1.6) is reduced to the equation for the control u:

$$Lu = F, \tag{2.7}$$

and the operator $L : X \to X$ is called the *control operator* (Shutyaev, 1995). The operator L plays a fundamental role in study of solvability of the original data assimilation problem and in the development of numerical algorithms to solve it.

We suppose that C is self-adjoint and non-negative:

$$\int\limits_0^T (C\varphi, \psi)dt = \int\limits_0^T (\varphi, C\psi)dt, \quad \int\limits_0^T (C\varphi, \varphi)dt \geq 0, \quad \forall \varphi, \psi.$$

The following statement is valid.

Lemma 1. *The operator L acts in X with domain of definition $D(L) = X$, it is self-adjoint and non-negative. If $\alpha > 0$, the operator L is coercive (positive definite):*

$$(Lv, v) \geq c(v, v), \quad c = const > 0, \ v \in X.$$

Proof. Let $v, w \in X$, and ψ be the solution to (2.1). We have

$$(Lv, w) = (\alpha v - \psi^*|_{t=0}, w) = \alpha(v, w) - (\psi^*|_{t=0}, w) =$$

$$= \alpha(v, w) + \int\limits_0^T (C\psi, \psi_1)dt = \alpha(v, w) + \int\limits_0^T (\psi, C\psi_1)dt = (v, Lw),$$

where ψ_1 is the solution to (2.1) with $v = w$. Hence, L is self-adjoint, and

$$(Lv, v) = \alpha(v, v) + \int\limits_0^T (C\psi, \psi)dt \geq 0,$$

that is, L is non-negative. Moreover, L is coercive if $\alpha > 0$.

Corollary. The following estimate is valid:

$$(Lv, v) \geq \mu_{\min}(v, v), \quad \forall v \in X, \tag{2.8}$$

where $\mu_{\min}$ is the lower spectrum bound of the operator L, and $\mu_{\min} \geq \alpha$.

Hence, the following solvability result holds.

Lemma 2. *For $\alpha > 0$ and $F \in X$, the control equation (2.7) has a unique solution $u \in X$, and*

$$\|u\| \leq \frac{1}{\mu_{\min}} \|F\|. \tag{2.9}$$

Proof. From (2.7)–(2.8),

$$\mu_{\min} \|u\|^2 \leq (Lu, u) = (F, u) \leq \|F\| \|u\|.$$

Hence, (2.9) holds.

Let $\alpha = 0$. In many cases, the operator L is *compact* (Shutyaev, 2001), that is, L maps every bounded set of X into a compact set. By this is meant that if $u_n \in X$ is a bounded sequence, Lu_n is a convergent in X sequence.

If L is compact, it has a complete orthonormal system in X of eigenfunctions v_n corresponding to the eigenvalues μ_k:

$$Lv_k = \mu_k v_k,$$

where $(v_k, v_l) = \delta_{kl}$, $k, l = 1, 2, \ldots$, and δ_{kl} is the Kronecker delta.

Example 1. Consider the finite-dimensional case when $X = \mathbf{R}^n$, $n \in \mathbf{N}$, and let $A(t) = A$ be a matrix independent of t. Then, from (2.1)–(2.3), we obtain the explicit form of the control operator:

$$L = \alpha I + \int_0^T e^{-A^* s} C e^{-As} ds,$$

where I is the identity matrix, and e^{-As}, $e^{-A^* s}$ are the matrix exponentials defined (as functions of matrices) by

$$e^{-At} = \sum_{k=1}^{\infty} \frac{(-At)^k}{k!}, \quad A^0 = I.$$

Example 2. Consider the case when $A = A^*$ is a self-adjoint operator independent of t and generating a basis consisting of the eigenfunctions ω_j corresponding to the eigenvalues λ_j:

$$A\omega_j = \lambda_j \omega_j, \quad (\omega_j, \omega_k) = \delta_{jk}.$$

Then, for any function $u \in X$, the following expansion is valid:

$$u = \sum_{j=1}^{\infty} u_j \omega_j, \quad u_j = (u, \omega_j).$$

For $C = E$ we have

$$L = \alpha I + \int\limits_0^T e^{-2As} ds,$$

and $v_k = w_k$ are the eigenfunctions of the control operator L corresponding to the eigenvalues

$$\mu_k = \begin{cases} \alpha + \dfrac{1 - e^{-2\lambda_k T}}{2\lambda_k}, & \lambda_k \neq 0 \\[2ex] \alpha + T, & \lambda_k = 0. \end{cases}$$

If $\alpha = 0, \lambda_k \to \infty$, then $\mu_k \to 0$, and L^{-1} will be unbounded. The equation $Lu = F$ is equivalent to

$$\sum_j u_j L\omega_j = \sum_j F_j \omega_j, \quad u_j = (u, \omega_j), \quad F_j = (F, \omega_j).$$

It is solvable if and only if

$$\sum_j \frac{F_j^2}{\mu_j^2} < \infty, \tag{2.10}$$

then,

$$u = \sum_j u_j \omega_j, \quad u_j = \frac{F_j}{\mu_j}.$$

The condition (2.10) is equivalent to

$$\sum_j \left(\frac{2\lambda_j}{1 - e^{-2\lambda_j T}} \right)^2 F_j^2 < \infty. \tag{2.11}$$

If the eigenvalues λ_j satisfy the condition $0 < \lambda_1 \leq \lambda_2 \leq \ldots \leq \lambda_j \leq \ldots, \lambda_j \to \infty$, then

$$c_1 \leq \frac{1 - e^{-2\lambda_j T}}{2} \leq c_2$$

with

$$c_1 = \frac{1 - e^{-2\lambda_1 T}}{2}, \quad c_2 = \frac{1}{2}.$$

In this case, the solvability condition is equivalent to

$$\sum_j \lambda_j^2 F_j^2 < \infty. \tag{2.12}$$

It means that F should be not only from X but much more regular.

The sufficient condition for $\hat{\varphi}, f$ to satisfy (2.12) is

$$\sum_j \lambda_j \int_0^T (\hat{\varphi}, w_j)^2 dt < \infty, \quad \sum_j \int_0^T (f, w_j)^2 dt < \infty.$$

3. Fundamental Control Functions

Consider the case that $C = I$ (the identity operator). We assume that the control operator L defined by (2.1)–(2.3) has a complete orthonormal system in X of eigenfunctions v_k corresponding to the eigenvalues $\mu_k > 0$:

$$Lv_k = \mu_k v_k, \tag{3.1}$$

where $(v_k, v_l)_X = \delta_{kl}$, $k, l = 1, 2, \ldots$, and δ_{kl} the Kronecker delta.

It is easily seen that the eigenvalue problem (3.1) is equivalent to the system:

$$\begin{cases} \dfrac{d\varphi_k}{dt} + A(t)\varphi_k &= 0, \ t \in (0, T), \\ \varphi_k|_{t=0} &= v_k, \end{cases} \tag{3.2}$$

$$\begin{cases} -\dfrac{d\varphi_k^*}{dt} + A^*(t)\varphi_k^* &= -\varphi_k, \ t \in (0, T) \\ \varphi_k^*|_{t=T} &= 0, \end{cases} \tag{3.3}$$

$$\alpha v_k - \varphi_k^*|_{t=0} = \mu_k v_k. \tag{3.4}$$

By the analogy with the Poincaré-Steklov operator theory, we say that the system of functions $\{\varphi_k, \varphi_k^*, v_k\}$ satisfying (3.2)–(3.4) is the system of *fundamental control functions*.

Let W be the Hilbert space of real-valued functions with the inner product and the norm:

$$(\varphi, \psi)_W = \int_0^T \left(\frac{d\varphi}{dt} + A(t)\varphi, \frac{d\psi}{dt} + A(t)\psi \right) dt + (\varphi|_{t=0}, \psi|_{t=0}),$$

$$\|\varphi\|_W = (\varphi, \varphi)_W^{1/2}.$$

We introduce the subspace $W_0 = \{\varphi \in W : \frac{d\varphi}{dt} + A(t)\varphi = 0\}$. It is easy to show that W_0 is closed in W. We consider also the subspace

$$W_1 = \left\{ \varphi \in W : -\frac{d\varphi}{dt} + A^*(t)\varphi \in W, \ \varphi|_{t=T} = 0 \right\},$$

putting

$$(\varphi, \psi)_{W_1} = \left(-\frac{d\varphi}{dt} + A^*(t)\varphi, -\frac{d\psi}{dt} + A^*(t)\psi\right)_W.$$

The following statement holds.

Lemma 3. *The functions* $\{\varphi_k\}, \{\varphi_k^*\}, \{v_k\}$ *defined by formulas* (3.2), (3.3), (3.4) *form complete orthonormal systems in* W_0, W_1, X, *respectively.*

Proof. Since $(v_k, v_l)_X = \delta_{kl}$, we have

$$(\varphi_k, \varphi_l)_W = \int_0^T \left(\frac{d\varphi_k}{dt} + A(t)\varphi_k, \frac{d\varphi_l}{dt} + A(t)\varphi_l\right) dt + (\varphi_k|_{t=0}, \varphi_l|_{t=0}) =$$

$$= (\varphi_k|_{t=0}, \varphi_l|_{t=0}) = (v_k, v_l) = \delta_{kl},$$

$$(\varphi_k^*, \varphi_k^*)_{W_1} = \left(-\frac{d\varphi_k^*}{dt} + A^*(t)\varphi_k^*, -\frac{d\varphi_l^*}{dt} + A^*(t)\varphi_l^*\right)_W = (\varphi_k, \varphi_l)_W = \delta_{kl}.$$

Let $\varphi \in W_0$ and $(\varphi, \varphi_k)_W = 0$ for all $k = 1, 2, \ldots$ Then

$$(\varphi, \varphi_k)_W = \int_0^T \left(\frac{d\varphi}{dt} + A(t)\varphi, \frac{d\varphi_k}{dt} + A(t)\varphi_k\right) dt + (\varphi|_{t=0}, \varphi_k|_{t=0}) =$$

$$= (\varphi|_{t=0}, v_k) = 0, \quad k = 1, 2, \ldots$$

Because of the completeness of $\{v_k\}$ in X we find that $\varphi|_{t=0} = 0$. Then, $\varphi = 0$, which leads to the completeness of $\{\varphi_k\}$ in W_0. Analogously, we show that $\{\varphi_k^*\}$ is complete in W_1.

Using the fundamental control functions, we can obtain the solution of the original data assimilation problem (1.4)–(1.6) in the explicit form. The equation (2.7) is equivalent to the system (1.4)–(1.6) and may be written as the following system:

$$\begin{cases} \dfrac{d\psi}{dt} + A(t)\psi = 0, \ t \in (0, T), \\ \psi|_{t=0} = u, \end{cases} \tag{3.5}$$

$$\begin{cases} -\dfrac{d\psi^*}{dt} + A^*(t)\psi^* = -\psi, \ t \in (0, T) \\ \psi^*|_{t=T} = 0, \end{cases} \tag{3.6}$$

$$\alpha u - \psi^*|_{t=0} = F, \tag{3.7}$$

where F is the right-hand side of (2.7).

Theorem. *The solution* ψ, ψ^*, u *of the system* (3.5)–(3.7) *may be represented in the form:*

$$\psi = \sum_k a_k \varphi_k, \quad \psi^* = \sum_k a_k \varphi_k^*, \quad u = \sum_k a_k v_k, \tag{3.8}$$

where $\varphi_k, \varphi_k^*, v_k$ are the fundamental control functions defined by (3.2)–(3.4), $a_k = (F, v_k)_X/\mu_k$, the series for ψ, ψ^*, u converging in W_0, W_1, X, respectively.

Proof. The system (3.5)–(3.7) is equivalent to the equation (2.7), which gives $\mu_k(u, v_k) = (F, v_k)$ due to (3.1). Hence, the solution u of (3.1) is represented as $u = \sum_k a_k v_k$, $a_k = (F, v_k)_X/\mu_k$. Then, it is easily seen that $\psi = \sum_k a_k \varphi_k$, $\psi^* = \sum_k a_k \varphi_k^*$ are the solutions to (3.5), (3.6), respectively. The completeness of the system $\{\varphi_k\}, \{\varphi_k^*\}, \{v_k\}$ gives the convergence of the corresponding series in W_0, W_1, X. This ends the proof.

From (3.8), we have the representation for the solution φ, φ^*, u of the system (1.4)–(1.6):

$$\varphi = \psi + \psi_1, \quad \varphi^* = \psi^* - \psi_1^*, \tag{3.9}$$

where ψ_1, ψ_1^* are the solutions to the problems:

$$\begin{cases} \dfrac{d\psi_1}{dt} + A(t)\psi_1 = f, \ t \in (0, T), \\ \psi_1|_{t=0} = 0, \end{cases} \tag{3.10}$$

$$\begin{cases} -\dfrac{d\psi_1^*}{dt} + A^*(t)\psi_1^* = \hat{\varphi} - \psi_1, \ t \in (0, T) \\ \psi_1^*|_{t=T} = 0. \end{cases} \tag{3.11}$$

Note that $F = \alpha\hat{\varphi}^\circ + \psi_1^*(0)$.

Example 3. Consider the case that $A(t) = A^*(t) = A$ is a t-independent self-adjoint operator such that A generates an orthonormal basis in X consisting of eigenfunctions w_k: $Aw_k = \lambda_k w_k$, $(w_k, w_j) = \delta_{kj}$, where $\lambda_k > 0$ are the corresponding eigenvalues of A, and δ_{kj} the Kronecker delta. Then, the eigenfunctions of the operator $L : X \to X$ coincide with the eigenfunctions w_k of the operator A, and the eigenvalues μ_k of H are defined by the formula

$$\mu_k = \alpha + \frac{1}{2\lambda_k}(1 - e^{-2\lambda_k T}).$$

It is easily seen that the fundamental control functions $\varphi_k, \varphi_k^*, v_k$ defined by (3.2)–(3.4) are

$$\varphi_k = e^{-\lambda_k t} w_k, \quad \varphi_k^* = \frac{1}{2\lambda_k} e^{\lambda_k t}(e^{-2\lambda_k T} - e^{-2\lambda_k t})w_k, \quad v_k = w_k.$$

Then, the solution φ, φ^*, u of the optimality system (1.4)–(1.6) has the form (3.9) with

$$\psi = \sum_k a_k e^{-\lambda_k t} w_k, \quad \psi^* = \sum_k \frac{a_k}{2\lambda_k} e^{\lambda_k t}(e^{-2\lambda_k T} - e^{-2\lambda_k t})w_k, \quad u = \sum_k a_k w_k,$$

where

$$a_k = \frac{(F, w_k)}{\mu_k} = \frac{2\lambda_k(F, w_k)}{2\alpha\lambda_k + 1 - e^{-2\lambda_k T}}.$$

References

Agoshkov, V.I. and Marchuk, G.I. (1993) On solvability and numerical solution of data assimilation problems, *Russ. J. Numer. Anal. Math. Modelling* **8**, 1–16.

Bellman, R. (1957)*Dynamic Programming*, Princeton Univ. Press, New Jersey.

Glowinski, R. and J.-L.Lions, J.-L. (1994) Exact and approximate controllability for distributed parameter systems, *Acta Numerica* **1**, 269–378.

Kravaris, C. and Seinfeld, J.H. (1985) Identification of parameters in distributed parameter systems by regularization, *SIAM J. Control and Optimization* **23**, 217.

Kurzhanskii, A.B. and Khapalov, A.Yu. (1991) An observation theory for distributed-parameter systems, *J. Math. Syst. Estimat. Control* **1**, 389–440.

Le Dimet, F.-X. and Talagrand, O. (1986) Variational algorithms for analysis and assimilation of meteorological observations: theoretical aspects, *Tellus* **38A**, 97–110.

Lions, J.-L. (1968)*Contrôle Optimal des Systèmes Gouvernés par des Équations aux Dérivées Partielles*, Dunod, Paris.

Lions, J.-L. (1997) On controllability of distributed systems, *Proc. Natl. Acad. Sci. USA* **94**, 4828–4835.

Marchuk, G.I. (1995) *Adjoint Equations and Analysis of Complex Systems*, Kluwer, Dordrecht.

Marchuk, G.I., Agoshkov, V.I., and Shutyaev, V.P. (1996) *Adjoint Equations and Perturbation Algorithms in Nonlinear Problems*, CRC Press Inc., New York.

Marchuk, G.I. and Penenko, V.V. (1978) Application of optimization methods to the problem of mathematical simulation of atmospheric processes and environment, in G.I.Marchuk (ed.), *Modelling and Optimization of Complex Systems: Proc. of the IFIP-TC7 Working conf.*, Springer, New York, pp. 240-252.

Marchuk, G.I. and Shutyaev, V.P. (1994) Iteration methods for solving a data assimilation problem, *Russ. J. Numer. Anal. Math. Modelling* **9**, 265–279.

Marchuk, G., Shutyaev, V., and Zalesny V. (2001) Approaches to the solution of data assimilation problems, in J.L.Menaldi, E.Rofman, and A.Sulem (eds.), *Optimal Control and Partial Differential Equations*, IOS Press, Amsterdam, pp.489–497.

Marchuk, G.I. and Zalesny, V.B. (1993) A numerical technique for geophysical data assimilation problem using Pontryagin's principle and splitting-up method, *Russ. J. Numer. Anal. Math. Modelling* **8**, 311–326.

Navon, I.M. (1986) A review of variational and optimization methods in meteorology, in Y.K.Sasaki (ed.), *Variational Methods in Geosciences*, Elsevier, New York, pp.29–34.

Pontryagin, L.S., Boltyanskii, V.G., Gamkrelidze, R.V., and Mischenko, E.F. (1962) *The Mathematical Theory of Optimal Processes*, John Wiley, New York.

Sasaki, Y.K. (1970) Some basic formalisms in numerical variational analysis, *Monthly Weather Review* **98**, 857–883.

Shutyaev, V.P. (1995) Some properties of the control operator in the problem of data assimilation and iterative algorithms, *Russ. J. Numer. Anal. Math. Modelling* **10**, 357–371.

Shutyaev, V.P. (2001) *Control Operators and Iterative Algorithms for Variational Data Assimilation Problems*, Nauka, Moscow.

Tikhonov, A.N. (1963) On the solution of ill-posed problems and the regularization method, *Dokl. Akad. Nauk SSSR* **151**, 501–504.

SOLVABILITY OF VARIATIONAL DATA ASSIMILATION PROBLEMS AND ITERATIVE ALGORITHMS

V. SHUTYAEV

Institute of Numerical Mathematics
Russian Academy of Sciences
Gubkina 8, 119991, GSP-1, Moscow, Russia

1. Statement of the Problem

Let H and X be real Hilbert spaces such that $X \subset H$, and H^*, X^* be the spaces dual to H, X, respectively. We assume that $H \equiv H^*$, $(\cdot, \cdot)_{L_2(0,T;H)} = (\cdot, \cdot)$, $\|\cdot\| = (\cdot, \cdot)^{1/2}$. Let us consider also the spaces $Y^0 = L_2(0, T; H)$, $Y = L_2(0, T; X)$, $Y^* = L_2(0, T; X^*)$ of functions $f(t)$ with the values in H, X, X^*, respectively, and the space

$$W = \{f \in L_2(0, T; X) : \frac{df}{dt} \in L_2(0, T; X^*)\},$$

$$\|f\|_W = (\|\frac{df}{dt}\|^2_{L_2(0,T; X^*)} + \|f\|^2_{L_2(0,T; X)})^{1/2}. \tag{1.1}$$

Let $A(t) : X \to X^*$ be a linear operator defined for any $t \in [0, T]$ and satisfied the inequalities:

$$(A(t)\varphi, \psi)_H \leq c_1 \|\varphi\|_X \|\psi\|_X, \quad c_1 = const > 0, \tag{1.2}$$

$$c_2 \|\varphi\|^2_X \leq (A(t)\varphi, \varphi)_H, \quad c_2 = const > 0, \quad \forall\, t \in [0, T], \quad \forall\, \varphi, \psi \in X. \tag{1.3}$$

Consider the following quasilinear evolution problem:

$$\begin{cases} \frac{d\varphi}{dt} + A(t)\varphi + \tau F(\varphi) = f(t), & t \in (0, T) \\ \varphi(0) = u, \end{cases} \tag{1.4}$$

where $f \in Y^*$, $u \in H$, $\tau \in [-\tau_0, \tau_0]$ is a parameter, $\tau_0 \in \mathbf{R}^+$, $F(\varphi)$ is a nonlinear Frechet differentiable operator, $F : Y \to Y^*$. Introduce a functional of $u \in H$ of the form:

$$S(u) = \frac{\alpha}{2} \|u - \widehat{\varphi}^\circ\|^2_H + \frac{1}{2} \|B\varphi - \widehat{\varphi}\|^2_Z, \tag{1.5}$$

65

R. Swinbank et al. (eds.), Data Assimilation for the Earth System, 65–74.
© 2003 *Kluwer Academic Publishers. Printed in the Netherlands.*

where $\alpha = const \geq 0$, Z is a Hilbert space (observational space) with the scalar product $(\cdot, \cdot)_Z$ and the norm $\| \cdot \|_Z = (\cdot, \cdot)_Z^{1/2}$, $B : Y \to Z$ is a linear continuous operator, $\widehat{\varphi} \in Z$, $\widehat{\varphi}^o \in H$. The function $\widehat{\varphi}$ is generally determined by *a priory* observational data. The weight coefficient α is often called a regularization parameter (Tikhonov, 1963).

Consider the following data assimilation problem: for given $f \in Y^*$, $\widehat{\varphi} \in Z$, find $u \in H$, $\varphi \in W$ such that

$$\begin{cases} \dfrac{d\varphi}{dt} + A(t)\varphi + \tau F(\varphi) &= f, \ t \in (0, T) \\ \varphi(0) &= u \\ S(u) &= \min_{\tilde{u} \in H} S(\tilde{u}). \end{cases} \qquad (1.6)$$

The problems of the form (1.6) were studied by L.S.Pontryagin (1962), J.-L.Lions (1968, 1997) and many others (see, e.g., Le Dimet and Talagrand (1986), Glowinski and Lions (1994), Agoshkov and Marchuk (1993), Marchuk and Zalesny (1993), Marchuk and Shutyaev (1994), Marchuk *et al.* (1995, 1996, 2001), Shutyaev (1994, 2001) etc.).

The necessary optimality condition (Lions, 1968) reduces the problem (1.6) to the system for finding the functions $\varphi, \varphi^* \in W$, $u \in H$, of the form:

$$\frac{d\varphi}{dt} + A(t)\varphi + \tau F(\varphi) = f, \ t \in (0, T); \quad \varphi(0) = u, \qquad (1.7)$$

$$-\frac{d\varphi^*}{dt} + A^*(t)\varphi^* + \tau(F'(\varphi))^*\varphi^* = C\widehat{\varphi} - K\varphi, \ t \in (0, T); \quad \varphi^*(T) = 0, \ (1.8)$$

$$\alpha(u - \widehat{\varphi}^o) - \varphi^*(0) = 0, \qquad (1.9)$$

where $(F'(\varphi))^* : Y \to Y^*$ is the operator adjoint to the Frechet derivative of F at the point $\varphi \in W$, $A^*(t) : Y \to Y^*$ is adjoint to $A(t)$, $K : Y \to Y^*$, $C : Z \to Y^*$ are linear bounded operators, $K = CB$, C is defined by the equality $(C\theta, \psi) = (\theta, B\psi)_Z \ \forall \theta \in Z, \psi \in Y$, and equations (1.7), (1.8) are considered in the space Y^*.

The solvability of the systems of the form (1.7)–(1.9) was studied by J.-L.Lions (1968, 1997) and other authors (see, e.g., Agoshkov and Marchuk (1993), Marchuk *et al.* (1996), Shutyaev (1994, 2001) etc.) In this paper, following Agoshkov and Marchuk (1993), Shutyaev (2001), we reduce the problem to the equation for the control function, study its properties in linear case, discuss the solvability of the optimality system, and present iterative algorithms to solve it.

2. Control Operator and Linear Problem

Consider the problem (1.7)–(1.9) for $\tau = 0$. The solutions of problems (1.7), (1.8) for $\tau = 0$ may by represented (Lions and Magenes, 1968) as

$$\varphi = G_0 u + G_1 f, \quad \varphi^* = G_1^{(T)}(C\widehat{\varphi} - K\varphi), \qquad (2.1)$$

where $G_0 : H \to W$, $G_1 : Y^* \to W$, $G_1^{(T)} : Y^* \to W$ are linear continuous operators. Eliminating φ, φ^* from (1.7)–(1.9) for $\tau = 0$, we come to the equation for the control u:

$$Lu = P, \qquad (2.2)$$

where the operator $L : H \to H$ and the right-hand side P are defined by

$$L = \alpha E + T_0 G_1^{(T)} K G_0, \quad P = \alpha \widehat{\varphi}^o + T_0 G_1^{(T)} C\widehat{\varphi} - T_0 G_1^{(T)} K G_1 f, \qquad (2.3)$$

E is the identity operator, $T_0 : W \to H$ is the trace operator: $T_0 \varphi = \varphi|_{t=0}$.

Consider the control operator L for $\alpha = 0$ and denote it by $\bar{L}$. Let $G_0 : H \to W$ be the operator from (2.1), where the element $G_0 u$ is defined as the solution of (1.7) for $\tau = 0$, $f = 0$. The following statement holds.

Lemma 2.1. *The operator* $\bar{L} : H \to H$ *is continuous, self-adjoint, and non-negative:*

$$(\bar{L}v, v)_H \geq 0 \quad \forall v \in H.$$

If the operator $BG_0 : H \to Z$ *is invertible, the operator* $\bar{L}$ *is positive:* $(\bar{L}v, v)_H > 0 \quad \forall v \in H, v \neq 0.$

Proof. Let $\rho \in H$ and $\varphi = G_0 \rho$. Then

$$\bar{L}\rho = T_0 G_1^{(T)} K\varphi.$$

Since (Lions and Magenes, 1968)

$$\|\varphi\|_W \leq c_1 \|\rho\|_H, \quad c_1 = const > 0,$$

and similarly for $\varphi^* = G_1^{(T)} K\varphi$

$$\|\varphi^*\|_W \leq c_2 \|K\varphi\|_{Y^*}, \quad c_2 = const > 0,$$

and by definition of K,

$$\|K\varphi\|_{Y^*} \leq \|CB\| \|\varphi\|_W,$$

then

$$\|\varphi^*\|_W \leq c_3 \|\rho\|_H, \quad c_3 = const > 0.$$

The imbedding of W into $C^0([0,T]; H)$ is continuous (Lions and Magenes, 1968), hence

$$\|T_0\, \varphi^*\|_H = \|\varphi^*(0)\|_H \le c_4\, \|\varphi^*\|_W\,, \quad c_4 = const > 0,$$

therefore, for $\bar{L}\rho = T_0\, \varphi^*$ we get the inequality

$$\|\bar{L}\rho\|_H \le c_3 c_4\, \|\rho\|_H\,,$$

which implies the continuity of $\bar{L}$. Obviously, $\bar{L}$ is self-adjoint. The positive definiteness of $\bar{L}$ follow from the equalities:

$$(\bar{L}\rho, \rho)_H = (T_0 G_1^{(T)} K G_0\, \rho, \rho)_H = (K\varphi, \varphi) = (B\varphi, B\varphi)_Z = \|BG_0\rho\|_Z^2.$$

The lemma is proved.

Corollary 2.1. *If the operator $BG_0 : H \to Z$ is invertible, then*
(I) *The range $R(\bar{L})$ of the operator $\bar{L}$ is dense in H.*
(II) *The equation $\bar{L}u = P$ is solvable uniquely and densely in H.*

Example 2.1. In the case when $Z = \mathbf{R}^n$, $n \in \mathbf{N}$, and the observational operator $B : Y \to Z$ is given by the formula $B\varphi = ((\varphi, p_1), ..., (\varphi, p_n))^T$, where $p_i \in Y^*, i = 1, ..., n$, the operators $C : Z \to Y$, $K : Y \to Y^*$ in (1.8) are defined by

$$C\theta = \sum_{i=1}^{n} \theta_i p_i, \quad K\varphi = \sum_{i=1}^{n} (\varphi, p_i) p_i, \tag{2.4}$$

where $\theta = (\theta_1, ..., \theta_n)^T \in Z$. Then $(K\varphi, \psi) = (K\psi, \varphi)$ and $(K\psi, \psi) = \sum_{i=1}^{n} (\psi, p_i)^2 \ge 0\ \forall \varphi, \psi \in Y$.

Example 2.2. In case of "complete observation", when $Z = Y^0, B = E$ (the identity operator), we have $C = E, K = E$, and the operator $\bar{L}$ is positive.

Lemma 2.2. *For the spectrum $\sigma(\bar{L})$ of the operator $\bar{L}$ the estimate*

$$0 \le \sigma(\bar{L}) \le \nu^2\, \|B\|^2 \tag{2.5}$$

holds with the constant ν from the inequality $\|\varphi\|_Y \le \nu\, \|u\|_H$, where $u \in H$, and $\varphi = G_0 u$ is the solution of the problem $\frac{d\varphi}{dt} + A(t)\varphi = 0$, $t \in (0,T)$; $\varphi(0) = u$.

Proof. To estimate the spectrum of the self-adjoint operator $\bar{L}$ consider $(\bar{L}u, u)$ for $u \in H$. Let $\varphi = G_0 u$, $\varphi^* = G_1^{(T)}\varphi$, then

$$(\bar{L}u, u)_H = (\varphi^*(0), u)_H = (K\varphi, \varphi) = \|B\varphi\|_Z^2 \le \|B\|^2\, \|\varphi\|_Y^2 \le \nu^2 \|B\|^2\, \|u\|_H^2.$$

Hence,

$$\sigma(\bar{L}) \leq \sup_{u \in H,\ u \neq 0} \frac{(\bar{L}u, u)}{(u, u)} \leq \nu^2 \|B\|^2.$$

This ends the proof.

Corollary 2.2. *For the spectrum $\sigma(L)$ of the operator L the estimates hold:*

$$m \leq \sigma(L) \leq M,$$

where

$$m = \alpha, \quad M = \alpha + \nu^2 \|B\|^2.$$

In some cases (when, for example, A is self-adjoint and independent of t), for ν we can put $\nu = 1$.

If $K = E$, for the spectrum $\sigma(L)$ of the operator L defined by (2.2) the following estimates hold (Shutyaev, 2001):

$$m \leq \sigma(L) \leq M, \tag{2.7}$$

where

$$m = \alpha + \int_0^T e^{-\int_0^t \lambda_{\max}(\tau)d\tau} dt, \quad M = \alpha + \int_0^T e^{-\int_0^t \lambda_{\min}(\tau)d\tau} dt,$$

and $\lambda_{\min}, \lambda_{\max}$ are the lower and upper bounds, respectively, of the spectrum of the operator $A + A^*$.

If $K = E$, and $A(t) = A : H \to H$ is independent of time and self-adjoint in H with the compact inverse, then the eigenvalues μ_k of the operator $\bar{L}$ are defined by the formula (Shutyaev, 2001):

$$\mu_k = \frac{1 - e^{-2\lambda_k T}}{2\lambda_k},$$

where λ_k are the eigenvalues of the operator A. Then in (2.7) $\lambda_{\min} = 2\lambda_1$, $\lambda_{\max} = \infty$, and m, M are given in the explicit form:

$$m = \alpha, \quad M = \alpha + \frac{1 - e^{-2\lambda_1 T}}{2\lambda_1}. \tag{2.8}$$

where λ_1 is the least eigenvalue of the operator A.

3. Solvability Results

It follows from Lemma 2.1 that for $\alpha > 0$ the operator $L : H \to H$ is coercive. Then, we come to the solvability theorems for linear and nonlinear problem (1.7)–(1.9):

Theorem 3.1. *Let $f \in Y^*$, $\widehat{\varphi} \in Z$. Then for $\alpha > 0$ the problem (1.7)–(1.9) for $\tau = 0$ has a unique solution $\varphi_0 \in W$, $\varphi_0^* \in W$, $u_0 \in H$, and the following estimate holds:*

$$\|\varphi_0\|_W + \|\varphi_0^*\|_W + \|u_0\|_H \leq c_0(\|\widehat{\varphi}^o\|_H + \|C\widehat{\varphi}\|_{Y^*} + \|f\|_{Y^*}), \quad c_0 = \text{const} > 0. \tag{3.1}$$

Theorem 3.2. *Let $u \in H$, $\widehat{\varphi} \in Z$ and for some $R > 0$ the inequalities*

$$\|F'(\xi)\|_{Y \to Y^*} \leq k_1, \quad \|F'(\xi) - F'(\eta)\|_{Y \to Y^*} \leq k_2\|\xi - \eta\|_W \tag{3.2}$$

are satisfied for any $\xi, \eta \in B(\varphi_0, R) = \{\varphi \in Y : \|\varphi - \varphi_0\|_W \leq R\}$, where $k_i = k_i(\varphi_0, R) = \text{const} > 0$. Then for $|\tau| \leq \tau_0$, with

$$\tau_0 = 1/c_0[k_1 + k_2(R + \|\varphi_0^*\|_W) + \frac{1}{R}(\|F(\varphi_0)\|_{Y^*} + k_1\|\varphi_0^*\|_W)]^{-1}, \tag{3.3}$$

the problem (1.2)–(1.4) has a unique solution $(\varphi, \varphi^*, u) \in W \times W \times H$.

Proof. Theorem 3.1 follows from Lemma 2.1 and the well-known results on solvability of linear optimal control problems (see, e.g., Lions, 1968; Agoshkov and Marchuk, 1993). To prove Theorem 3.2, consider the problem for the remainders $\tilde{\varphi} = \varphi - \varphi_0$, $\tilde{\varphi}^* = \varphi^* - \varphi_0^*$, $\tilde{u} = u - u_0$, where $(\varphi_0, \varphi_0^*, u_0)$ is the solution to the problem (1.7)–(1.9) for $\tau = 0$. The problem for $\tilde{\varphi}, \tilde{\varphi}^*, \tilde{u}$ reads:

$$\frac{d\tilde{\varphi}}{dt} + A(t)\tilde{\varphi} + \tau F(\varphi_0 + \tilde{\varphi}) = 0, \ t \in (0, T); \quad \tilde{\varphi}(0) = \tilde{u}, \tag{3.4}$$

$$-\frac{d\tilde{\varphi}^*}{dt} + A^*(t)\tilde{\varphi}^* + \tau(F'(\varphi_0 + \tilde{\varphi}))^*(\varphi_0^* + \tilde{\varphi}^*) = -K\tilde{\varphi}, \ t \in (0, T); \quad \tilde{\varphi}^*(T) = 0, \tag{3.5}$$

$$\alpha\tilde{u} - \tilde{\varphi}^*(0) = 0. \tag{3.6}$$

Consider the following iterative process:

$$\frac{d\tilde{\varphi}^{(n+1)}}{dt} + A(t)\tilde{\varphi}^{(n+1)} + \tau F(\tilde{\varphi}^{(n)} + \varphi_0) = 0, \ t \in (0, T); \quad \tilde{\varphi}^{(n+1)}(0) = \tilde{u}^{(n+1)}, \tag{3.7}$$

$$-\frac{d\tilde{\varphi}^{*(n+1)}}{dt} + A^*(t)\tilde{\varphi}^{*(n+1)} + \tau(F'(\tilde{\varphi}^{(n)} + \varphi_0))^*(\tilde{\varphi}^{*(n)} + \varphi_0^*) = -K\tilde{\varphi}^{(n+1)}, \tag{3.8}$$

$$\alpha\tilde{u}^{(n+1)} - \tilde{\varphi}^{*(n+1)}(0) = 0 \tag{3.9}$$

for $\|\tilde{\varphi}^{(0)}\|_W + \|\tilde{\varphi}^{*(0)}\|_W \leq R$ with $\tilde{\varphi}^{*(n+1)}(T) = 0$. Since (for a fixed n) $\tilde{\varphi}^{(n+1)}$, $\tilde{\varphi}^{*(n+1)}$, $\tilde{u}^{(n+1)}$ is the solution of the linear problem, then, in view of (3.1), it is easily seen that

$$\|\tilde{\varphi}^{(n+1)}\|_W + \|\tilde{\varphi}^{*(n+1)}\|_W + \|\tilde{u}^{(n+1)}\|_H \leq k|\tau|(\|\tilde{\varphi}^{(n)}\|_W + \|\tilde{\varphi}^{*(n)}\|_W) + f_0,$$

where

$$k = c_0(k_1 + k_2(R + \|\varphi_0^*\|_W)), \quad f_0 = c_0|\tau|(\|F(\varphi_0)\|_{Y^*} + k_1\|\varphi_0^*\|_W).$$

By successive use of the last inequality, we get

$$\|\tilde{\varphi}^{(n)}\|_W + \|\tilde{\varphi}^{*(n)}\|_W + \|\tilde{u}^{(n)}\|_H \leq (k|\tau|)^n(\|\tilde{\varphi}^{(0)}\|_W + \|\tilde{\varphi}^{*(0)}\|_W)+$$

$$+\frac{1 - (k|\tau|)^n}{1 - k|\tau|}f_0 \leq (k|\tau|)^n R + \frac{1 - (k|\tau|)^n}{1 - k|\tau|}f_0 \leq R \qquad (3.10)$$

if $|\tau| \leq \tau_0$. Then, consider the problem for $\tilde{\varphi}^{(n+1)} - \tilde{\varphi}^{(n)}$, $\tilde{\varphi}^{*(n+1)} - \tilde{\varphi}^{*(n)}$, $\tilde{u}^{(n+1)} - \tilde{u}^{(n)}$. This leads to the estimate:

$$\|\tilde{\varphi}^{(n+1)} - \tilde{\varphi}^{(n)}\|_W + \|\tilde{\varphi}^{*(n+1)} - \tilde{\varphi}^{*(n)}\|_W + \|\tilde{u}^{(n+1)} - \tilde{u}^{(n)}\|_H \leq$$

$$\leq k|\tau|(\|\tilde{\varphi}^{(n)} - \tilde{\varphi}^{(n-1)}\|_W + \|\tilde{\varphi}^{*(n)} - \tilde{\varphi}^{*(n-1)}\|_W),$$

which implies

$$\tilde{\varphi}^{(n)} \to \tilde{\varphi}, \quad \tilde{\varphi}^{*(n)} \to \tilde{\varphi}^*, \quad \tilde{u}^{(n)} \to \tilde{u} \text{ as } n \to \infty, \text{ for } |\tau| \leq \tau_0,$$

where $\tilde{\varphi}, \tilde{\varphi}^*, \tilde{u}$ is the solution to the problem (3.4)–(3.6), and the convergence rate estimate holds:

$$\|\tilde{\varphi}^{(n)} - \tilde{\varphi}\|_W + \|\tilde{\varphi}^{*(n)} - \tilde{\varphi}^*\|_W + \|\tilde{u}^{(n)} - \tilde{u}\|_H \leq c\frac{(k|\tau|)^n}{1 - k|\tau|} \qquad (3.11)$$

with $c = const > 0$. It is easily seen that for $|\tau| \leq \tau_0$ this solution is unique and satisfies the condition $\|\tilde{\varphi}\|_W + \|\tilde{\varphi}^*\|_W + \|\tilde{u}\|_H \leq R$. Thus, under the hypotheses of Theorem, there exists a unique solution of the problem (1.7)–(1.9). Theorem is proved.

Remark 3.1. If the operator $F(\varphi)$ is analytic, then the functions (φ, φ^*, u) are represented as the series in the powers of τ:

$$\varphi = \varphi_0 + \sum_{i=1}^{\infty}\tau^i\varphi_i, \quad \varphi^* = \varphi_0^* + \sum_{i=1}^{\infty}\tau^i\varphi_i^*, \quad u = u_0 + \sum_{i=1}^{\infty}\tau^iu_i,$$

convergent for $|\tau| < \tau_0$ in W, W, H, respectively, where $\varphi_i, \varphi_i^*, u_i$ may be found by the small parameter method (Marchuk et al., 1996).

4. Iterative Algorithms

To solve (1.7)–(1.9) one may use the successive approximation method (3.7)–(3.9). Each step of this method involves a linear data assimilation problem of the form (1.7)–(1.9) for $\tau = 0$. To solve it we consider a class of iterative algorithms:

$$\frac{d\varphi^k}{dt} + A(t)\varphi^k = f, \ t \in (0, T); \quad \varphi^k(0) = u^k, \tag{4.1}$$

$$-\frac{d\varphi^{*k}}{dt} + A^*(t)\varphi^{*k} = C\widehat{\varphi} - K\varphi^k, \ t \in (0, T); \quad \varphi^{*k}(T) = 0, \tag{4.2}$$

$$u^{k+1} = u^k - \alpha_{k+1} B_k(\alpha(u^k - \widehat{\varphi}^o)\varphi^{*k}|_{t=0}) + \beta_{k+1} C_k(u^k - u^{k-1}), \tag{4.3}$$

where $B_k, C_k : H \to H$ are some operators, and $\alpha_{k+1}, \beta_{k+1}$ the iterative parameters. Let $\gamma = \nu^2 \|B\|^2$ with ν defined in (2.5). We introduce the following notations:

$$\tau_{opt} = 2(2\alpha + \gamma)^{-1}, \ \ \theta = (2\alpha + \gamma)\gamma^{-1}, \tag{4.4}$$

$$\tau_k = 2(2\alpha + \gamma - \gamma \cos \omega_k \pi)^{-1}, \ \ k = 1, 2, \ldots, s, \tag{4.5}$$

$$\alpha_{k+1} = \begin{cases} 2(2\alpha + \gamma)^{-1}, & k = 0 \\ 4\gamma^{-1} \dfrac{T_k(\theta)}{T_{k+1}(\theta)}, & k > 0 \end{cases};$$

$$\beta_{k+1} = \begin{cases} 0, & k = 0 \\ \dfrac{T_{k-1}(\theta)}{T_{k+1}(\theta)}, & k > 0, \end{cases} \tag{4.6}$$

$$e_k = \begin{cases} 0, & k = 0 \\ p_k \|\xi^k\|_H^2 / \|\xi^{k-1}\|_H^2, & k > 0, \end{cases} \tag{4.7}$$

$$p_{k+1} = \alpha + (K\eta^k, \eta^k) / \|\xi^k\|_H^2 - e_k, \ \ k = 0, 1, \ldots, \tag{4.8}$$

where $\omega_k = (2i - 1)/2s$, T_k is the k-th degree Chebyshev polynomial of the first kind, $\xi^k = \alpha u^k - \varphi^{*k}(0)$, and η^k is the solution of the problem $\dfrac{d\eta^k}{dt} + A\eta^k = 0, \ t \in (0, T); \ \eta^k(0) = \xi^k$.

Theorem 4.1. (I) *If $\alpha_{k+1} = \tau$, $B_k = E$, $\beta_{k+1} = 0$, $0 < \tau < 2/(\alpha + \gamma)$, then the iterative process* (4.1)–(4.3) *is convergent. For $\tau = \tau_{opt}$ defined by* (4.4) *the following convergence rate estimates are valid:*

$$\|\varphi - \varphi^k\|_W \leq c_1 q_k, \ \ \|\varphi^* - \varphi^{*k}\|_W \leq c_2 q_k, \ \ \|u - u^k\|_H \leq c_3 q_k, \tag{4.9}$$

where $q_k = 1/\theta^k$, θ is given by (4.4), and the constants c_1, c_2, c_3, c_4 do not depend on the number of iterations and on the functions $\varphi, \varphi^k, \varphi^*, \varphi^{*k}, u, u^k$.

(II) If $B_k = E$, $\beta_{k+1} = 0$, and $\alpha_{k+1} = \tau_k$, where the parameters τ_k are defined by (4.5) and repeated cyclically with the period s, then the error in the iterative process (4.1)–(4.3) is suppressed after each cycle of the length s. After $k = ls$ iterations the error estimates (4.9) are valid with $q_k = (T_s(\theta))^{-l}$.

(III) If $B_k = C_k = E$ and $\alpha_{k+1}, \beta_{k+1}$ are defined by (4.6), then the error in the algorithm (4.1)–(4.3) is suppressed for each $k \geq 1$, and the estimates (4.9) hold for $q_k = (T_k(\theta))^{-1}$.

(IV) If $B_k = C_k = E$ and $\alpha_{k+1} = 1/p_{k+1}$, $\beta_{k+1} = e_k/p_{k+1}$, where e_k, p_{k+1} are defined by (4.7), (4.8), then the iterative process (4.1)–(4.3) is convergent, and the convergence rate estimates (4.9) are valid with $q_k = (T_k(\theta))^{-1}$.

Proof. It is not difficult to show (Marchuk and Shutyaev, 1994) that the iterative process (4.1)–(4.3) is equivalent to the following iterative algorithm

$$u^{k+1} = u^k - \alpha_{k+1}B_k(Lu^k - P) + \beta_{k+1}C_k(u^k - u^{k-1}) \qquad (4.10)$$

for solving the control equation (2.2).

According to Lemma 2.3, the bounds of the spectrum of the control operator L are given by

$$m \overset{\text{def}}{=} \inf_{u \in H,\, u \neq 0} \frac{(Lu, u)}{(u, u)} \geq \alpha, \quad M \overset{\text{def}}{=} \sup_{u \in H,\, u \neq 0} \frac{(Lu, u)}{(u, u)} \leq \alpha + \nu^2 \|B\|^2. \quad (4.11)$$

Thus, for $\alpha > 0$ for solving the equation $Lu = P$ we may use the well-known iterative algorithms with optimal choice of parameters. The theory of these methods is well developed (Marchuk and Lebedev, 1986). Taking into account the explicit form of the bounds for m and M from (4.11) and applying for the equation $Lu = P$ the simple iterative method, the Chebyshev acceleration methods (s-cyclic and two-step ones), and the conjugate gradient method in the form (4.10), we arrive at the conclusions of Theorem, using the well-known convergence results (Marchuk and Lebedev, 1986) for these methods.

In case of complete observation, when $Z = Y^0$, $B = E, K = E$, for the spectrum of the operator L the estimates (2.7) are valid, and in the formulas for iterative parameters (4.4)–(4.8) we may take

$$\gamma = \int_0^T e^{-\int_0^t \lambda_{\min}(\tau)d\tau}\, dt.$$

If, moreover, the operator A is self-adjoint and independent of t, we can put, due to (2.8), $\gamma = (1 - e^{-2\lambda_1 T})/(2\lambda_1)$.

The numerical analysis of the above-formulated iterative algorithms has been done in (Parmuzin and Shutyaev, 1999) for the data assimilation problem with a linear parabolic state equation.

In case $\alpha_k = 1/\alpha$, $B_k = E$, $\beta_k = 0$, the iterative algorithm (4.1)–(4.3) coincides with the Krylov-Chernousko method (Krylov and Chernousko, 1962).

References

Agoshkov, V.I. and Marchuk, G.I. (1993) On solvability and numerical solution of data assimilation problems, *Russ. J. Numer. Anal. Math. Modelling* **8**, 1–16.

Glowinski, R. and J.-L.Lions, J.-L. (1994) Exact and approximate controllability for distributed parameter systems, *Acta Numerica* **1**, 269–378.

Kravaris, C. and Seinfeld, J.H. (1985) Identification of parameters in distributed parameter systems by regularization, *SIAM J. Control and Optimization* **23**, 217.

Krylov, I.A. and Chernousko, F.L. (1962) On a successive approximation method for solving optimal control problems, *Zh. Vychisl. Mat. Mat. Fiz.* **2**, 1132–1139.

Kurzhanskii, A.B. and Khapalov, A.Yu. (1991) An observation theory for distributed-parameter systems, *J. Math. Syst. Estimat. Control* **1**, 389-440.

Le Dimet, F.-X. and Talagrand, O. (1986) Variational algorithms for analysis and assimilation of meteorological observations: theoretical aspects, *Tellus* **38A**, 97-110.

Lions, J.-L. (1968) *Contrôle Optimal des Systèmes Gouvernés par des Équations aux Dérivées Partielles*, Dunod, Paris.

Lions, J.-L. (1997) On controllability of distributed systems, *Proc. Natl. Acad. Sci. USA* **94**, 4828–4835.

Lions, J.-L. and Magenes, E. (1968) *Problémes aux Limites non Homogenes et Applications*, Dunod, Paris.

Marchuk, G.I. (1995) *Adjoint Equations and Analysis of Complex Systems*, Kluwer, Dordrecht.

Marchuk, G.I., Agoshkov, V.I., and Shutyaev, V.P. (1996) *Adjoint Equations and Perturbation Algorithms in Nonlinear Problems*, CRC Press Inc., New York.

Marchuk, G.I. and Lebedev, V.I. (1986) *Numerical Methods in the Theory of Neutron Transport*, Harwood Academic Publishers, New York.

Marchuk, G.I. and Shutyaev, V.P. (1994) Iteration methods for solving a data assimilation problem, *Russ. J. Numer. Anal. Math. Modelling* **9**, 265–279.

Marchuk, G., Shutyaev, V., and Zalesny V. (2001) Approaches to the solution of data assimilation problems, in J.L.Menaldi, E.Rofman, and A.Sulem (eds.), *Optimal Control and Partial Differential Equations*, IOS Press, Amsterdam, pp.489–497.

Marchuk, G.I. and Zalesny, V.B. (1993) A numerical technique for geophysical data assimilation problem using Pontryagin's principle and splitting-up method, *Russ. J. Numer. Anal. Math. Modelling* **8**, 311–326.

Parmuzin, E.I. and Shutyaev, V.P. (1999) Numerical analysis of iterative methods for solving evolution data assimilation problems, *Russ. J. Numer. Anal. Math. Modelling* **14**, 275–289.

Pontryagin, L.S., Boltyanskii, V.G., Gamkrelidze, R.V., and Mischenko, E.F. (1962) *The Mathematical Theory of Optimal Processes*, John Wiley, New York.

Shutyaev, V.P. (1994) On a class of insensitive control problems, *Control and Cybernetics* **23**, 257–266.

Shutyaev, V.P. (2001) *Control Operators and Iterative Algorithms for Variational Data Assimilation Problems*, Nauka, Moscow.

Tikhonov, A.N. (1963) On the solution of ill-posed problems and the regularization method, *Dokl. Akad. Nauk SSSR* **151**, 501–504.

FUNDAMENTAL CONTROL FUNCTIONS
AND ERROR ANALYSIS

V. SHUTYAEV

Institute of Numerical Mathematics
Russian Academy of Sciences
Gubkina 8, 119991, GSP-1, Moscow, Russia

1. Statement of the Problem

Consider mathematical model of a physical process that is described by the evolution problem

$$\begin{cases} \dfrac{\partial \varphi}{\partial t} = F(\varphi) + f, & t \in (0, T) \\ \varphi|_{t=0} = u, \end{cases} \tag{1.1}$$

where $\varphi = \varphi(t)$ is the unknown function belonging for any t to the Hilbert space X, $u \in X$, F is a nonlinear operator mapping X into X. Let $Y = L_2(0, T; X)$, $\| \cdot \|_Y = (\cdot, \cdot)_Y^{1/2}$, $f \in Y$. Let us introduce the functional

$$S(u) = \frac{\alpha}{2} \|u - u_0\|_X^2 + \frac{1}{2} \|C\varphi - \varphi_{obs}\|_{Y_{obs}}^2, \tag{1.2}$$

where $\alpha = const \geq 0$, $u_0 \in X$, $\varphi_{obs} \in Y_{obs}$ are prescribed functions (observational data), Y_{obs} is a Hilbert space (observational space), $C : Y \to Y_{obs}$ a linear continuous operator.

Consider the following data assimilation problem with the aim to identify the initial condition: find u and φ such that they satisfy (1.1), and on the set of solutions to equation (1.1), the functional (1.2) takes the minimum value, i.e.

$$\begin{cases} \dfrac{\partial \varphi}{\partial t} = F(\varphi) + f, & t \in (0, T) \\ \varphi|_{t=0} = u \\ S(u) = \inf_v S(v). \end{cases} \tag{1.3}$$

R. Swinbank et al. (eds.), Data Assimilation for the Earth System, 75–84.
© *2003 Kluwer Academic Publishers. Printed in the Netherlands.*

The necessary optimality condition reduces the problem (1.3) to the following system:

$$\begin{cases} \dfrac{\partial \varphi}{\partial t} &= F(\varphi) + f, \quad t \in (0, T) \\ \varphi|_{t=0} &= u, \end{cases} \tag{1.4}$$

$$\begin{cases} -\dfrac{\partial \varphi^*}{\partial t} - (F'(\varphi))^* \varphi^* &= -C^*(C\varphi - \varphi_{obs}), \quad t \in (0, T) \\ \varphi^*|_{t=T} &= 0, \end{cases} \tag{1.5}$$

$$\alpha(u - u_0) - \varphi^*|_{t=0} = 0 \tag{1.6}$$

with the unknowns φ, φ^*, and u, where $(F'(\varphi))^*$ is the adjoint to the Frechet derivative of F, and C^* is the adjoint to C defined by $(C\varphi, \psi)_{Y_{obs}} = (\varphi, C^*\psi)_Y$, $\varphi \in Y, \psi \in Y_{obs}$.

Suppose that $u_0 = \bar{u} + \xi_1$, $\varphi_{obs} = C\bar{\varphi} + \xi_2$, $f = \bar{f} + \xi_3$, where $\xi_1 \in X$, $\xi_2 \in Y_{obs}$, $\xi_3 \in Y$, and $\bar{\varphi}$ is the solution to the problem (1.1) with $u = \bar{u}, f = \bar{f}$:

$$\begin{cases} \dfrac{\partial \bar{\varphi}}{\partial t} &= F(\bar{\varphi}) + \bar{f}, \quad t \in (0, T) \\ \bar{\varphi}|_{t=0} &= \bar{u}. \end{cases} \tag{1.7}$$

The solution $\bar{\varphi}$ may be treated as "exact", and the functions ξ_1, ξ_2, ξ_3 may be treated as the errors of the input data u_0, φ_{obs}, f, respectively, ξ_2 being the observational error, and ξ_3 the model error. In this paper, having supposed that the solution of the problem (1.4)–(1.6) exists, we study the influence of the errors ξ_1, ξ_2, ξ_3 on the optimal solution u.

2. Error Equation

The system (1.4)–(1.6) with the three unknowns φ, φ^*, u may be treated as an operator equation of the form

$$\mathcal{F}(U, U_d) = 0, \tag{2.1}$$

where $U = (\varphi, \varphi^*, u)$, $U_d = (u_0, \varphi_{obs}, f)$.

The following equality holds:

$$\mathcal{F}(\bar{U}, \bar{U}_d) = 0, \tag{2.2}$$

with $\bar{U} = (\bar{\varphi}, \bar{\varphi}^*, \bar{u})$, $\bar{U}_d = (\bar{u}, C\bar{\varphi}, \bar{f})$, $\bar{\varphi}^* = 0$. From (2.1)–(2.2), we get

$$\mathcal{F}(U, U_d) - \mathcal{F}(\bar{U}, \bar{U}_d) = 0. \tag{2.3}$$

Let $\delta U = U - \bar{U}$, $\delta U_d = U_d - \bar{U}_d$. Then (2.3) gives

$$\mathcal{F}(\bar{U} + \delta U, \bar{U}_d + \delta U_d) - \mathcal{F}(\bar{U}, \bar{U}_d) = 0. \tag{2.4}$$

From (2.4), with an accuracy of the second order in $\delta U, \delta U_d$, we obtain

$$\mathcal{F}'_U(\bar{U}, \bar{U}_d)\delta U + \mathcal{F}'_{U_d}(\bar{U}, \bar{U}_d)\delta U_d = 0, \qquad (2.5)$$

where $\mathcal{F}'_U, \mathcal{F}'_{U_d}$ are the Gateâux derivatives with respect to U and U_d.

Let $\delta\varphi = \varphi - \bar{\varphi}$, $\delta u = u - \bar{u}$; then $\delta U = (\delta\varphi, \varphi^*, \delta u)$, $\delta U_d = (\xi_1, \xi_2, \xi_3)$. By calculating the derivatives $\mathcal{F}'_U, \mathcal{F}'_{U_d}$, it is easily seen that equation (2.5) is equivalent to the system:

$$\begin{cases} \dfrac{\partial \delta\varphi}{\partial t} - F'(\bar{\varphi})\delta\varphi &= \xi_3, \quad t \in (0, T), \\ \delta\varphi|_{t=0} &= \delta u, \end{cases} \qquad (2.6)$$

$$\begin{cases} -\dfrac{\partial \varphi^*}{\partial t} - (F'(\bar{\varphi}))^*\varphi^* &= (F''(\bar{\varphi})\delta\varphi)^*\bar{\varphi}^* - C^*(C\delta\varphi - \xi_2), \\ \varphi^*|_{t=T} &= 0, \end{cases} \qquad (2.7)$$

$$\alpha(\delta u - \xi_1) - \varphi^*|_{t=0} = 0, \qquad (2.8)$$

where $\bar{\varphi}$ is the solution of the original problem (1.7).

The problem (2.6)–(2.8) is a linear data assimilation problem; for $\bar{\varphi}^* = 0$ it is equivalent to the following minimization problem: find u and φ such that

$$\begin{cases} \dfrac{\partial \varphi}{\partial t} - F'(\bar{\varphi})\varphi &= \xi_3, \quad t \in (0, T) \\ \varphi|_{t=0} &= u \\ S_1(u) &= \inf_v S_1(v), \end{cases} \qquad (2.9)$$

where

$$S_1(u) = \frac{\alpha}{2}\|u - \xi_1\|_X^2 + \frac{1}{2}\|C\varphi - \xi_2\|_{Y_{obs}}^2. \qquad (2.10)$$

Consider the Hessian $H(u)$ of the functional (1.2) which is defined by the successive solution of the following problems:

$$\begin{cases} \dfrac{\partial \varphi}{\partial t} &= F(\varphi) + f, \quad t \in (0, T) \\ \varphi|_{t=0} &= u, \end{cases}$$

$$\begin{cases} -\dfrac{\partial \varphi^*}{\partial t} - (F'(\varphi))^*\varphi^* &= -C^*(C\varphi - \varphi_{obs}), \quad t \in (0, T) \\ \varphi^*|_{t=T} &= 0, \end{cases}$$

$$\begin{cases} \dfrac{\partial \psi}{\partial t} - F'(\varphi)\psi &= 0, \ t \in (0, T), \\ \psi|_{t=0} &= v, \end{cases}$$

$$\begin{cases} -\dfrac{\partial \psi^*}{\partial t} - (F'(\varphi))^*\psi^* &= (F''(\varphi)\psi)^*\varphi^* - C^*C\psi, \ t \in (0, T) \\ \psi^*|_{t=T} &= 0, \end{cases}$$

$$H(u)v = \alpha v - \psi^*|_{t=0}.$$

For $u = \bar{u}$, $\xi_2 = \xi_3 = 0$, the the Hessian $H(\bar{u})$ coincides with the Hessian H of the functional (2.10); it is defined by the successive solutions of the following problems:

$$\begin{cases} \dfrac{\partial \psi}{\partial t} - F'(\bar{\varphi})\psi &= 0, \ t \in (0, T), \\ \psi|_{t=0} &= v, \end{cases} \tag{2.11}$$

$$\begin{cases} -\dfrac{\partial \psi^*}{\partial t} - (F'(\bar{\varphi}))^*\psi^* &= -C^*C\psi, \ t \in (0, T) \\ \psi^*|_{t=T} &= 0, \end{cases} \tag{2.12}$$

$$Hv = \alpha v - \psi^*|_{t=0}. \tag{2.13}$$

Let us introduce the operator R_2 acting on the functions $g \in Y_{obs}$ according to the formula

$$R_2 g = \theta^*|_{t=0}, \tag{2.14}$$

where θ^* is the solution to the adjoint problem

$$\begin{cases} -\dfrac{\partial \theta^*}{\partial t} - (F'(\bar{\varphi}))^*\theta^* &= C^*g, \ t \in (0, T) \\ \theta^*|_{t=T} &= 0. \end{cases} \tag{2.15}$$

We introduce also the operator $R_3 : Y \to X$ defined successively by the formulas:

$$\begin{cases} \dfrac{\partial \theta_1}{\partial t} - F'(\bar{\varphi})\theta_1 &= h, \ h \in Y, \\ \theta_1|_{t=0} &= 0, \end{cases} \tag{2.16}$$

$$\begin{cases} -\dfrac{\partial \theta_1^*}{\partial t} - (F'(\bar{\varphi}))^*\theta_1^* &= -C^*C\theta_1, \ t \in (0, T) \\ \theta_1^*|_{t=T} &= 0, \end{cases} \tag{2.17}$$

$$R_3 h = \theta_1^*|_{t=0}. \tag{2.18}$$

From (2.11)–(2.18) we conclude that the system (2.6)–(2.8) is equivalent to the one equation for δu:

$$H\delta u = R_1 \xi_1 + R_2 \xi_2 + R_3 \xi_3, \tag{2.19}$$

where $R_1 = \alpha E$, and E is the identity operator in X.

For $\alpha > 0$, the operator H is coercive, H^{-1} exists, and the error δu of the optimal solution depends on the errors ξ_1, ξ_2, ξ_3 linearly and continuously. The influence of the errors ξ_1, ξ_2, ξ_3 on the value of δu is determined by the operators $H^{-1}R_1, H^{-1}R_2, H^{-1}R_3$, respectively. The values of the norms of these operators may be considered as an influence criteria: the less is the norm of the operator $H^{-1}R_i$, the less impact on δu is given by the corresponding error ξ_i. This criteria may be used also to choose the regularization parameter α (Tikhonov *et al.*, 1995; Morozov, 1987).

3. Fundamental Control Functions for Error Analysis

Consider the case that $Y_{obs} = Y$, $C = E$ (the identity operator). We assume that the operator H defined by (2.11)–(2.13) is positive (i.e. the inequality holds: $(Hv, v)_X > 0$, $v \neq 0$) and has a complete orthonormal system in X of eigenfunctions v_k corresponding to the eigenvalues μ_k:

$$Hv_k = \mu_k v_k, \tag{3.1}$$

where $(v_k, v_l)_X = \delta_{kl}$, $k, l = 1, 2, \ldots$, and δ_{kl} the Kronecker delta.

It is easily seen that the eigenvalue problem (3.1) is equivalent to the system:

$$\begin{cases} \dfrac{\partial \varphi_k}{\partial t} - F'(\bar\varphi)\varphi_k &= 0, \ t \in (0, T), \\ \varphi_k|_{t=0} &= v_k, \end{cases} \tag{3.2}$$

$$\begin{cases} -\dfrac{\partial \varphi_k^*}{\partial t} - (F'(\bar\varphi))^*\varphi_k^* &= -\varphi_k, \ t \in (0, T) \\ \varphi_k^*|_{t=T} &= 0, \end{cases} \tag{3.3}$$

$$\alpha v_k - \varphi_k^*|_{t=0} = \mu_k v_k. \tag{3.4}$$

We say that the system of functions $\{\varphi_k, \varphi_k^*, v_k\}$ satisfying (1.4)–(1.6) is the system of *fundamental control functions*. It can be easily proved that the functions $\{\varphi_k\}, \{\varphi_k^*\}, \{v_k\}$ defined by formulas (3.2)–(3.4) form complete orthonormal systems in the corresponding spaces.

Using the fundamental control functions, we can obtain the solution of the error equation (2.19) in the explicit form. The equation (2.19) is equivalent to the system (2.6)–(2.8) and may be written as the following system:

$$\begin{cases} \dfrac{\partial \psi}{\partial t} - F'(\bar\varphi)\psi &= 0, \ t \in (0, T), \\ \psi|_{t=0} &= \delta u, \end{cases} \tag{3.5}$$

$$\begin{cases} -\dfrac{\partial \psi^*}{\partial t} - (F'(\bar\varphi))^*\psi^* &= -\psi, \ t \in (0, T) \\ \psi^*|_{t=T} &= 0, \end{cases} \tag{3.6}$$

$$\alpha\delta u - \psi^*|_{t=0} = P, \tag{3.7}$$

where $P = R_1\xi_1 + R_2\xi_2 + R_3\xi_3$ is the right-hand side of (2.19).

The following theorem holds (Shutyaev, 2001).

Theorem 3.1. *The solution $\psi, \psi^*, \delta u$ of the system (3.5)–(3.7) may be represented in the form:*

$$\psi = \sum_k a_k\varphi_k, \quad \psi^* = \sum_k a_k\varphi_k^*, \quad \delta u = \sum_k a_k v_k, \tag{3.8}$$

where $\varphi_k, \varphi_k^*, v_k$ are the fundamental control functions defined by (3.2)–(3.4), $a_k = (P, v_k)_X / \mu_k$.

From (3.8), we have the representation for the Fourier coefficients $(\delta u)_k$ of the error δu:

$$(\delta u)_k = (\delta u, v_k)_X = a_k = \frac{1}{\mu_k}(R_1\xi_1 + R_2\xi_2 + R_3\xi_3, v_k)_X. \qquad (3.9)$$

Note that

$$(R_1\xi_1, v_k)_X = \alpha(\xi_1, v_k)_X. \qquad (3.10)$$

By definition of R_2, R_3,

$$(R_2\xi_2, v_k)_X = (\theta^*|_{t=0}, v_k)_X, \quad (R_3\xi_3, v_k)_X = (\theta_1^*|_{t=0}, v_k)_X,$$

where θ^* is the solution of (2.15) for $g = \xi_2$, and θ_1, θ_1^* are the solutions of (2.16)–(2.17) for $h = \xi_3$. From (2.15) and (3.2) we get

$$(\theta^*|_{t=0}, v_k)_X = (\xi_2, \varphi_k)_Y.$$

Hence,

$$(R_2\xi_2, v_k)_X = (\xi_2, \varphi_k)_Y. \qquad (3.11)$$

Analogously, from (2.17) and (3.2),

$$(\theta_1^*|_{t=0}, v_k)_X = (-\theta_1, \varphi_k)_Y.$$

Further, (2.16) and (3.3) give

$$(\theta_1, -\varphi_k)_Y = (\theta_1|_{t=0}, \varphi_k^*|_{t=0})_X + (\xi_3, \varphi_k^*)_Y = (\xi_3, \varphi_k^*)_Y.$$

Hence,

$$(R_3\xi_3, v_k)_X = (\xi_3, \varphi_k^*)_Y. \qquad (3.12)$$

From (3.9)–(3.12) we obtain the expression for the Fourier coefficients $(\delta u)_k$ of the error δu of the optimal solution through the errors ξ_1, ξ_2, ξ_3:

$$(\delta u)_k = \frac{\alpha}{\mu_k}(\xi_1, v_k)_X + \frac{1}{\mu_k}(\xi_2, \varphi_k)_Y + \frac{1}{\mu_k}(\xi_3, \varphi_k^*)_Y, \qquad (3.13)$$

where $\{\varphi_k, \varphi_k^*, v_k\}$ are the fundamental control functions defined by (3.2)–(3.4).

In more general case, when $Y_{obs} \neq Y$, $C \neq E$, we may also define the control functions $\varphi_k, \varphi_k^*, v_k$ as the solutions to the system:

$$\begin{cases} \dfrac{\partial \varphi_k}{\partial t} - F'(\bar{\varphi})\varphi_k &= 0, \ t \in (0, T), \\ \varphi_k|_{t=0} &= v_k, \end{cases} \qquad (3.14)$$

$$\begin{cases} -\dfrac{\partial \varphi_k^*}{\partial t} - (F'(\bar{\varphi}))^* \varphi_k^* = -C^*C\varphi_k, \quad t \in (0, T) \\ \varphi_k^*|_{t=T} = 0, \end{cases} \quad (3.15)$$

$$\alpha v_k - \varphi_k^*|_{t=0} = \mu_k v_k. \quad (3.16)$$

It may be easily verified that in this case the error relationship (3.13) changes to

$$(\delta u)_k = \frac{\alpha}{\mu_k}(\xi_1, v_k)_X + \frac{1}{\mu_k}(\xi_2, C\varphi_k)_{Y_{obs}} + \frac{1}{\mu_k}(\xi_3, \varphi_k^*)_Y. \quad (3.17)$$

From (3.13), (3.17), it is seen that the fundamental control functions play a role of "sensitivity functions"; they are the weight-functions for the corresponding errors ξ_1, ξ_2, ξ_3 in the representations (3.13), (3.17). Note that the fundamental control functions $\{\varphi_k, \varphi_k^*, v_k\}$ do not depend on the structure of the errors ξ_1, ξ_2, ξ_3 and may be calculated beforehand for each k in need.

4. Singular Vectors and Error Sensitivity

Consider the error equation (2.19). Under the hypotheses of Theorem 3.1, we may rewrite (2.19) as

$$\delta u = H^{-1}R_1\xi_1 + H^{-1}R_2\xi_2 + H^{-1}R_3\xi_3. \quad (4.1)$$

Suppose that the errors ξ_1, ξ_2, ξ_3 do not correlate and the following relation is satisfied:

$$\|\delta u\|_X^2 = \|T_1\xi_1\|_X^2 + \|T_2\xi_2\|_X^2 + \|T_3\xi_3\|_X^2, \quad (4.2)$$

where $T_i = H^{-1}R_i$. From (4.2),

$$\|\delta u\|_X^2 = (T_1^*T_1\xi_1, \xi_1)_X + (T_2^*T_2\xi_2, \xi_2)_{Y_{obs}} + (T_3^*T_3\xi_3, \xi_3)_Y, \quad (4.3)$$

where $T_1^* : X \to X$, $T_2^* : X \to Y_{obs}$, $T_3^* : X \to Y$ are the adjoints to $T_i, i = 1, 2, 3$.

Each summand in (4.3) determines the impact given by the corresponding error ξ_i. We have

$$(T_1^*T_1\xi_1, \xi_1)_X \le \|T_1^*T_1\|\|\xi_1\|_X^2, \quad (T_2^*T_2\xi_2, \xi_2)_{Y_{obs}} \le \|T_2^*T_2\|\|\xi_2\|_{Y_{obs}}^2,$$

$$(T_3^*T_3\xi_3, \xi_3)_Y \le \|T_3^*T_3\|\|\xi_3\|_Y^2, \quad (4.4)$$

and the i-th inequality becomes an equality when ξ_i is the singular vector of T_i corresponding to the largest singular value $\sigma_{max}^2 = \|T_i^*T_i\|$. The values $r_i = \sqrt{\|T_i^*T_i\|}$ may be considered as sensitivity coefficients which clearly demonstrate the measure of influence of the corresponding error upon the optimal solution. The higher the relative sensitivity coefficient, the more effectual is the error in question.

As above, we assume that the Hessian H defined by (2.11)–(2.13) is positive and has a complete orthonormal system in X of eigenfunctions v_k corresponding to the eigenvalues μ_k: $Hv_k = \mu_k v_k$, $(v_k, v_l)_X = \delta_{kl}$.

Consider the operator T_1. Since $T_1 = H^{-1}R_1 = \alpha H^{-1} = T_1^*$, the singular vectors of T_1 are the eigenvectors v_i of the Hessian H, and the corresponding sensitivity coefficient is equal to

$$r_1 = \sqrt{\|T_1^*T_1\|} = \frac{\alpha}{\mu_{\min}}, \qquad (4.5)$$

where $\mu_{\min}$ is the lower spectrum bound of H.

For the operator $T_2 : Y_{obs} \to X$ the following statement is valid (Le Dimet, Ngnepieba, and Shutyaev, 2002).

Lemma 4.1. *The singular values σ_k^2 and the corresponding orthonormal (right) singular vectors $w_k \in Y_{obs}$ of the operator T_2 are defined by the formulas:*

$$\sigma_k^2 = \frac{\mu_k - \alpha}{\mu_k^2}, \quad w_k = \frac{1}{\sqrt{\mu_k - \alpha}}C\varphi_k, \qquad (4.6)$$

where μ_k are the eigenvalues of the Hessian H, and φ_k are the fundamental control functions defined by (3.14). The left singular vectors of T_2 coincide with the eigenvectors v_k of H:

$$T_2 T_2^* v_k = \sigma_k^2 v_k, \quad k = 1, 2, \ldots$$

Corollary 4.1. The sensitivity coefficient $r_2 = \sqrt{\|T_2^*T_2\|}$ is defined by the formula:

$$r_2 = \max_k \frac{\sqrt{\mu_k - \alpha}}{\mu_k}. \qquad (4.7)$$

The equality $(T_2^*T_2\xi_2, \xi_2)_{Y_{obs}} = r_2^2\|\xi_2\|_{Y_{obs}}^2$ holds if $\xi_2 = w_{k_0}$, where w_{k_0} is the singular vector of T_2 corresponding to the largest singular value $\sigma_{k_0}^2$.

Consider now the operator $T_3 = H^{-1}R_3$. To determine the sensitivity coefficient $r_3 = \sqrt{\|T_3^*T_3\|}$, we need to derive R_3^*. For $h \in Y, p \in X$, we have from (2.16)–(2.18):

$$(R_3h, p)_X = (\theta_1^*|_{t=0}, p)_X = -(C^*C\theta_1, \phi)_Y = -(C\theta_1, C\phi)_{Y_{obs}},$$

where θ_1, θ_1^* are the solutions to (2.16)–(2.17), and ϕ is the solution to (2.11) for $v = p$. Further,

$$(R_3h, p)_X = -(\theta_1, C^*C\phi)_Y = (h, \phi^*)_Y$$

and $R_3^* p = \phi^*$, where ϕ^* is the solution to the adjoint problem:

$$\begin{cases} -\dfrac{\partial \phi^*}{\partial t} - (F'(\bar{\varphi}))^* \phi^* &= -C^* C\phi, \ \ t \in (0, T) \\ \phi^*|_{t=T} &= 0. \end{cases} \qquad (4.8)$$

The operator $R_3 R_3^* : X \to X$ may be defined as follows: for given $p \in X$ find ϕ as the solution of (2.11) for $v = p$, find ϕ^* as the solution of (4.9), and for $h = \phi^*$ find θ_1, θ_1^* as the solutions of (2.16)–(2.17); then, put $R_3 R_3^* = \theta_1^*|_{t=0}$.

Therefore, the operator $T_3 T_3^* = H^{-1} R_3 R_3^* H^{-1}$ is defined by the successive solutions of the following problems (for given $v \in X$):

$$Hp = v, \qquad (4.9)$$

$$\begin{cases} \dfrac{\partial \phi}{\partial t} - F'(\bar{\varphi})\phi &= 0, \ t \in (0, T), \\ \phi|_{t=0} &= p, \end{cases} \qquad (4.10)$$

$$\begin{cases} -\dfrac{\partial \phi^*}{\partial t} - (F'(\bar{\varphi}))^* \phi^* &= -C^* C\phi, \ \ t \in (0, T) \\ \phi^*|_{t=T} &= 0. \end{cases} \qquad (4.11)$$

$$\begin{cases} \dfrac{\partial \theta_1}{\partial t} - F'(\bar{\varphi})\theta_1 &= \phi^*, \ \ t \in (0, T) \\ \theta_1|_{t=0} &= 0, \end{cases} \qquad (4.12)$$

$$\begin{cases} -\dfrac{\partial \theta_1^*}{\partial t} - (F'(\bar{\varphi}))^* \theta_1^* &= -C^* C\theta_1, \ \ t \in (0, T) \\ \theta_1^*|_{t=T} &= 0, \end{cases} \qquad (4.13)$$

$$Hw = \theta_1^*|_{t=0}, \qquad (4.14)$$

then

$$T_3 T_3^* v = w, \qquad (4.15)$$

and for the sensitivity coefficient r_3 we have

$$r_3 = \sqrt{\|T_3 T_3^*\|}. \qquad (4.16)$$

From the formulas (4.5), (4.7), (4.16), one can derive the typical behaviour of the sensitivity coefficients r_1, r_2, r_3, depending on the parameter α. The sensitivity coefficient r_1 is small and r_2, r_3 are large if α goes to zero and μ_{min} is close to zero. If α goes to 1, the coefficient r_1 also is close to 1, and r_2, r_3 are decreasing, being less than r_1. Thus, with α increasing, the output regularization error increases, whereas the sensitivity to the observation and model errors decreases. If α goes to zero, the output regularization error vanishes, however, the sensitivity to the observation and model errors is increasing (it is usually due to the fact that the Hessian of the cost functional becomes ill-conditioned with μ_{min} close to zero).

84

In most cases, $r_3 < r_2$, that is, the optimal solution is more sensitive to variations of the observation errors then to the model errors.

Thus, the sensitivity of the optimal solution to the input errors is determined by the value of the sensitivity coefficients which are the norms of the specific response operators relating the error of the input to the error of the optimal initial-value function. The maximum error growth for the output is given by the singular vectors of the corresponding response operator. The singular vectors are the fundamental control functions which form complete orthonormal systems in specific functional spaces and may be used for error analysis.

References

Agoshkov, V.I. and Marchuk, G.I. (1993) On solvability and numerical solution of data assimilation problems, *Russ. J. Numer. Anal. Math. Modelling* **8**, 1–16.

Cacuci, D.G. (1981) Sensitivity theory for nonlinear systems: II.Extensions to additional classes of responses, *J. Math. Phys* **22**, 2803–2812.

Chavent, G. (1983) Local stability of the output least square parameter estimation technique, *Math. Appl. Comp.* **2**, 3–22.

Dontchev, A.L. (1983) *Perturbations, Approximations and Sensitivity Analysis of Optimal Control Systems.* (Lecture Notes in Control and Information Sciences; 52), Springer, Berlin.

Gejadze, I.Yu. and Shutyaev, V.P. (1999) An optimal control problem of initial data restoration, *Comp. Math. Math. Phys.* **39**, 1416–1425.

Kravaris, C. and Seinfeld, J.H. (1985) Identification of parameters in distributed parameter systems by regularization, *SIAM J. Control and Optimization* **23**, 217.

Kurzhanskii, A.B. and Khapalov, A.Yu. (1991) An observation theory for distributed-parameter systems, *J. Math. Syst. Estimat. Control* **1**, 389-440.

Le Dimet, F.X., Navon, I.M., and Daescu, D.N. (2002) Second-order information in data assimilation, *Monthly Weather Review* **130**, 629–648.

Le Dimet, F.-X., Ngnepieba P., and Shutyaev, V.P. On error analysis in data assimilation problems. *Russ. J. Numer. Anal. Math. Modelling* (2002), **17** (1), 71–97.

Lions, J.-L. (1988) *Contrôllabilité Exacte Perturbations et Stabilisation de Systèmes Distribués*, Masson, Paris.

Marchuk, G.I. (1995) *Adjoint Equations and Analysis of Complex Systems*, Kluwer, Dordrecht.

Marchuk, G.I., Agoshkov, V.I., and Shutyaev, V.P. (1996) *Adjoint Equations and Perturbation Algorithms in Nonlinear Problems*, CRC Press Inc., New York.

Morozov, V.A. (1987) *Regular Methods for Solving the Ill-Posed Problems*, Nauka, Moscow.

Navon, I.M. (1995) Variational data assimilation, optimal parameter estimation and sensitivity analysis for environmental problems, in Atluri, Yagawa, and Cruse (eds.) *Computational Mechanics'95. V.1*, Springer, New York, pp.740–746.

Ngodock, H.E. (1995) *Assimilation de Données et Analyse de Sensibilité: une Application à la Circulation Océanique. Thèse de l'Université Joseph Fourier*, UJF, Grenoble.

Shutyaev, V.P. (1994) On a class of insensitive control problems, *Control and Cybernetics* **23**, 257–266.

Shutyaev, V.P. (2001) *Control Operators and Iterative Algorithms for Variational Data Assimilation Problems*, Nauka, Moscow.

Tikhonov, A.N., Leonov, A.S., Yagola, A.G. (1995) *Nonlinear Inverse Problems*, Nauka, Moscow.

A *POSTERIORI* VALIDATION OF ASSIMILATION ALGORITHMS

O. TALAGRAND
Laboratoire de Météorologie Dynamique / CNRS
École Normale Supérieure, Paris, France

1. Introduction

The theory of statistical linear estimation, upon which most of presently existing assimilation algorithms are based, has been succinctly described in the chapter *Bayesian Estimation. Optimal Interpolation. Statistical Linear Estimation* (which will hereafter be referred to as Part I). We recall eq. (4.1) of that chapter, which links the data vector z, belonging to data space $\mathcal{D}$, with dimension m, to the state vector x to be determined, belonging to state space $\mathcal{S}$, with dimension n

$$z = \Gamma x + \zeta \tag{1.1}$$

We have assumed the data error ζ to be centred [$E(\zeta)=0$], and to have covariance matrix $E(\zeta\zeta^{\mathrm{T}}) = S$. Assuming that ζ is centred is actually assuming that the expectation $E(\zeta)$ is known. If that expectation happened to be different from 0, it would suffice to subtract it from the data vector in order to obtain a centred error.

Statistical linear estimation intends at determining the *BLUE*, defined by eq. (4.2a) of Part I. On the face of it, that equation requires the *a priori* specification of the first- and second-order statistical moments of the error vector (the first-order moment being necessary for centring the error). That raises an obvious question. How can these moments be determined in the first place? And, if some estimate of these moments has been obtained, is it possible to validate it and to improve on it, for instance through appropriate statistical processing of the continuous flow of new observations? Other questions that naturally arise are the following. How to objectively evaluate the quality of an assimilation system? Is it possible to objectively determine if an assimilation is 'optimal', especially in the sense of least-variance which is at the basis of the theory of the *BLUE*? More generally, assimilation is a costly process, and it is legitimate to try and make the best of it. Is it possible to obtain more from assimilation than its primary purpose, namely as accurate as possible an estimate of the state of the flow, and of the associated estimation error? And, if yes, how is it possible to obtain more? This chapter discusses some of these questions, within the limits of statistical linear estimation, and provides at least a number of partial answers.

Starting from the description (1.1) of the data, the *BLUE* of the state vector x is equal to

$$x^{a} = (\Gamma^{\mathrm{T}} S^{-1} \Gamma)^{-1} \Gamma^{\mathrm{T}} S^{-1} z \tag{1.2}$$

R. Swinbank et al. (eds.), Data Assimilation for the Earth System, 85–95.

The matrix $A = (\Gamma^T S^{-1} \Gamma)^{-1} \Gamma^T S^{-1}$ is a left-inverse of Γ ($A\Gamma = I_m$, where I_m is the unit matrix of order m). This means in particular that the assimilation scheme (1.2), if implemented on exact data ($\zeta=0$), will produce the exact state vector. It can be shown that any left-inverse A' of Γ, whether it defines the *BLUE* or not, is of the form

$$A' = (\Gamma^T S'^{-1} \Gamma)^{-1} \Gamma^T S'^{-1} \qquad (1.3)$$

where S' is an ($m \times m$) symmetric positive definite matrix.

We will in this chapter consider assimilation schemes which, without necessarily being optimal (one question to be studied is precisely whether it is possible to determine if a given scheme is optimal), are built on a left-inverse (1.3) of Γ. In the background-observation description of the data (eqs 4.4a-b of Part I), this is equivalent to saying that the estimated state vector x^a is of the form

$$x^a = x^b + K[y - Hx^b] \qquad (1.4)$$

where the gain matrix K can be any $n \times p$ matrix ($p=m-n$).

2. Evaluation of the quality of an assimilation system

Considering first objective evaluation of the quality of assimilated fields, it cannot be achieved by comparison with data that have been used in the assimilation itself: it is intuitively obvious, and it can be proved, that the fit to any particular piece of data can be arbitrarily reduced by decreasing the corresponding assumed error variance. Objective evaluation can be done, if at all, by comparison with data that have not been used in the assimilation. Let

$$w = p^T x + \gamma \qquad (2.1)$$

be such a scalar piece of data. In this equation, p is a vector belonging to the dual of state space, and γ is an error, which we assume to be centred. Comparison between w and the analogue $w^a = p^T x^a$ of $p^T x$ in the assimilation yields the difference

$$w - w^a = p^T(x - x^a) + \gamma = -p^T A'\zeta + \gamma \qquad (2.2)$$

where use has been made of the assumption that the assimilation operator A' is a left-inverse of the observation operator Γ. Looking first at the expectation $E(w-w^a)$, a non-zero value, under the assumption that γ is centred, would mean that the expectation $E(x-x^a)$ is non zero, *i.e.* that a bias in the data error ζ has not been properly taken into account. If the difference $w-w^a$ is centred (possibly as a consequence of an *a posteriori* unbiasing), the variance $E[(w-w^a)^2]$ will depend on the covariance between the errors ζ and γ. If these errors are uncorrelated, the variance reads

$$E[(w-w^a)^2] = p^T A'SA'^T p + E[\gamma^2] \qquad (2.3)$$

The second term on the right-hand side is independent of the assimilation system, and the variance $E[(w-w^a)^2]$ therefore provides an objective measure of the quality of an assimilation system (strictly speaking, of the quality of the estimate $w^a = p^T x^a$). The smaller the variance, the better the quality of the assimilation. Tests of the form (2.3) are commonly performed, but they are entirely based on the assumptions that the observation error γ is centred, and uncorrelated with the error ζ.

If the errors ζ and γ are correlated, objective evaluation is still possible, but the mutual correlation must be taken into account. The observation w is then to be compared to its own *BLUE* from the data vector z. That *BLUE* is no more equal to $p^T x^a$, and can be shown to be equal to

$$w^a = p^T x^a + c^T S^{-1}(z - \Gamma x^a) \tag{2.4}$$

where c is the covariance vector $E(\gamma \zeta)$.

3. Evaluation of the optimality of an assimilation system

Statistical linear estimation, as described in Part I, is intended at producing the linear combination of the data that fits best, in the least-variance sense, the real meteorological fields. In the backgound-observation decomposition of the data, the analysed state can be expressed, among other forms, by eq. (4.9a) of Part I, which we rewrite here

$$x^a = x^b + BH^T [HBH^T + R]^{-1} [y - Hx^b] \tag{3.1}$$

This equation expresses that the increment $x^a - x^b$ is the orthogonal projection, in the sense of covariance, of the difference $x - x^b$ onto the space spanned by the innovation vector defined by eq. (4.7) of Part I, *viz.*, $d \equiv y - Hx^b$. As a consequence, the estimation error $x - x^a$ is necessarily orthogonal to the innovation vector

$$E[(x - x^a)d^T] = 0 \tag{3.2}$$

as can be verified directly from eq. (3.1). Condition (3.2) characterizes the *BLUE* of x from the data z, in the sense that an unbiased estimate x^a of x is the *BLUE* of x if and only if it verifies condition (3.2).

Considering now an observation of form (2.1), optimality of the estimate x^a is equivalent to the two conditions

$$E(w - w^a) = 0 \tag{3.3a}$$

$$E[(w - w^a)d^T] = 0 \tag{3.3b}$$

where w^a is the *BLUE* of w from the data z, given by the general expression (2.4). Eqs (3.3) define an objective criterion for the optimality of an estimate of w. That criterion is absolutely general. It is often expressed by saying that a sequential assimilation process is optimal (*i.e.*, is the exact Kalman filter) if, and only if, the sequence of innovation vectors is unbiased and uncorrelated (Kailath, 1968).

Although eqs (3.3) provide an objective and practically usable criterion for optimality of an assimilation process, it has not been used much so far in meteorology and oceanography. The only example seems to be a study by Daley (1992), who computed the correlation of the innovation sequence for the sequential assimilation system that was then in use at the Canadian Meteorological Centre (that system is described by Mitchell *et al.*, 1990). Daley found significantly non-zero correlations, reaching values of more than 0.4 for the 500-hPa geopotential innovation, at time-lag 12 hours. It is reasonable to assume that operational assimilation systems have significantly improved in the more than ten years that have passed since, and it would be very instructive to perform similar diagnostics on present systems.

It is seen that in all cases, objective evaluation of the quality or of the optimality of estimated fields requires the *a priori* knowledge of the expectation of the errors affecting the verifying observations, and of the covariance between those errors and the data errors. Those quantities cannot however be objectively estimated, at least not on the basis of the data and the observations.

4. The innovation and data-minus-analysis vectors

As already mentioned, a basic question is whether the first- and second-order statistical moments of the data error, whose knowledge is in principle required for estimating the *BLUE*, can be objectively determined from the data themselves. Consider the case when two observations of a scalar quantity x are available in the form

$$z_1 = x + \zeta_1 \tag{4.1a}$$

$$z_2 = x + \zeta_2 \tag{4.1b}$$

The only combination of those data that is independent of x is the difference $z_1 - z_2 = \zeta_1 - \zeta_2$. Assume statistics accumulated on the data have shown that difference to have expectation m and variance σ^2, *viz.*,

$$E[z_1 - z_2] = m \tag{4.2a}$$

$$E[(z_1 - z_2)^2] = m^2 + \sigma^2 \tag{4.2b}$$

What can be obtained from those statistics concerning the *BLUE* x^a of x? Assume the errors ζ_1 and ζ_2 have respective expectations

$$E(\zeta_1) = m_1 \quad ; \quad E(\zeta_2) = m_2 \tag{4.3a}$$

and covariance matrix

$$E(\zeta'_1 \ \zeta'_2) = \begin{pmatrix} s_1^2 & cs_1 s_2 \\ cs_1 s_2 & s_2^2 \end{pmatrix} \qquad |c| < 1 \tag{4.3b}$$

where the prime (') denotes a centred variable. Eqs (4.2) imply

$$m_1 - m_2 = m \tag{4.4a}$$

$$s_1^2 + s_2^2 - 2c\, s_1\, s_2 = \sigma^2 \tag{4.4b}$$

The *BLUE* x^a of x, and the associated estimation error variance p^a, are given by eqs (4.2) of Part I which, taking into account the presence of the biases m_1 and m_2, read here

$$x^a = (1/2 + \alpha)\, z_1 + (1/2 - \alpha)\, z_2 - \mu \tag{4.5a}$$

$$p^a = s_1^2\, s_2^2\, (1 - c^2) / (s_1^2 + s_2^2 - 2c\, s_1\, s_2) \tag{4.5b}$$

where

$$\alpha = (1/2)\, (s_2^2 - s_1^2) / (s_1^2 + s_2^2 - 2c\, s_1\, s_2) \tag{4.6a}$$

$$\mu = (1/2 + \alpha)\, m_1 + (1/2 - \alpha)\, m_2 \tag{4.6b}$$

The three quantities α, μ, $p^a > 0$ entirely define the estimation scheme and the associated error, and the question is whether conditions (4.4) restrict the values that can be taken by those quantities. For any m, $\sigma^2 > 0$, α, μ, $p^a > 0$, eqs (4.4), (4.5b) and (4.6), considered as equations for the unknowns m_1, m_2, s_1, s_2, and c, possess the solution

$$m_1 = \mu - (\alpha - 1/2)\, m \quad ; \quad m_2 = \mu - (\alpha + 1/2)\, m \tag{4.7a}$$

$$s_1 = [(\alpha + 1/2)^2 \sigma^2 + p^a]^{1/2} \quad ; \quad s_2 = [(\alpha - 1/2)^2 \sigma^2 + p^a]^{1/2} \tag{4.7b}$$

$$c = (s_1^2 + s_2^2 - \sigma^2) / (2\, s_1\, s_2) \tag{4.7c}$$

Any estimation scheme (4.5) is therefore compatible with any innovation statistics (4.2). The knowledge of the latter is totally useless in that it does not restrict in any way the *BLUE* (4.5a), nor even the variance p^a (4.5b) of the associated estimation error.

This result is not due to some particular feature of the data (4.1), but is absolutely general. To see that, we first note that objective information on the error vector ζ (eq. 1.1) can be obtained only to the extent that some components of the data vector z can be mutually combined to produce quantities that are independent of x, *i.e.* to the extent that the estimation problem (1.1) is overdetermined. The combinations of the components of z that are independent of x make up the innovation vector, *viz.*,

$$d = y - Hx^b = -H\zeta^b + \varepsilon \tag{4.8}$$

where ζ^b and ε are respectively the background and observation errors (eqs 4.4 of Part I). The *BLUE* is then given by eq. (3.1), where we note that the matrix $HBH^T + R$ is the covariance matrix of the innovation vector (see eqs 4.4c and 4.5 of Part I). A basic problem becomes clearly apparent here. The matrix $HBH^T + R$ can be determined from accumulation of statistics on the data, but the same is not true of the matrix BH^T that appears in front of it in eq. (3.1). Determining the *BLUE* requires knowing, at least to some extent, how the background and observation errors contribute individually to the innovation vector. That obviously cannot be known from the innovation vector alone. Clearly, consistency between *a priori* assumed and *a posteriori* observed statistics of the innovation vector cannot be a sufficient condition for optimality of the estimation process.

Consider the difference

$$\delta \equiv z - \Gamma x^a \tag{4.9}$$

i.e. the *a posteriori* difference between the data vector and the analogue of the data in the analysed fields. That vector will be called the Data-minus-Analysis (briefly, DmA) difference. For an assimilation system of form (1.4), it reads

$$\delta = \begin{pmatrix} x^b - x^a \\ y - Hx^a \end{pmatrix} = \begin{pmatrix} -Kd \\ (I_p - HK)d \end{pmatrix} \tag{4.10}$$

For given gain matrix K, the DmA difference is a linear invertible function of the innovation vector d. It is therefore exactly equivalent to perform statistics on either one of those two vectors.

It has been shown in Part I, and discussed in detail in the chapter *Variational Assimilation. Adjoint Equations*, that the *BLUE* can be obtained by minimization of the scalar objective function

$$\mathcal{J}(\xi) \equiv (1/2) [\Gamma\xi - z]^T S^{-1} [\Gamma\xi - z] \tag{4.11}$$

which is defined on state space $\mathcal{S}$. Given any two vectors u and v in data space $\mathcal{D}$, the quantity

$$(1/2) u^T S^{-1} v \tag{4.12}$$

is easily verified to be invariant in a change of linear coordinates. It therefore defines a scalar product on data space, called the *Mahalanobis scalar product* associated with the covariance matrix S. Now, saying that the *BLUE* x^a minimises the objective function (4.11) means that Γx^a is the point in the image space $\Gamma(\mathcal{S})$ that lies closest, in the sense of the Mahalanobis distance (4.12), to the data vector z. Γx^a is therefore the orthogonal projection, in the S-Mahalanobis sense, of z onto $\Gamma(\mathcal{S})$. The *BLUE* is thus obtained by first projecting the data vector z onto $\Gamma(\mathcal{S})$, and then taking the inverse of the projection through Γ. The determinacy condition $\text{rank}\Gamma = n$ (see Part I) insures that the inverse is uniquely defined.

Let us decompose the data space $\mathcal{D}$ into the image space $\Gamma(\mathcal{S})$ and the space $\perp\Gamma(\mathcal{S})$ orthogonal to $\Gamma(\mathcal{S})$ according to the S-Mahalanobis scalar product. In that decomposition, the covariance matrix S reads

$$S = \begin{pmatrix} S_1 & 0 \\ 0 & S_2 \end{pmatrix} \tag{4.13}$$

where S_1 and S_2 are positive definite matrices with respective dimensions $n \times n$ and $p \times p$. The data matrix Γ reads

$$\Gamma = \begin{pmatrix} \Gamma_1 \\ 0 \end{pmatrix} \tag{4.14}$$

where Γ_1 is an $(n \times n)$-matrix which, because of the determinacy condition, is invertible. Denoting by ζ_1 and ζ_2 the projections of the error vector ζ onto the space $\Gamma(\mathcal{S})$ and the orthogonal space $\perp\Gamma(\mathcal{S})$, the components of the data vector z read

$$z_1 = \Gamma_1 x + \zeta_1 \tag{4.15a}$$

$$z_2 = \zeta_2 \tag{4.15b}$$

For the sake of generality, we now assume that the error vector may have nonzero expectation. The projection onto $\Gamma(\mathcal{S})$ of the unbiased data vector is equal to $z_1 - E(\zeta_1)$, and the *BLUE* x^a is equal to:

$$x^a = \Gamma_1^{-1}[z_1 - E(\zeta_1)] \tag{4.16a}$$

As for the corresponding estimation error covariance matrix, it is equal to

$$P^a = \Gamma_1^{-1} S_1 \Gamma_1^{-T} \tag{4.16b}$$

as can be seen for instance from eq. (2.13b) of Part I. Considering now the DmA difference (4.9), it is orthogonal to $\Gamma(\mathcal{S})$ and, considered as a vector of $\perp\Gamma(\mathcal{S})$, equal to ζ_2. It therefore has expectation $E(\zeta_2)$ and covariance matrix S_2. These quantities are independent of the quantities $E(\zeta_1)$ and S_1 which determine the *BLUE* and the associated error. This shows that any expectation and covariance matrix of the DmA difference are compatible with any estimation scheme of form (4.16), and generalizes the result obtained on data (4.1). The knowledge of the statistics of the DmA difference, or equivalently of the innovation vector, is totally useless for determining the *BLUE* or the associated error covariance matrix. As said above, consistency between the observed and specified statistics of the innovation vector is not a sufficient condition for optimality of the estimate. More than that, it is not even a necessary condition.

This is true of course in the absence of any knowledge on the data error other than the innovation vector. For example, if it is known that the observational errors ζ_1 and ζ_2 in data (4.1) are statistically uncorrelated, then the parameter α must be comprised between $-1/2$ and $+1/2$, and the estimation error variance p^a is at most equal to $\sigma^2/2$. But the conclusion is that, in order to be able to draw inferences from statistics of the innovation vector, it is necessary to make independent hypotheses. This is actually very commonly done, sometimes implicitly. A systematic bias in the innovation vector, at least in the components of the innovation vector obtained by comparison with well-calibrated observations, is usually interpreted as resulting from a bias in the background. Several authors (Hollingsworth and Lönnberg, 1986, Daley, 1993) have studied the horizontal correlation of the innovation vector obtained from radiosonde observations. It is reasonable to assume that radiosonde observation errors are horizontally uncorrelated. In these conditions, a direct estimate of the horizontal correlation of the forecast error can be obtained. In addition, if the observation and forecast errors are supposed to be uncorrelated, the residual obtained by extrapolating the covariance to zero horizontal distance provides an estimate of observation error variance. One conclusion from these studies is that the 6-hour forecast error is typically of the same magnitude as the observational error.

But independent hypotheses are always necessary. If one wants to use the statistics of the innovation vector, or of any quantity derived from the innovation vector, for drawing inferences that can be useful for the assimilation, one must necessarily use independent hypotheses that cannot be objectively validated, at least not on the basis of the innovation vector.

5. Various diagnostics of consistency

The number of statistical diasgnostics that can in principle be implemented on either the innovation or the DmA vector is in practice unlimited. Critical aspects are the statistical significance of the diagnostics on the one hand, and whether or not independent appropriate information is available for usefully exploiting any observed inconsistency between the *a priori* assumed and *a posteriori* observed statistics. Many *adaptive* schemes have been defined for progressively adjusting the expectation or covariance parameters on the basis of observed statistical inconsistencies. In the context of Kalman filtering, and independently of meteorological or oceanographical applications, one can mention the early works of Mehra (1972) and Godbole (1974). For more recent meteorological and oceanographical applications, see, *e.g.*, Blanchet *et al.* (1997) or Dee *et al.* (1999). We briefly describe here a number of basic diagnostics.

One first remark is that, in a consistent system, the innovation and the DmA vectors have zero expectation. If they are observed to have non-zero expectation, it necessarily means that a systematic bias in the data has not been properly taken into account. This fact has been systematically exploited by Dee and Da Silva (1998) who, assuming any bias to come from the background, have developed algorithms for constantly correcting the latter.

Using eq. (1.2), the covariance matrix of the DmA difference is shown to be equal to

$$E(\delta \delta^{T}) = S - \Gamma P^{a} \Gamma^{T} \tag{5.1}$$

where P^{a} is, as before, the covariance matrix of the estimation error. The term subtracted on the right-hand side is positive definite [it is actually the covariance matrix of the vector $\Gamma(x^{a}-x)$]. This means that the variance of any component of the DmA difference is less than the variance of the corresponding component of the error. Optimally assimilated fields statistically fit the data to within the accuracy of the latter (Hollingsworth and Lönnberg, 1989, have called *efficient* an assimilation system that possesses that particular property). If an unbiased assimilation system does not fit the data to within their assumed accuracy, that means that error covariance matrix S has been misspecified (the misspecification may of course be in the variance of the error affecting the ill-fitted data).

Let us consider a perfect assimilation system (*i.e.*, a system that exactly estimates the state vector). The DmA difference will be equal to the data error, *i.e.*, the second term on the right-hand side of eq. (5.1) will be zero. This means that if the intrinsic quality of a statistically optimal (but imperfect) system increases (as a result for instance of an increase of the number of assimilated data), the fit of the assimilated fields to the data must *increase*, and asymptote to the variance of the data error. This confirms, if need be, that the closeness of the fit of the estimated fields to the data cannot be a measure of the quality of those fields.

Consider a set of observations affected by uncorrelated observations. The corresponding submatrix of S is therefore diagonal. Eq. (5.1) shows by continuity that the DmA difference must be *negatively* correlated at short spatial distance. Hollingsworth and Lönnberg (1989) describe an example of a positive short-distance correlation of the DmA difference, which is necessarily the sign of a misspecification somewhere in the matrix S.

The objective function (4.11) assumes at its minimum x^a the value

$$\mathcal{J}(x^a) = (1/2) [\Gamma x^a - z]^T S^{-1} [\Gamma x^a - z] \tag{5.2}$$

It is the squared S-Mahalanobis norm of the DmA difference. Using the background-observation decomposition, standard matrix manipulations lead to

$$\mathcal{J}(x^a) = (1/2) d^T [HBH^T + R]^{-1} d \tag{5.3}$$

As already mentioned, the matrix $HBH^T + R$ is the covariance matrix $E(dd^T)$ of the innovation d, and $\mathcal{J}(x^a)$ is therefore the squared Mahalanobis norm of d with respect to its own covariance matrix. Writing eq. (5.3) in the basis of the principal components of d, in which $E(dd^T)$ is the unit matrix I_p, it is easily seen that, on statistical expectation

$$E[\mathcal{J}(x^a)] = p/2 \tag{5.4}$$

Eq. (5.4) provides a very simple diagnostic of global consistency of an assimilation algorithm. A number of authors (Ménard and Chang, 2000, Talagrand and Bouttier, 2000, Cañizares *et al.*, 2001, see also Bennett, 2002) have implemented this diagnostic on various systems. A rather general conclusion is that the expectation $E[\mathcal{J}(x^a)]$ is often significantly smaller (by a factor of typically between 1/2 and 2/3) than the theoretical value $p/2$. As seen from eq. (5.3), this means that the assumed innovation covariance matrix $HBH^T + R$ is larger than it should be. This probably results from an overestimation of the background error covariance matrix B, and is actually not surprising. Given the choice between overestimating and underestimating the background error, the safer choice is overestimation. Underestimation will produce an analysed state that is too close to the background, and may eventually lead to a progressive divergence of the analysed states from the reality, as it is described by the observations. It is therefore not surprising that systems that have been 'tuned' over long periods of time to produce a matrix B that itself produces reasonable forecasts in all circumstances tend to overestimate the background error. But, if the interpretation given here is correct, there is clearly a possibility for improvement in the specification of the matrix B, particularly by making it more dependent on the current state of the flow.

The objective function (4.11) will most often be the sum of a number of independent terms, *viz.*,

$$\mathcal{J}(\xi) = \Sigma_{k=1,\ldots,K} \mathcal{J}_k(\xi) \tag{5.5}$$

where

$$\mathcal{J}_k(\xi) \equiv (1/2) [\Gamma_k \xi - z_k]^{\mathrm{T}} S_k^{-1} [\Gamma_k \xi - z_k] \tag{5.6}$$

In this equation, z_k is an m_k–dimensional component of the data vector z ($\Sigma_k m_k = m$), and the rest of the notation is obvious. It can be shown (Talagrand, 1999, Desroziers and Ivanov, 2001) that the expectation of the k-th term at the minimum is equal to

$$E[\mathcal{J}_k(x^a)] = (1/2) [m_k - \mathrm{tr}(\Gamma_k^{\mathrm{T}} S_k^{-1} \Gamma_k P^a)] \tag{5.7}$$

This formula includes eq. (5.4) as a particular case. Desroziers and Ivanov (2001) have defined an economical algorithm for computing the matrix trace on the right-hand side of eq. (5.7). They have shown that, if the observation error is supposed to be uncorrelated in space and uncorrelated with the background error, eq. (5.7) can be used for estimating the observation and background error variances (see also Chapnik et al., 2003). This is in essence a systematic extension of the already mentioned works by Hollingsworth and Lönnberg (1986), and Daley (1993).

The sum over k of the traces on the right-hand side of eq. (5.7) is equal to n, and it results that the quantity $\mathrm{tr}(\Gamma_k^{\mathrm{T}} S_k^{-1} \Gamma_k P^a)/n$ measures the relative contribution of the subset z_k of the data vector to the overall accuracy of the assimilation system. This potentially defines a very powerful tool, for instance in so-called Observing System Simulation Experiments, intended at estimating the usefulness of hypothetical systems of observations. Eq. (5.7) (like actually all equations in Part I and in this chapter) is valid in any system of coordinates in data and state spaces, and the measure $\mathrm{tr}(\Gamma_k^{\mathrm{T}} S_k^{-1} \Gamma_k P^a)/n$ is absolutely intrinsic. In particular, given any subset z' of the data vector, its contribution to the objective function is independent of whether, in the coordinates used in data space, the errors affecting z' are correlated or not with the errors affecting the other components of z. Eq. (5.7) is therefore valid, and can be used as a diagnostic tool, for any subset of data in any basis in data space.

We have presented a short overview of the theory and practical implementation of a posteriori validation, and possible improvement, of assimilation algorithms. The approach that has been described here is explicitly intended at determining the first- and second-order statistical moments of the data error as such. A different approach is to determine the parameters of the estimation process, independently of any a priori statistical significance, as those that minimize the a posteriori estimation error, as it is measured against data assumed to be independent. The difficulty there is to define an efficient numerical algorithm which, starting from the observed estimation error, proceeds back to the estimation parameters in order to further decrease the estimation error. Hoang et al. (1997, 2001), using the power of the adjoint approach, have defined such an algorithm, which they call Reduced-Order Adaptive Filtering (ROAC). In essence, the ROAC uses the adjoint of the assimilation process itself in order to determine the values of the estimation parameters (actually, of a reduced set of those parameters) that statistically minimize the estimation error. The ROAC has been shown to work very efficiently in a number of experiments performed with simulated data. It may constitute a very useful alternative, or rather complement, to the general approach described above.

References

Bennett, A. F. (2002) *Inverse Modeling of the Ocean and Atmosphere*, Cambridge University Press, Cambridge, United Kingdom.

Blanchet, I., Frankignoul, C. and Cane, M. A. (1997) A Comparison of Adaptive Kalman Filters for a Tropical Pacific Ocean Model, *Mon. Wea. Rev.* **125**, 40-58.

Cañizares, R., Kaplan, A., Cane, M. A., Chen, D. and Zebiak, S. E. (2001) Use of data assimilation via linear low-order models for the initialization of El Niño - Southern Oscillation predictions, *J. Geophys. Res.*, **106** (C12) 30,947-30,959.

Chapnik, B., Desroziers, G., Rabier, F. and Talagrand, O. (2003) Properties and first application of an error statistics tuning method in variational assimilation, submitted for publication in *Q. J. R. Meteorol. Soc.*.

Daley, R. (1992) The Lagged Innovation Covariance: A Performance Diagnostic for Atmospheric Data Assimilation, *Mon. Wea. Rev.* **120**, 178-196.

Daley, R. (1993) Estimating observation error statistics for atmospheric data assimilation, *Ann. Geophysicae* **11**, 634-647.

Dee, D. P. and Da Silva, A. M. (1998) Data assimilation in the presence of forecast bias, *Q. J. R. Meteorol. Soc.* **124**, 269-295.

Dee, D. P., Gaspari, G., Redder, C., Rukhovets, L. and Da Silva, A. M. (1999) Maximum-Likelihood Estimation of Forecast and Observation Error Covariance Parameters. Part II: Applications, *Mon. Wea. Rev.* **127**, 1835-1849.

Desroziers, G. and Ivanov, S. (2001) Diagnosis and adaptive tuning of information error parameters in a variational assimilation, *Q. J. R. Meteorol. Soc.* **127**, 1433-1452.

Godbole, S. S. (1974) Kalman Filtering with No A Priori Information About Noise - White Noise Case : Identification of Covariances, *IEEE Trans. Automat. Contr.* **AC-19**, 561-563.

Hoang, H. S., Baraille, R. and Talagrand, O. (2001) On the design of a stable adaptive filter for state estimation in high dimensional systems, *Automatica* **37**, 341-359.

Hoang S., Baraille, R., Talagrand, O., Carton, X. and De Mey, P. (1997) Adaptive filtering: application to satellite data assimilation in oceanography, *Dyn. Atmos. Oceans*, **27**, 257-281.

Hollingsworth, A. and Lönnberg, P. (1986) The statistical structure of short-range forecast errors as determined from radiosonde data. Part I: The wind field, *Tellus* **38A**, 111-136.

Hollingsworth, A. and Lönnberg, P. (1989) The Verification of Objective Analyses: Diagnostic of Analysis System Performance, Meteorol. Atmos. Phys. 40, 3-27.

Kailath, T. (1968) An Innovations Approach to Least-Squares Estimation. Part I: Linear Filtering in Additive White Noise, *IEEE Trans. Automat. Contr.* **AC-13**, 646-655.

Mehra, R. K. (1972) Approaches to adaptive filtering, *IEEE Trans. Automat. Contr.* **15**, 693-698.

Ménard, R. and Chang, L.-P. (2000) Assimilation of Stratospheric Chemical Tracer Observations Using a Kalman Filter. Part II: χ^2-Validated Results and Analysis of Variance and Correlation Dynamics. *Mon. Wea. Rev.* **128**, 2672–2686.

Mitchell, H., Charette, C., Chouinard, C. and Brasnett, B. (1990) Revised interpolation statistics for the Canadian data assimilation procedure: Their derivation and application, *Mon. Wea. Rev.* **118**, 1591-1614.

Talagrand, O. (1999) A posteriori evaluation and verification of analysis and assimilation algorithms, in Proceedings of Workshop on *Diagnosis of Data Assimilation Systems* (November 1998). ECMWF, Reading, United Kingdom, pp. 17-28.

Talagrand, O. and Bouttier, F. (2000) Internal Diagnostics of Data Assimilation Systems, in Proceedings of Seminar on *Diagnosis of Models and Data Assimilation Systems* (September 1999). ECMWF, Reading, United Kingdom, pp. 407-409.

INTRODUCTION TO INITIALIZATION

PETER LYNCH

Met Éireann,
Glasnevin Hill, Dublin 9, Ireland.

1. Introduction

The spectrum of atmospheric motions is vast, encompassing phenomena having periods ranging from seconds to millennia. The motions of interest to the forecaster have timescales greater than a day, but the mathematical models used for numerical prediction describe a broader span of dynamical features than those of direct concern. For many purposes these higher frequency components can be regarded as *noise* contaminating the motions of meteorological interest. The elimination of this noise is achieved by adjustment of the initial fields, a process called *initialization*. In this chapter, the fundamental equations are examined and the causes of spurious oscillations are elucidated. The history of methods of eliminating high-frequency noise is recounted and various initialization methods are described. The normal mode initialization method is described, and illustrated by application to a simple mechanical system, the swinging spring.

1.1. RICHARDSON'S FORECAST

The story of Lewis Fry Richardson's forecast, made about eighty years ago, is well known. Richardson forecast the change in surface pressure at a point in central Europe, using the mathematical equations. He described his methods and results in his book *Weather Prediction by Numerical Process* (Richardson, 1922). His results implied a change in surface pressure of 145 hPa in 6 hours. As Sir Napier Shaw remarked, "the wildest guess ... would not have been wider of the mark ...". Yet, Richardson claimed that his forecast was "... a fairly correct deduction from a somewhat unnatural initial distribution"; he ascribed the unrealistic value of pressure tendency to errors in the observed winds, leading to a spuriously large value of the calculated divergence. This large tendency reflects the fact that the atmosphere can support motions with a great range of timescales.

R. Swinbank et al. (eds.), Data Assimilation for the Earth System, 97–111.

1.2. THE SPECTRUM OF ATMOSPHERIC MOTIONS

The natural oscillations of the atmosphere fall into two groups (see, *e.g.*, Kasahara, 1976). The solutions of meteorological interest have low frequencies and are close to geostrophic balance. They are called rotational or vortical modes, since their vorticity is greater than their divergence; if divergence is ignored, these modes reduce to the Rossby-Haurwitz waves. There are also very fast gravity-inertia wave solutions, with phase speeds of hundreds of metres per second and large divergence. For typical conditions of large scale atmospheric flow (when the Rossby and Froude numbers are small) the two types of motion are clearly separated and interactions between them are weak. The high frequency gravity-inertia waves may be locally significant in the vicinity of steep orography, where there is strong thermal forcing or where very rapid changes are occurring; but overall they are of minor importance and may be regarded as undesirable noise.

1.3. THE PROBLEM OF INITIALIZATION.

A subtle and delicate state of balance exists in the atmosphere between the wind and pressure fields, ensuring that the fast gravity waves have much smaller amplitude than the slow rotational part of the flow. Observations show that the pressure and wind fields in regions not too near the equator are close to a state of geostrophic balance and the flow is quasi-nondivergent. The bulk of the energy is contained in the slow rotational motions and the amplitude of the high frequency components is small. The existence of this geostrophic balance is a perennial source of interest; it is a consequence of the forcing mechanisms and dominant modes of hydrodynamic instability and of the manner in which energy is dispersed and dissipated in the atmosphere. For a review of balanced flow, see McIntyre (2003). The gravity-inertia waves are instrumental in the process by which the balance is maintained, but the nature of the sources of energy ensures that the low frequency components predominate in the large scale flow. The atmospheric balance is subtle, and difficult to specify precisely. It is *delicate* in that minor perturbations may disrupt it but *robust* in that local imbalance tends to be rapidly removed through radiation of gravity-inertia waves in a process known as geostrophic adjustment.

When the primitive equations are used for numerical prediction the forecast may contain spurious large amplitude high frequency oscillations. These result from anomalously large gravity-inertia waves which occur because the balance between the mass and velocity fields is not reflected faithfully in the analysed fields. High frequency oscillations of large amplitude are engendered, and these may persist for a considerable time unless strong dissipative processes are incorporated in the forecast model.

One of the long-standing problems in numerical weather prediction has been to overcome the problems associated with high frequency motions. This is achieved by the process known as *initialization*, the principal aim of which is to define the initial fields in such a way that the gravity inertia waves remain small throughout the forecast. If the fields are not initialized the spurious oscillations which occur in the forecast can lead to various problems. In particular, new observations are checked for accuracy against a short-range forecast. If this forecast is noisy, good observations may be rejected or erroneous ones accepted. Thus, *initialization is essential for satisfactory data assimilation.* Another problem occurs with precipitation forecasting. A noisy forecast has unrealistically large vertical velocity. This interacts with the humidity field to give hopelessly inaccurate rainfall patterns. To avoid this *spin-up*, we must control the gravity wave oscillations.

2. Scale-analysis of the Shallow Water Equations.

Considerable insight into the problem of initialization is achieved by consideration of the scale properties of the linear shallow water equations. We introduce characteristic scales for the dependent variables, and examine the relative sizes of the terms in the equations. Let $L = 10^6$ m represent the length scale and $V = 10\,\mathrm{m\,s^{-1}}$ the velocity scale. An advective time-scale $T = L/V = 10^5$ s is assumed. P represents the scale of pressure variations. For simplicity we take $f = 10^{-4}\,\mathrm{s^{-1}}$ and density $\rho_0 = 1\,\mathrm{kg\,m^{-3}}$ to be constants. The linear rotational shallow water equations are:

$$\underbrace{\frac{\partial u}{\partial t}}_{V/T} \; - \; \underbrace{fv}_{fV} \; + \; \underbrace{\frac{1}{\rho_0}\frac{\partial p}{\partial x}}_{P/L} \; = \; 0 \qquad (2.1)$$

$$\underbrace{\frac{\partial v}{\partial t}}_{V/T} \; + \; \underbrace{fu}_{fV} \; + \; \underbrace{\frac{1}{\rho_0}\frac{\partial p}{\partial y}}_{P/L} \; = \; 0 \qquad (2.2)$$

$$\underbrace{\frac{1}{\rho_0}\frac{\partial p}{\partial t}}_{P/T} \; + \; \underbrace{gH\left[\frac{\partial u}{\partial x} + \frac{\partial v}{\partial y}\right]}_{gHV/L} \; = \; 0 \qquad (2.3)$$

The scale of each term in the equations is indicated. The ratio of the velocity tendencies to the Coriolis terms is the Rossby number,

$$\mathrm{Ro} \equiv \frac{1}{fT} = \frac{V}{fL} = 10^{-1},$$

a small parameter. For balance in the momentum equations, we must have $P = LfV = 10^3$ Pa. Then the sizes of the terms are as indicated here:

$$\underbrace{\frac{\partial u}{\partial t}}_{10^{-4}} \; - \; \underbrace{fv}_{10^{-3}} \; + \; \underbrace{\frac{1}{\rho_0}\frac{\partial p}{\partial x}}_{10^{-3}} \; = \; 0 \qquad (2.4)$$

$$\underbrace{\frac{\partial v}{\partial t}}_{10^{-4}} + \underbrace{fu}_{10^{-3}} + \underbrace{\frac{1}{\rho_0}\frac{\partial p}{\partial y}}_{10^{-3}} = 0. \tag{2.5}$$

To the lowest order of approximation, the tendency terms are negligible and there is geostrophic balance between the Coriolis and pressure terms.

2.1. SCALING THE DIVERGENCE

The vorticity is the same scale as each of its components:

$$\zeta = \frac{\partial v}{\partial x} - \frac{\partial u}{\partial y} \sim \frac{V}{L} = 10^{-5}\,\mathrm{s}^{-1}.$$

Due to the cancellation between the two terms in the divergence, one might expect it to scale an order of magnitude smaller than each of its terms:

$$\delta = \frac{\partial u}{\partial x} + \frac{\partial v}{\partial y} \sim \mathrm{Ro}\frac{V}{L} = 10^{-6}\,\mathrm{s}^{-1}$$

This is generally appropriate but here we are considering barotropic motions of large vertical scale. If we assume this magnitude for the divergence, and take $g = 10\,\mathrm{m\,s}^{-2}$ and $H = 10^4$ m (the scale height of the atmosphere is approximately 10 km), the terms of the continuity equation scale as follows:

$$\underbrace{\frac{1}{\rho_0}\frac{\partial p}{\partial t}}_{10^{-2}} + \underbrace{gH\left[\frac{\partial u}{\partial x} + \frac{\partial v}{\partial y}\right]}_{10^{-1}} = 0, \tag{2.6}$$

which is *impossible*, as there is nothing to balance the second term. We recall that the divergence term arises through vertical integration:

$$g\int \delta\,dz \approx gH\left[\frac{\partial u}{\partial x} + \frac{\partial v}{\partial y}\right].$$

There is a strong tendency in the atmosphere towards cancellation between convergence at low levels and divergence at higher levels and vice-versa, called the Dines mechanism. Thus, we assume

$$\int \delta\,dz \sim \mathrm{Ro}\,\delta H\,, \qquad \text{so that} \qquad g\int \delta\,dz \sim \mathrm{Ro}^2 gH\frac{V}{L} = 10^{-2}\,.$$

The terms of the continuity equation are now brought into balance.

2.2. THE EFFECT OF DATA ERRORS

Suppose there is a 10% error Δv in the v-component of the wind observation at a point. The scales of the terms are as before:

$$\underbrace{\frac{\partial u}{\partial t}}_{10^{-4}} - \underbrace{f(v + \Delta v)}_{10^{-3}} + \underbrace{\frac{1}{\rho_0}\frac{\partial p}{\partial x}}_{10^{-3}} = 0 \tag{2.7}$$

However, the error in the tendency is $\Delta(\partial u/\partial t) \sim f\Delta v \sim 10^{-4}$, comparable in size to the tendency itself: the signal-to-noise ratio is 1. The forecast may be qualitatively reasonable, but will be quantitatively invalid.

A similar conclusion is reached for a 10% error in the pressure gradient. However, if the spatial scale Δx of the pressure error is small (say, $\Delta x \sim L/10$) the error in its gradient is correspondingly large:

$$\frac{\partial p}{\partial x} \sim \frac{P}{L}, \qquad \text{but} \qquad \Delta\frac{\partial p}{\partial x} \sim \frac{\Delta p}{\Delta x} \sim \frac{P}{L} \sim \frac{\partial p}{\partial x},$$

so that the error in the wind tendency is now

$$\Delta\frac{\partial u}{\partial t} \sim \frac{1}{\rho_0}\frac{\partial p}{\partial x} \sim 10^{-3} \gg \frac{\partial u}{\partial t}.$$

The forecast will be qualitatively incorrect, indeed unreasonable.

Now consider the continuity equation. The pressure tendency has scale

$$\frac{\partial p}{\partial t} \sim 10^{-2}\,\mathrm{Pa\,s}^{-1} \approx 1\,\mathrm{hPa} \text{ in 3 hours}.$$

If there is a 10% error in the wind, the resulting error in divergence is $\Delta\delta \sim \Delta v/L \sim 10^{-6}$. The error is larger than the divergence itself! As a result, the pressure tendency is unrealistic. Worse still, if the wind error is of small spatial scale, the divergence error is correspondingly greater:

$$\Delta\delta \sim \Delta\frac{\partial v}{\partial x} \sim \frac{\Delta v}{\Delta x} \sim \frac{V}{L} \sim 10^{-5} \sim 10^2\delta.$$

Clearly, this implies a pressure tendency *two orders of magnitude larger* than the correct value. Instead of the value $\partial p/\partial t \sim 1$ hPa in 3 hours we get a change of order 100 hPa in 3 hours. This is strikingly reminiscent of Richardson's result.

3. Early Initialization Methods

3.1. THE FILTERED EQUATIONS

The first computer forecast was made in 1950 by Charney, Fjørtoft and Von Neumann. In order to avoid Richardson's error, they modified the prediction equations in such a way as to eliminate the high frequency solutions. This process is known as filtering. The basic filtered system is the set of quasi-geostrophic equations. These equations were used in operational forecasting for a number of years. However, they involve approximations which are not always valid, and this can result in poor forecasts. A more accurate filtering of the primitive equations leads to the *balance equations*. This system is more complicated to solve than the quasi-geostrophic system, and has not been widely used.

3.2. STATIC INITIALIZATION

Hinkelmann (1951) investigated the problem of noise in numerical integrations and concluded that if the initial winds were geostrophic, high frequency oscillations would occur but would remain small in amplitude. He later succeeded in integrating the primitive equations, using a very short timestep, with geostrophic initial winds (Hinkelmann, 1959). Forecasts made with the primitive equations were soon shown to be clearly superior to those using the quasi-geostrophic system. However, the use of geostrophic initial winds had a huge disadvantage: the valuable information contained in the observations of the wind field was completely ignored. Moreover, the remaining noise level is not tolerable in practice. Charney (1955) proposed that a better estimate of the initial wind field could be obtained by using the nonlinear balance equation. This equation — part of the balance system — is a diagnostic relationship between the pressure and wind fields. It implies that the wind is nondivergent. It was later argued by Phillips (1960) that a further improvement would result if the divergence of the initial field were set equal to that implied by quasi-geostrophic theory. Each of these steps represented some progress, but the noise problem still remained essentially unsolved.

3.3. DYNAMIC INITIALIZATION

Another approach, called dynamic initialization, uses the forecast model itself to define the initial fields. The dissipative processes in the model can damp out high frequency noise as the forecast procedes. We integrate the model first forward and then backward in time, keeping the dissipation active all the time. We repeat this forward-backward cycle many times until we finally obtain fields, valid at the initial time, from which the high frequency components have been damped out. The forecast starting from these fields is noise-free. However, the procedure is expensive in computer time, and damps the meteorologically significant motions as well as the gravity waves, so it is no longer popular. Digital filtering initialization, described in another chapter of this book, is essentially a refinement of dynamic initialization. Because it used a highly selective filtering technique, is is computationally more efficient than the older method.

3.4. VARIATIONAL INITIALIZATION

An elegant initialization method based on the calculus of variations was introduced by Sasaki (1958). We consider the simplest case: given an analysis of the mass and wind fields, how can they be modified so as to impose geostrophic balance? This problem can be be formulated as the minimization of an integral representing the deviation of the resulting fields from balance. The variation of the integral leads to the Euler-Lagrange equations, which yield diagnostic relationships for the new mass and wind fields in terms of the incoming analysis. Although the method was not widely used, the variational method is now at the centre of modern data assimilation practice.

4. Atmospheric Normal Mode Oscillations

The solutions of the model equations can be separated, by a process of spectral analysis, into two sets of components or linear normal modes, slow rotational components or Rossby modes, and high frequency gravity modes. We assume that the amplitude of the motion is so small that all nonlinear terms can be neglected. The horizontal structure is then governed by a system equivalent to the linear shallow water equations which describe the small-amplitude motions of a shallow layer of incompressible fluid. These equations were first derived by Laplace in his discussion of tides in the atmosphere and ocean, and are called the Laplace Tidal Equations. The simplest means of deriving the linear shallow water equations from the primitive equations is to assume that the vertical velocity vanishes identically.[1]

4.1. THE LAPLACE TIDAL EQUATIONS

Let us assume that the motions under consideration can be described as small perturbations about a state of rest, in which the temperature is a constant, T_0, and the pressure $\bar{p}(z)$ and density $\bar{\rho}(z)$ vary only with height. The basic state variables satisfy the gas law and are in hydrostatic balance: $\bar{p} = \mathcal{R}\bar{\rho}T_0$ and $d\bar{p}/dz = -g\bar{\rho}$. The variations of mean pressure and density follow immediately:

$$\bar{p}(z) = p_0 \exp(-z/H), \qquad \bar{\rho}(z) = \rho_0 \exp(-z/H),$$

where $H = p_0/g\rho_0 = \mathcal{R}T_0/g$ is the scale-height of the atmosphere. We consider only motions for which the vertical component of velocity vanishes identically, $w \equiv 0$. Let u, v, p and ρ denote variations about the basic state, each of these being a small quantity. The horizontal momentum, continuity and thermodynamic equations become

$$\frac{\partial \bar{\rho}u}{\partial t} - f\bar{\rho}v + \frac{\partial p}{\partial x} = 0 \tag{4.1}$$

$$\frac{\partial \bar{\rho}v}{\partial t} + f\bar{\rho}u + \frac{\partial p}{\partial y} = 0 \tag{4.2}$$

$$\frac{\partial \rho}{\partial t} + \nabla\cdot\bar{\rho}\mathbf{V} = 0 \tag{4.3}$$

$$\frac{1}{\gamma\bar{p}}\frac{\partial p}{\partial t} - \frac{1}{\bar{\rho}}\frac{\partial \rho}{\partial t} = 0 \tag{4.4}$$

---

[1] This assumption precludes the possibility of studying internal modes for which $w \neq 0$. A more general derivation, based on a separation of the horizontal and vertical dependencies of the variables, is presented in Daley, (1991, Ch. 9).

Density can be eliminated from the continuity equation, Eq. 4.3, by means of the thermodynamic equation, Eq. 4.4. Now let us assume that the horizontal and vertical dependencies of the perturbation quantities are separable:

$$\left\{ \begin{array}{c} \bar{\rho}u \\ \bar{\rho}v \\ p \end{array} \right\} = \left\{ \begin{array}{c} U(x,y,t) \\ V(x,y,t) \\ P(x,y,t) \end{array} \right\} Z(z) \ . \tag{4.5}$$

The momentum and continuity equations can then be written

$$\frac{\partial U}{\partial t} - fV + \frac{\partial P}{\partial x} = 0 \tag{4.6}$$

$$\frac{\partial V}{\partial t} + fU + \frac{\partial P}{\partial y} = 0 \tag{4.7}$$

$$\frac{\partial P}{\partial t} + (gh)\nabla \cdot \mathbf{V} = 0 \tag{4.8}$$

where $\mathbf{V} = (U, V)$ is the momentum vector and $h = \gamma H = \gamma \mathcal{R} T_0/g$. This is a set of three equations for the three dependent variables U, V, and P. They are mathematically isomorphic to the Laplace tidal equations with a mean depth h. The quantity h is called the equivalent depth. There is no dependence in this system on the vertical coordinate z.

The vertical structure follows from the hydrostatic equation, together with the relationship $p = (\gamma g H)\rho$ implied by the thermodynamic equation. It is determined by

$$\frac{dZ}{dz} + \frac{Z}{\gamma H} = 0 \ , \tag{4.9}$$

the solution of which is $Z = Z_0 \exp(-z/\gamma H)$, where Z_0 is the amplitude at $z = 0$. If we set $Z_0 = 1$, then U, V and P give the momentum and pressure fields at the earth's surface. These variables all decay exponentially with height. It follows from Eq. 4.5 that u and v actually increase with height as $\exp(\kappa z/H)$, but the kinetic energy decays. Solutions with more general vertical structures, and with non-vanishing vertical velocity, are discussed in Daley, (1991, Ch. 9).

4.2. VORTICITY AND DIVERGENCE

We examine the solutions of the Laplace Tidal Equations in some enlightening limiting cases. Holton (1975) gives a more extensive analysis, including treatments of the equatorial and mid-latitude β-plane approximations. By means of the Helmholtz Theorem, a general horizontal wind field $\mathbf{V}$ may be partitioned into rotational and divergent components

$$\mathbf{V} = \mathbf{V}_\psi + \mathbf{V}_\chi = \mathbf{k} \times \nabla \psi + \nabla \chi \ .$$

The stream function ψ and velocity potential χ are related to the vorticity and divergence by the Poisson equations $\nabla^2 \psi = \zeta$ and $\nabla^2 \chi = \delta$.

It is straightforward to derive equations for the vorticity and divergence tendencies. Together with the continuity equation, they are

$$\frac{\partial \zeta}{\partial t} + f\delta + \beta v = 0 \tag{4.10}$$

$$\frac{\partial \delta}{\partial t} - f\zeta + \beta u + \nabla^2 P = 0 \tag{4.11}$$

$$\frac{\partial P}{\partial t} + gh\delta = 0. \tag{4.12}$$

These equations are completely equivalent to Eqs. 4.6–4.8; no additional approximations have yet been made. However, the vorticity and divergence forms enable us to examine various simple approximate solutions.

4.3. ROSSBY-HAURWITZ MODES

If we suppose that the solution is quasi-nondivergent, $i.e.$, we assume $|\delta| \ll |\zeta|$, the wind is given approximately in terms of the stream function $(u, v) \approx (-\psi_y, \psi_x)$, the vorticity equation becomes

$$\nabla^2 \psi_t + \beta\psi_x = O(\delta), \tag{4.13}$$

and we can ignore the right-hand side. Assuming the stream function has the wave-like structure of a spherical harmonic, we substitute the expression $\psi = \psi_0 Y_n^m(\lambda, \phi) \exp(-i\nu t)$ in the vorticity equation and immediately deduce an expression for the frequency:

$$\nu = \nu_R \equiv -\frac{2\Omega m}{n(n+1)}. \tag{4.14}$$

This is the celebrated dispersion relation for Rossby-Haurwitz waves (Haurwitz, 1940). If we ignore sphericity (the β-plane approximation) and assume harmonic dependence $\psi(x, y, t) = \psi_0 \exp[i(kx + \ell y - \nu t)]$, then (4.13) has the dispersion relation

$$c = \frac{\nu}{k} = -\frac{\beta}{k^2 + \ell^2},$$

which is the expression for phase-speed found by Rossby (1939). The Rossby or Rossby-Haurwitz waves are, to the first approximation, non-divergent waves which travel westward, the phase speed being greatest for the waves of largest scale. They are of relatively low frequency — Eq. 4.14 implies that $|\nu| \leq \Omega$ — and the frequency decreases as the spatial scale decreases.

To the same degree of approximation, we may write the divergence equation, Eq. 4.11, as

$$\nabla^2 P - f\zeta - \beta\psi_y = O(\delta). \tag{4.15}$$

Ignoring the right-hand side, we get the *linear balance equation*

$$\nabla^2 P = \nabla \cdot f\nabla\psi, \tag{4.16}$$

a diagnostic relationship between the geopotential and the stream function. This also follows immediately from the assumption that the wind is both non-divergent ($\mathbf{V} = \mathbf{k} \times \nabla\psi$) and geostrophic ($f\mathbf{V} = \mathbf{k} \times \nabla P$). If variations of f are ignored, we can assume $P = f\psi$. The wind and pressure are in approximate geostrophic balance for Rossby-Haurwitz waves.

4.4. GRAVITY WAVE MODES

If we assume now that the solution is quasi-irrotational, *i.e.* that $|\zeta| \ll |\delta|$, then the wind is given approximately by $(u, v) \approx (\chi_x, \chi_y)$ and the divergence equation becomes

$$\nabla^2\chi_t + \beta\chi_x + \nabla^2 P = O(\zeta)$$

with the right-hand side negligible. Using the continuity equation to eliminate P, we get

$$\nabla^2\chi_{tt} + \beta\chi_{xt} - gh\nabla^4\chi = 0\,.$$

Seeking a solution $\chi = \chi_0 Y_n^m(\lambda, \phi)\exp(-i\nu t)$, we find that

$$\nu^2 + \left(-\frac{2\Omega m}{n(n+1)}\right)\nu - \frac{n(n+1)gh}{a^2} = 0\,. \tag{4.17}$$

The coefficient of the second term is just the Rossby-Haurwitz frequency ν_R found in Eq. 4.14 above, so that

$$\nu = \pm\sqrt{\nu_G^2 + (\tfrac{1}{2}\nu_R)^2} - \frac{1}{2}\nu_R\,, \qquad \text{where} \qquad \nu_G \equiv \sqrt{\frac{n(n+1)gh}{a^2}}\,,$$

Noting that $|\nu_G| \gg |\nu_R|$, it follows that

$$\nu_\pm \approx \pm\nu_G\,,$$

the frequency of pure gravity waves. There are then two solutions, representing waves travelling eastward and westward with equal speeds. The frequency increases approximately linearly with the total wavenumber n.

5. Normal Mode Initialization

The model equations can be written schematically in the form

$$\dot{\mathbf{X}} + i\mathbf{L}\mathbf{X} + \mathbf{N}(\mathbf{X}) = 0 \tag{5.1}$$

with $\mathbf{X}$ the state vector, $\mathbf{L}$ a matrix and $\mathbf{N}$ a nonlinear vector function. If $\mathbf{L}$ is diagonalized, the system separates into two subsystems, for the low and high frequency components:

$$\dot{\mathbf{Y}} + i\Lambda_Y\mathbf{Y} + \mathbf{N}_Y(\mathbf{Y}, \mathbf{Z}) = 0 \tag{5.2}$$

$$\dot{\mathbf{Z}} + i\Lambda_Z\mathbf{Z} + \mathbf{N}_Z(\mathbf{Y}, \mathbf{Z}) = 0 \tag{5.3}$$

where **Y** and **Z** are the coefficients of the LF and HF components of the flow, referred to colloquially as the *slow* and *fast* components respectively, and Λ_Y and Λ_Z are diagonal matrices of eigenfrequencies for the two types of modes.

Let us suppose that the initial fields are separated into slow and fast parts, and that the latter are removed so as to leave only the Rossby waves. It might be hoped that this process of "linear normal mode initialization" (imposing the condition **Z** = **0** at $t = 0$) would ensure a noise-free forecast. However, the results of the technique are disappointing: the noise is reduced initially, but soon reappears; the forecasting equations are nonlinear, and the slow components interact nonlinearly in such a way as to generate gravity waves. The problem of noise remains: the gravity waves are small to begin with, but they grow rapidly (see Daley, Ch. 9).

To control the growth of HF components, Machenhauer (1977) proposed setting their initial rate-of-change to zero, in the hope that they would remain small throughout the forecast. Baer (1977) proposed a somewhat more general method, using a two-timing perturbation technique. The forecast, starting from initial fields modified so that **Ż** = **0** at $t = 0$ is very smooth and the spurious gravity wave oscillations are almost completely removed. The method takes account of the nonlinear nature of the equations, and is referred to as nonlinear normal mode initialization. The method is comprehensively reviewed in Daley (1991). Rather than considering the full complexity of an atmospheric model, we will illustrate linear normal mode initialization (LNMI) and nonlinear normal mode initialization (NNMI) by application to a simple mechanical system.

6. Initialization and the Swinging Spring

The procedure of linear and nonlinear normal mode initialization can be clearly illustrated by applying the method to the equations of the elastic pendulum or 'swinging spring'. This system comprises a heavy bob suspended by a light elastic spring. The bob is free to move in a vertical plane. The oscillations of this system are of two types, distinguished by their physical restoring mechanisms. For an appropriate choice of parameters, the elastic oscillations have much higher frequency than the rotation or libration of the bob. We consider the elastic oscillations to be analogues of the high frequency gravity waves in the atmosphere. Similarly, the low frequency rotational motions are considered to correspond to the rotational or Rossby-Haurwitz waves.

6.1. THE DYNAMICAL EQUATIONS

Let ℓ_0 be the unstretched length of the spring, k its elasticity or stiffness and m the mass of the bob. At equilibrium the elastic restoring force is balanced by the weight: $k(\ell - \ell_0) = mg$. Polar coordinates $q_r = r$ and $q_\theta = \theta$ are used, and the radial and angular momenta are $p_r = m\dot{r}$ and

$p_\theta = mr^2\dot\theta$. The Hamiltonian is, in this case, the sum of kinetic, elastic potential and gravitational potential energy:

$$H = \frac{1}{2m}\left(p_r^2 + \frac{p_\theta^2}{r^2}\right) + \frac{1}{2}k(r - \ell_0)^2 - mgr\cos\theta.$$

The (canonical) dynamical equations may now be written explicitly

$$
\begin{aligned}
\dot\theta &= p_\theta/mr^2 \\
\dot p_\theta &= -mgr\sin\theta \\
\dot r &= p_r/m \\
\dot p_r &= p_\theta^2/mr^3 - k(r - \ell_0) + mg\cos\theta.
\end{aligned}
$$

(If the Hamiltonian formalism is unfamiliar, the equations may be derived by considering the forces on the bob). These equations may also be written symbolically in vector form

$$\dot{\mathbf{X}} + \mathbf{L}\mathbf{X} + \mathbf{N}(\mathbf{X}) = 0$$

where $\mathbf{X} = (\theta, p_\theta, r, p_r)^{\mathrm{T}}$, $\mathbf{L}$ is the matrix of coefficients of the linear terms and $\mathbf{N}$ is a nonlinear vector function.

Let us now suppose that the amplitude of the motion is small, so that $|r'| = |r - \ell| \ll \ell$ and $|\theta| \ll 1$. The state vector $\mathbf{X}$ comprises two sub-vectors:

$$\mathbf{X} = \begin{pmatrix} \mathbf{Y} \\ \mathbf{Z} \end{pmatrix}, \qquad \text{where} \quad \mathbf{Y} = \begin{pmatrix} \theta \\ p_\theta \end{pmatrix} \quad \text{and} \quad \mathbf{Z} = \begin{pmatrix} r' \\ p_r \end{pmatrix},$$

and the linear dynamics of these components evolve independently. We call the motion described by $\mathbf{Y}$ the rotational component and that described by $\mathbf{Z}$ the elastic component. The rotational equations may be written

$$\ddot\theta + (g/\ell)\theta = 0$$

which is the equation for a simple pendulum having oscillatory solutions with frequency $\sqrt{g/\ell}$. The remaining two equations yield

$$\ddot r' + (k/m)r' = 0,$$

the equations for elastic oscillations with frequency $\sqrt{k/m}$. We define the rotational and elastic frequencies and their ratio by

$$\omega_{\mathrm{R}} = \sqrt{\frac{g}{\ell}}, \qquad \omega_{\mathrm{E}} = \sqrt{\frac{k}{m}}, \qquad \epsilon \equiv \left(\frac{\omega_{\mathrm{R}}}{\omega_{\mathrm{E}}}\right).$$

It is easily shown that $\epsilon < 1$, so the rotational frequency is always less than the elastic. We assume that the parameters are such that $\epsilon \ll 1$. In this case the linear normal modes are clearly distinct: the rotational mode has low frequency (LF) and the elastic mode has high frequency (HF).

6.2. LINEAR AND NONLINEAR INITIALIZATION

For small amplitude motions, for which the nonlinear terms are negligible, the LF and HF oscillations are completely independent of each other and evolve without interaction. We can suppress the HF component completely by setting its initial amplitude to zero:

$$\mathbf{Z} = (r', p_r)^{\mathrm{T}} = \mathbf{0} \quad \text{at} \quad t = 0 \,.$$

This procedure is called linear initialization. When the amplitude is large, nonlinear terms are no longer negligible and the LF and HF motions interact. It is clear from the equations that linear initialization will not ensure permanent absence of HF motions: the nonlinear LF terms generate radial momentum. To achieve better results, we set the initial *tendency* of the HF components to zero:

$$\dot{\mathbf{Z}} = (\dot{r}, \dot{p}_r)^{\mathrm{T}} = \mathbf{0} \quad \text{at} \quad t = 0 \,,$$

This procedure is called nonlinear initialization. For the spring, we can deduce explicit expressions for the initial conditions:

$$r(0) = r_{\mathrm{B}} \equiv \frac{\ell(1 - \epsilon^2(1 - \cos\theta))}{1 - (\dot{\theta}/\omega_{\mathrm{E}})^2} \,, \qquad p_r(0) = 0 \,.$$

Thus, given arbitrary initial conditions $\mathbf{X} = (\theta, p_\theta, r, p_r)^{\mathrm{T}}$, we replace $\mathbf{Z} = (r, p_r)^{\mathrm{T}}$ by $\mathbf{Z}_{\mathrm{B}} = (r_{\mathrm{B}}, 0)^{\mathrm{T}}$. The rotational component $\mathbf{Y} = (\theta, p_\theta)^{\mathrm{T}}$ remains unchanged.

6.3. A NUMERICAL EXAMPLE

In Figure 1 we show the results of two integrations of the spring equations. The upper panels show the evolution and spectrum of the slow variable θ; the lower panels are for the fast variable r. Dotted curves are for linear initialization and solid curves for nonlinear initialization. The parameter values are $m = 1$, $g = \pi^2$, $k = 100\pi^2$ and $\ell = 1$ (all SI units), so that $\epsilon = 0.1$ and the periods of the swinging and springing motions are respectively $\tau_{\mathrm{R}} = 2\mathrm{s}$ and $\tau_{\mathrm{E}} = 0.2\mathrm{s}$. The initial conditions are vanishing velocity ($\dot{r} = \dot{\theta} = 0$), with $\theta(0) = 1$ and $r(0) \in \{1, 0.99540\}$. The equations are integrated over a period of 6 seconds For the slow variable, the curves are indistinguishable. The spectrum has a clear peak at a frequency of 0.5 cycles per second (Hz). For the fast variable, the linearly initialized evolution has high frequency noise (dotted curve, lower left panel). This is confirmed in the spectrum: there is a sharp peak at 5 Hz. When nonlinearly initialized, this peak is removed: only the peak at 1 Hz remains. This is the 'balanced fast motion'.

110

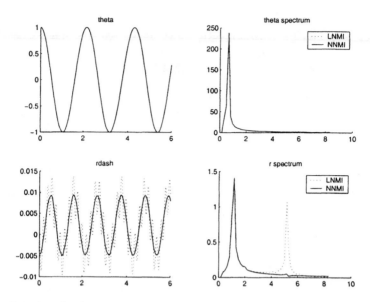

Figure 1. Solution of swinging spring equations for linear (LNMI) and nonlinear (NNMI) initialization. See text for details.

It can be understood physically: the centrifugal effect stretches the spring twice for each pendular swing: the result is a component of r with a period of one second. The radial variation does not disappear for balanced motion, but it is of low frequency. It is said to be 'slaved' (or, better, enslaved) to the slow motion.

References

Baer, F, 1977: Adjustment of initial conditions required to suppress gravity oscillations in non-linear flows. *Beit. Phys. Atmos.*, **50**, 350–366.

Charney, J.G., 1955: The use of the primitive equations of motion in numerical prediction. *Tellus*, **7**, 22–26

Charney, J.G., R. Fjørtoft and J. von Neumann, 1950: Numerical integration of the barotropic vorticity equation. *Tellus*, **2**, 237–254.

Daley, Roger, 1991: *Atmospheric Data Analysis*. Cambridge Univ. Press, 457pp.

Haurwitz, B., 1940: The motion of atmospheric disturbances on the spherical earth. *J. Marine Res.*, **3**, 254–267.

Hinkelmann, K., 1951: Der Mechanismus des meteorologischen Lärmes. *Tellus*, **3**, 285–296.

Hinkelmann, K., 1959: Ein numerisches Experiment mit den primitiven Gleichungen. In *The Atmosphere and the Sea in Motion*. Rossby Memorial Volume, Rockerfeller Institute Press, 486–500

Holton, J.R., 1975: *The Dynamic Meteorology of the Stratosphere and Mesosphere*. Met. Monogr., Vol. 15, No. 37. Amer. Meteor. Soc., Boston. 218pp.

Kasahara, A., 1976: Normal modes of ultralong waves in the atmosphere. *Mon Weather Rev*, **104**, 669–690.

Lynch, Peter, 1985: Initialization using Laplace Transforms. *Q. J. Roy. Meteor. Soc.*, **111**, 243–258

Machenhauer B, 1977: On the dynamics of gravity oscillations in a shallow water model, with applications to normal mode initialization. *Beitr. Phys. Atmos.*, **50**, 253–271

McIntyre, M E, 2003: Balanced Flow. In *Encyclopedia of Atmospheric Sciences*, Ed. J R Holton, J Pyle and J A Curry. Academic Press. 6 Vols. ISBN 0-12-227090-8.

Phillips, N A, 1960: On the problem of initial data for the primitive equations. *Tellus*, **12**, 121–126

Richardson, L.F., 1922: *Weather Prediction by Numerical Process*. Cambridge Univ. Press, 236 *pp*. Reprinted by Dover Publications, New York, 1965.

Rossby, C.G., *et al.*, 1939: Relations between variations in the intensity of the zonal circulation of the atmosphere and the displacements of the semipermanent centers of action. *J. Marine Res.*, **2**, 38–55.

Sasaki, Y., 1958: An objective method based on the variational method. *J. Met. Soc. Japan*, **36**, 77–88.

DIGITAL FILTER INITIALIZATION

PETER LYNCH
Met Éireann,
Glasnevin Hill, Dublin 9, Ireland.

1. Introduction

The requirement to modify meteorological analyses to avoid spurious high frequency oscillations in numerical forecasts has been known from the beginning of numerical weather prediction. The most popular method of initialization up to recently was normal mode initialization, or NMI (Machenhauer, 1977). This has been used in many NWP centres, and has performed satisfactorily. Its most natural context is for global models, for which the horizontal structure of the normal modes corresponds to the Hough functions, the eigenmodes of the Laplace Tidal Equations. For limited area models, normal modes can also be derived, but the lateral boundaries force the introduction of simplifying assumptions.

Recently, an alternative method of initialization, called digital filter initialization (DFI), was introduced by Lynch and Huang (1992). It was generalised to allow for diabatic effects by Huang and Lynch (1993). The latter paper also discussed the use of an optimal filter. A much simpler filter, the Dolph-Chebyshev filter, which is a special case of the optimal filter, was applied to the initialization problem by Lynch (1997). A more efficient formulation of DFI was presented by Lynch, Giard and Ivanovici (1997).

2. Advantages of DFI

The method of digital filter initialization, which is based on ideas from digital signal processing, has significant advantages over alternative methods, and is now in use operationally at several major weather prediction centres. Some of the principal advantages of DFI compared to available alternatives are:

1. No need to compute or store normal modes;
2. No need to separate vertical modes;
3. Complete compatibility with model discretisation;
4. Applicable to exotic grids on arbitrary domains;

R. Swinbank et al. (eds.), Data Assimilation for the Earth System, 113–126.

5. No iterative numerical procedure which may diverge;
6. Ease of implementation and maintenance, due to simplicity of scheme;
7. Applicable to all prognostic model variables;
8. Applicable to non-hydrostatic models.

The first advantage becomes more pronounced as the number of degrees of freedom of the model increases. The second advantage over NMI is that the latter method requires the introduction of an auxiliary geopotential variable, and partitioning of its changes between the temperature and surface pressure involves an *ad hoc* assumption. Advantage 3 eliminates discretization errors due to grid disparities. Advantage 4 facilitates the use of stretched or irregular model grids. Advantage 5 means that all the vertical modes can be initialized effectively. The sixth advantage is that the simplicity of the method makes it easy to implement and maintain. The seventh is that additional prognostic model variables, such as cloud water, rain water, turbulent kinetic energy, etc., are processed in the same way as the standard mass and wind variables; thus, DFI produces initial fields for these variables which are compatible with the basic dynamical fields. Last but not least, DFI filters the additional prognostic variables in non-hydrostatic models in a manner identical to the basic variables. The DFI method is thus immediately suitable for non-hydrostatic models (Bubnová, *et al.*, 1995). This is not the case for NMI. It must be pointed out that DFI is significantly more demanding of computational time than NMI. However, the significant benefits justify this additional cost.

3. The Primitive Notion of Filtering

The concept of filtering has a rôle in virtually every field of study, from topology to theology, seismology to sociology. The process of filtering involves the *selection* of those components of an assemblage having some particular property, and the removal or elimination of those components which lack it. A filter is any device or contrivance designed to carry out such a selection. It may be represented by a simple system diagram, having an input with both desired and undesired components, and an output comprising only the former.

$$Good/Bad/Ugly \implies \boxed{\textbf{Filter}} \longrightarrow Good$$

We are primarily concerned with filters as used in signal processing. The selection principle for these is generally based on the frequency of the signal components. There are a number of ideal types — lowpass, highpass, bandpass and bandstop — corresponding to the range of frequencies which pass through the filter and those which are rejected. In many cases the input consists of a low-frequency (LF) signal contaminated by high-frequency (HF) noise, and the information in the signal can be isolated by using a lowpass filter which rejects the noise. Such a situation is typical for the application to meteorology discussed below.

Filter theory originated from the need to design electronic circuits with precise frequency-selective characteristics, for radio and telecommunications. These analog filters were constructed from capacitors and inductors, and acted on continuous time signals. More recently, discrete time signal processing has assumed prominence, and the technique and theory of *digital filtering* has evolved. Digital filters may be implemented in hardware using integrated circuits, but are more commonly realized in software: the input is processed by a program designed to perform the required selection and compute the output.

4. Nonrecursive and Recursive Digital Filters

Given a discrete function of time, $\{x_n\}$, a *nonrecursive* digital filter is defined by

$$y_n = \sum_{k=-N}^{N} a_k x_{n-k}. \tag{1}$$

The output y_n at time $n\Delta t$ depends on both past and future values of x_n, but not on other output values. A *recursive* digital filter is defined by

$$y_n = \sum_{k=K}^{N} a_k x_{n-k} + \sum_{k=1}^{L} b_k y_{n-k} \tag{2}$$

where K, L and N are integers, with L and N positive. The output y_n at time $n\Delta t$ in this case depends on past and present values of the input (for $K = 0$), and also on previous output values (occasionally, future input values are also used ($K < 0$), in which case the recursive filter is *non-causal*). Recursive filters are more powerful than non-recursive ones, but can also be more problematical, as the feedback of the output can give rise to instability. The response of a nonrecursive filter to an impulse $\delta(n)$ is zero for $|n| > N$, giving rise to the alternative name *finite impulse response* or FIR filter. Since the response of a recursive filter to this input can persist indefinitely, it is known as an *infinite impulse response* or IIR filter.

The frequency response of a recursive filter is easily found: let $x_n = \exp(in\theta)$ and assume an output of the form $y_n = H(\theta)\exp(in\theta)$; substituting into Eq. 2, the transfer function $H(\theta)$ is

$$H(\theta) = \frac{\sum_{k=K}^{N} a_k e^{-ik\theta}}{1 - \sum_{k=1}^{L} b_k e^{-ik\theta}}. \tag{3}$$

For nonrecursive filters the denominator reduces to unity. This equation gives the response once the filter coefficients a_k and b_k have been specified.

However, what is really required is the opposite: to derive coefficients (and as few as possible) which will yield the desired response function. This *inverse problem* has no unique solution, and a great variety of techniques have been developed.

Only the most elementary design techniques for non-recursive filters will be considered below. Recursive filters generally have superior performance to nonrecursive filters with the same total number of coefficients. Numerous accounts of recursive digital filters are available in publications on digital signal processing (*e.g.*, Oppenheim and Schafer, 1989). For a review in the meteorological literature, see Raymond (1988), where another class of filter, the implicit filter, is also discussed.

5. Design of Nonrecursive Filters

Consider a function of time, $f(t)$, with low and high frequency components. To filter out the high frequencies one may proceed as follows:

[1] Calculate the Fourier transform $F(\omega)$ of $f(t)$;
[2] Set the coefficients of the high frequencies to zero;
[3] Calculate the inverse transform.

Step [2] may be performed by multiplying $F(\omega)$ by an appropriate weighting function $H_c(\omega)$. Typically, $H_c(\omega)$ is a step function

$$H_c(\omega) = \begin{cases} 1, & |\omega| \leq |\omega_c|; \\ 0, & |\omega| > |\omega_c|, \end{cases} \tag{4}$$

where ω_c is a cutoff frequency. These three steps are equivalent to a convolution of $f(t)$ with $h(t) = \sin(\omega_c t)/\pi t$, the inverse Fourier transform of $H_c(\omega)$. This follows from the convolution theorem

$$\mathcal{F}\{(h * f)(t)\} = \mathcal{F}\{h\} \cdot \mathcal{F}\{f\} = H_c(\omega) \cdot F(\omega) \tag{5}$$

Thus, to filter $f(t)$ one calculates

$$f^\star(t) = (h * f)(t) = \int_{-\infty}^{+\infty} h(\tau) f(t - \tau) d\tau. \tag{6}$$

For simple functions $f(t)$, this integral may be evaluated analytically. In general, some method of approximation must be used.

Suppose now that f is known only at discrete moments $t_n = n\Delta t$, so that the sequence $\{\cdots, f_{-2}, f_{-1}, f_0, f_1, f_2, \cdots\}$ is given. For example, f_n could be the value of some model variable at a particular grid point at time t_n. The shortest period component which can be represented with a time step Δt is $\tau_N = 2\Delta t$, corresponding to a maximum frequency, the so-called Nyquist frequency, $\omega_N = \pi/\Delta t$. The sequence $\{f_n\}$ may be regarded as the Fourier coefficients of a function $F(\theta)$:

$$F(\theta) = \sum_{n=-\infty}^{\infty} f_n e^{-in\theta}, \tag{7}$$

where $\theta = \omega \Delta t$ is the *digital frequency* and $F(\theta)$ is periodic, $F(\theta) = F(\theta + 2\pi)$. High frequency components of the sequence may be eliminated by multiplying $F(\theta)$ by a function $H_d(\theta)$ defined by

$$H_d(\theta) = \begin{cases} 1, & |\theta| \le |\theta_c|; \\ 0, & |\theta| > |\theta_c|, \end{cases} \tag{8}$$

where the cutoff frequency $\theta_c = \omega_c \Delta t$ is assumed to fall in the Nyquist range $(-\pi, \pi)$ and $H_d(\theta)$ has period 2π. This function may be expanded:

$$H_d(\theta) = \sum_{n=-\infty}^{\infty} h_n e^{-in\theta} \quad ; \quad h_n = \frac{1}{2\pi} \int_{-\pi}^{\pi} H_d(\theta) e^{in\theta} d\theta. \tag{9}$$

The values of the coefficients h_n follow immediately from Eqs. 8 and 9:

$$h_n = \frac{\sin n\theta_c}{n\pi}. \tag{10}$$

Let $\{f_n^\star\}$ denote the low frequency part of $\{f_n\}$, from which all components with frequency greater than θ_c have been removed. Clearly,

$$H_d(\theta) \cdot F(\theta) = \sum_{n=-\infty}^{\infty} f_n^\star e^{-in\theta}.$$

The convolution theorem for Fourier series now implies that $H_d(\theta) \cdot F(\theta)$ is the transform of the convolution of $\{h_n\}$ with $\{f_n\}$:

$$f_n^\star = (h * f)_n = \sum_{k=-\infty}^{\infty} h_k f_{n-k}. \tag{11}$$

This enables the filtering to be performed directly on the given sequence $\{f_n\}$. It is the discrete analogue of Eq. 6. In practice the summation must be truncated at some finite value of k. Thus, an approximation to the low frequency part of $\{f_n\}$ is given by

$$f_n^\star = \sum_{k=-N}^{N} h_k f_{n-k}. \tag{12}$$

Comparing Eq. 12 with Eq. 1, it is apparent that the finite approximation to the discrete convolution is formally identical to a nonrecursive digital filter.

As is well known, truncation of a Fourier series gives rise to Gibbs oscillations. These may be greatly reduced by means of an appropriately defined "window" function. The response of the filter is improved if h_n is multiplied by the Lanczos window

$$w_n = \frac{\sin(n\pi/(N+1))}{n\pi/(N+1)}.$$

The transfer function $H(\theta)$ of a filter is defined as the function by which a pure sinusoidal oscillation is multiplied when subjected to the filter. For symmetric coefficients, $h_k = h_{-k}$, it is real, implying that the phase is not altered by the filter. Then, if $f_n = \exp(in\theta)$, one may write $f_n^\star = H(\theta) \cdot f_n$, and $H(\theta)$ is easily calculated by substituting f_n in Eq. 12:

$$H(\theta) = \sum_{k=-N}^{N} h_k e^{-ik\theta} = \left[h_0 + 2 \sum_{k=1}^{N} h_k \cos k\theta \right]. \qquad (13)$$

The transfer functions for a windowed and unwindowed filter are shown in Lynch and Huang (1992, Fig. 2). The use of the window decreases the Gibbs oscillations in the stop-band $|\theta| > |\theta_c|$. However, it also has the effect of widening the pass-band beyond the nominal cutoff. For a fuller discussion of windowing see *e.g.* Hamming (1989) or Oppenheim and Schafer (1989).

One of the simplest design methods for nonrecursive filters is the expansion of the desired filtering function, $H(\theta)$, as a Fourier series, and the application of a suitable window function to improve the transfer characteristics. That is the method employed above. An alternative method called frequency sampling fits the desired frequency response by making a selection of values and calculating the inverse discrete Fourier transform to obtain the filter coefficients. A more sophisticated method uses the Chebyshev alternation theorem to obtain a filter whose maximum error in the pass- and stop-bands is minimized. This method yields a filter meeting required specifications with fewer coefficients that the other methods. The design of nonrecursive and recursive filters is outlined in Hamming (1989), where several methods are described, and fuller treatments may be found in Oppenheim and Schafer (1989).

6. Application of a Nonrecursive Digital Filter to Initialization

An initialization scheme using a nonrecursive digital filter has been developed by Lynch and Huang (1992) for the HIRLAM model. The value chosen for the cutoff frequency corresponded to a period $\tau_c = 6$ hours. With the time step $\Delta t = 6$ minutes used in the model, this corresponds to a (digital) cutoff frequency $\theta_c = \pi/30$. The coefficients were derived by Fourier expansion of a step-function, truncated at $N = 30$, with application of a Lanczos window, and are given by

$$h_n = \left[\frac{\sin(n\pi/(N+1))}{n\pi/(N+1)} \right] \left(\frac{\sin(n\theta_c)}{n\pi} \right).$$

The frequency response was depicted in Lynch and Huang (1992, Fig. 2). The central lobe of the coefficient function spans a period of six hours, from $t = -3$ hours to $t = +3$ hours. The summation in Eq. 1 was calculated over this range, with the coefficients normalized to have unit sum over the span.

Thus, the application of the technique involved computation equivalent to sixty time steps, or a six hour adiabatic integration.

The uninitialized fields of surface pressure, temperature, humidity and winds were first integrated forward for three hours, and running sums of the form

$$f_F^\star(0) = \frac{1}{2}h_0 f_0 + \sum_{n=1}^{N} h_{-n} f_n, \tag{14}$$

where $f_n = f(n\Delta t)$, were calculated for each field at each gridpoint and on each model level. These were stored at the end of the three hour forecast. The original fields were then used to make a three hour 'hindcast', during which running sums of the form

$$f_B^\star(0) = \frac{1}{2}h_0 f_0 + \sum_{n=-1}^{-N} h_{-n} f_n \tag{15}$$

were accumulated for each field, and stored as before. The two sums were then combined to form the required summations:

$$f^\star(0) = f_F^\star(0) + f_B^\star(0). \tag{16}$$

These fields correspond to the application of the digital filter Eq. 1 to the original data, and will be referred to as the filtered data.

In the foregoing, only the amplitudes of the transfer functions have been discussed. Since these functions are complex, there is also a phase change induced by the filters. Space prohibits further discussion here; however, it is essential that the phase characteristics of a filter be studied before it is considered for use. Ideally, the phase-error should be as small as possible for the low frequency components which are meteorologically important. The error in the high frequency stop-band is unimportant. It is salutary to recall that phase-errors are amongst the most prevalent and pernicious problems faced by the forecaster.

7. The Dolph-Chebyshev Filter

We now consider a particularly simple filter, having explicit expressions for its impulse response coefficients. The details of the Dolph-Chebyshev filter are presented in Lynch (1997). We will confine the present discussion to the definition and principal properties of the filter; further information may be found in the reference cited.

7.1. DEFINITION OF THE FILTER

The function to be described is constructed using Chebyshev polynomials, defined by the equations

$$T_n(x) = \begin{cases} \cos(n \cos^{-1} x), & \text{if } |x| \leq 1; \\ \cosh(n \cosh^{-1} x), & \text{if } |x| > 1. \end{cases}$$

Clearly, $T_0(x) = 1$ and $T_1(x) = x$. From the definition, the following recurrence relation follows immediately:

$$T_n(x) = 2xT_{n-1}(x) - T_{n-2}(x), \quad n \geq 2.$$

The main relevant properties of these polynomials are given in Lynch (1997).

Now consider the function defined in the frequency domain by

$$H(\theta) = \frac{T_{2M}\left(x_0 \cos\left(\theta/2\right)\right)}{T_{2M}(x_0)}$$

where $x_0 > 1$. Let θ_s be such that $x_0 \cos(\theta_s/2) = 1$. As θ varies from 0 to θ_s, $H(\theta)$ falls from 1 to $r = 1/T_{2M}(x_0)$. For $\theta_s \leq \theta \leq \pi$, $H(\theta)$ oscillates in the range $\pm r$. The form of $H(\theta)$ is that of a low-pass filter with a cut-off at $\theta = \theta_s$. By means of the definition of $T_n(x)$ and basic trigonometric identities, $H(\theta)$ can be written as a *finite expansion*

$$H(\theta) = \sum_{n=-M}^{+M} h_n \exp(-in\theta).$$

The coefficients $\{h_n\}$ may be evaluated from the inverse Fourier transform

$$h_n = \frac{1}{N}\left[1 + 2r \sum_{m=1}^{M} T_{2M}\left(x_0 \cos\frac{\theta_m}{2}\right) \cos m\theta_n\right],$$

where $|n| \leq M$, $N = 2M + 1$ and $\theta_m = 2\pi m/N$ (Antoniou, 1993). Since $H(\theta)$ is real and even, h_n are also real and $h_{-n} = h_n$. The weights $\{h_n : -M \leq n \leq +M\}$ define the Dolph-Chebyshev or, for short, Dolph filter.

In the HIRLAM model, the filter order $N = 2M + 1$ is determined by the time step Δt and forecast span T_S. The desired frequency cut-off is specified by choosing a value for the cut-off period, τ_s. Then $\theta_s = 2\pi\Delta t/\tau_s$ and the parameters x_0 and r are given by

$$\frac{1}{x_0} = \cos\frac{\theta_s}{2}, \quad \frac{1}{r} = \cosh\left(2M\cosh^{-1}x_0\right).$$

The ripple ratio r is a measure of the maximum amplitude in the stop-band $[\theta_s, \pi]$: $r = $ [side-lobe amplitude/main-lobe amplitude]. The Dolph filter has minimum ripple-ratio for given main-lobe width and filter order.

7.2. AN EXAMPLE OF THE DOLPH FILTER

Let us suppose components with period less than three hours are to be eliminated ($\tau_s = 3\,\text{h}$) and the time step is $\Delta t = \frac{1}{2}\,\text{h}$. Then $\theta_s = 2\pi\Delta t/\tau_s \approx 1.05$. It can be shown that a filter of order $N = 7$, or span $T = 2M\Delta t = 3\,\text{h}$, attenuates high frequency components by more than $20\,\text{dB}$ (the attenuation in the stop-band is $\delta = 20\log_{10}|H(\theta)|$). This level of damping implies that

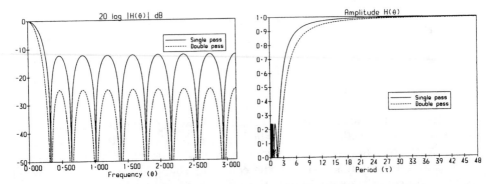

Figure 1. Frequency response for Dolph filter with span $T_S = 2$h, order $N = 2M + 1 = 17$ and cut-off $\tau_s = 3$h. Results for single and double application are shown. Left: Logarithmic response (dB) as a function of frequency. Right: Amplitude response as a function of period.

the amplitudes of high-frequency components are reduced by at least 90% and their energy by at least 99%, which is found to be adequate in practice.

The DFI procedure employed in the HIRLAM model involves a double application of the filter. Thus, we examine both the frequency response $H(\theta)$ and its square, $H(\theta)^2$, as the effect of a second pass through the filter is to square the frequency response. The parameters chosen for the DFI tests below are span $T_S = 2$ h, cut-off period $\tau_s = 3$ h and time step $\Delta t = 450$ s $= \frac{1}{8}$ h. So, $M = 8$, $N = 17$ and $\theta_s = 2\pi\Delta t/\tau_s \approx 0.26$. The response and square response are shown in Fig. 1 (left panel). The ripple ratio has the value $r = 0.241$. A single pass attenuates high frequencies (components with $|\theta| > |\theta_s|$) by at least 12.4dB. For a double pass, the minimum attenuation is about 25dB, more than adequate for elimination of HF noise. For ease of visualisation, the response is also plotted as a function of period in Fig. 1 (right panel). From this it is clear that the amplitudes of components with periods less than two hours are reduced to less than 5% of their original value. At the same time, components with periods greater than one day are substantially left unchanged. It is crucial for an initialization scheme that it does not distort the meteorologically significant components of the flow: the filter described here has the required property.

In Lynch (1997, Appendix) it is proved that the Dolph window is an *optimal* filter whose pass-band edge, θ_p, is the solution of the equation $H(\theta) = 1 - r$. Since an optimal filter is, by construction, the best possible solution to minimizing the maximum deviation from the ideal in the pass- and stop-bands, the Dolph filter shares this property provided the equivalence holds. However, note the essential distinction: for the general optimal filter, θ_p can be freely chosen; for the Dolph window, it is determined by the other parameters. The algorithm for the optimal filter is complex, involving about one thousand lines of code; calculation of the Dolph filter coefficients is trivial by comparison.

8. Implementation in HIRLAM

8.1. OUTLINE OF THE METHOD

The digital filter initialization is performed by applying the filter to time series of model variables. The coefficients of the Dolph filter, $\{h_n : -M \leq n \leq +M\}$, are real and symmetric: $h_{-n} = h_n$. Thus, the phase response is such that the output is valid at the centre of the span. If we had a model integration centred on the initial time, $t = 0$, the filter would produce output valid at that time. However, as the model contains irreversible physical processes, it is not possible to integrate it backwards in time. The solution is to apply the filter in two stages: in the first, a backward integration from $t = 0$ to $t = -T_S$ is performed, with all irreversible physics switched off. The filter output is calculated by accumulating sums of the form

$$\bar{x} = \sum_{n=0}^{n=N} h_{n-M} x_n \, ,$$

where x is a particular prognostic variable at a particular grid point and level (the same sum is accumulated for all prognostic variables). The output $\bar{x}$ is valid at time $t = -\frac{1}{2}T_S$. In the second stage, a forward integration is made from $t = -\frac{1}{2}T_S$ to $t = +\frac{1}{2}T_S$, starting from the output of the first stage. Once again, the filter is applied by accumulating sums formally identical to those of the first stage. But now the output is valid at the centre of the interval $[-\frac{1}{2}T_S, +\frac{1}{2}T_S]$, i.e., at $t = 0$. The output of the second pass of the filter is the initialized data. The values of the prognostic variables at the lateral boundaries are left unchanged during the digital filtering process.

An adiabatic backward integration followed by a diabatic forward integration will not return the variables to their initial values. This deviation, called the diabatic discrepancy, is reduced by the method of Lynch, et al. (1997), but not completely eliminated. Some work towards developing a boundary filter, relying solely on a forward integration, is reported in Lynch and McGrath (2001).

Complete technical details of the original implementation of DFI in the HIRLAM model may be found in Lynch, et al., (1999). A reformulation of the implementation, with further testing and evaluation, is presented in Huang and Yang, 2002.

8.2. INITIALIZATION EXAMPLE

A detailed case study based on the first implementation in HIRLAM was carried out to check the effect of the initialization on the initial fields and on the forecast, and to examine the efficacy of DFI in eliminating high frequency noise. The digital filter initialization was compared to the reference

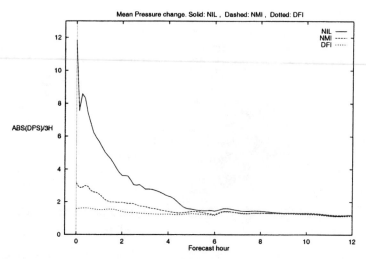

Figure 2. Mean absolute surface pressure tendency for three forecasts. Solid: uninitialized analysis (NIL). Dashed: Normal mode initialization (NMI). Dotted: Digital filter initialization (DFI). Units are hPa/3 hours.

implicit normal mode initialization (NMI) scheme, and to forecasts with no initialization (NIL). Forecasts starting from the analysis valid at 1200 UTC on 10 February, 1999 were compared.

We first checked the effect of DFI on the analysis and forecast fields. The maximum change in surface pressure is 2.2hPa, with an *rms* change of about 0.5hPa. The changes to the other analysed variables are in general comparable in size to analysis errors, and considerably smaller in magnitude than typical changes brought about by the analysis itself: the *rms* change in surface pressure from first-guess to analysis is about 1hPa. The *rms* and maximum differences between the uninitialized 24 hour forecast (NIL) and the filtered forecast (DFI) for all prognostic variables were examined. When we compare these values to the differences at the initial time they are seen to be generally smaller. The changes made by DFI are to the high frequency components; since these are selectively damped during the course of the forecast, the two forecasts are very similar. After 24 hours the maximum difference in surface pressure is less than 1hPa and the *rms* difference is only 0.1hPa.

The basic measure of noise is the mean absolute value of the surface pressure tendency

$$N_1 = \left(\frac{1}{\text{NMAX}} \right) \sum_{n=1}^{\text{NMAX}} \left| \frac{\partial p_s}{\partial t} \right| .$$

For well balanced fields this quantity has a value of about 1 hPa/3h. For uninitialized fields it can be an order of magnitude larger. In Figure 2 we plot the value of N_1 for three forecasts. The solid line represents the forecast

 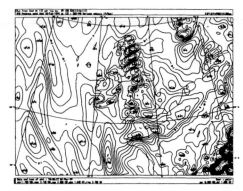

Figure 3. Vertical velocity at 500 hPa over western Europe and the eastern North Atlantic. (Left) Uninitialized analysis (NIL); (Right) after digital filtering (DFI).

from uninitialized data: we see that the value of N_1 at the beginning of the forecast is about 12 hPa/3h. This large value reflects the lack of an effective multivariate balance in the analysis. It takes about six hours to fall to a reasonable value. The dashed line is for a forecast starting from data initialized using the implicit normal mode method (NMI). The starting value is about 3 hPa/3h, falling to about 1.5 hPa/3h after twelve hours. The final graph (the dotted line) is for the digitally filtered data (DFI). The initial value of N_1 is now about 1.5, and remains more-or-less constant throughout the forecast. It is clear from this measure that DFI is more effective in removing high frequency noise than NMI.

The measure N_1 indicates the noise in the vertically integrated divergence field. However, even when this is small, there may be significant activity in the internal gravity wave modes. To see this, we look at the vertical velocity field at 500 hPa for the NIL and DFI analyses. The left panel in Fig. 3 shows the uninitialized vertical velocity field, zoomed in over western Europe and the eastern North Atlantic. There is clearly substantial gravity wave noise in this field. In fact, the field is physically quite unrealistic. The right panel shows the DFI vertical velocity. It is much smoother; the spurious features have been eliminated and the large values with small horizontal scales which remain are clearly associated with the Scottish Highlands, the Norwegian Mountains and the Alps. Comparison with the NMI method (see Lynch, *et al.*, 1999, for details) indicates that DFI is more effective than NMI in dealing with internal gravity wave noise. It is noteworthy that stationary mountain waves are unaffected by digital filtering, since they have zero frequency. This is a desirable characteristic of the DFI scheme.

8.3. BENEFITS FOR THE DATA ASSIMILATION CYCLE

In Lynch, *et al.*, 1999, a parallel test of data for one of the FASTEX intensive observing periods showed that the DFI method resulted in slightly improved scores compared to NMI. As it is not usual for an initialization scheme to yield significant improvements in forecast accuracy, some discussion is merited. We cannot demonstrate beyond question the reason for this improvement. However, the comparative results showed up some definite defects in the implicit normal mode initialization as implemented in the reference HIRLAM model. It was clear that the NMI scheme did not eliminate imbalance at lower model levels. Moreover, although the noise level indicated by the parameter N_1 fell to a reasonable level in six hours, there was still internal gravity wave noise, not measured by this parameter. Any noise in the six hour forecast will be carried through to the next analysis cycle, and will affect the quality control and assimilation of new observational data. It is believed that the DFI scheme, with its superior ability to establish atmospheric balance, results in improved assimilation of data and consequently in a reduction of forecast errors.

9. Digital Filtering as a Constraint in 4DVAR

We conclude with a remark on the application of a digital filter as a weak constraint in four-dimensional variational assimilation (4DVAR). The idea is that if the state of the system is noise-free at a particular time, *i.e.*, is close to the slow manifold, it will remain noise-free, since the slow manifold is an invariant subset of phase-space. We consider a sequence of values $\{x_0, x_1, x_2, \cdots x_N\}$ and form the filtered value

$$\bar{x} = \sum_{n=0}^{N} h_n x_n. \tag{17}$$

The evolution is constrained, so that the value at the mid-point in time is close to this filtered value, by addition of a term

$$J_c = \frac{1}{2}\mu ||x_{N/2} - \bar{x}||^2$$

to the cost function to be minimized (μ is an adjustable parameter). It is straightforward to derive the adjoint of the filter operator. Gauthier and Thépaut (2001) applied such a constraint to the 4DVAR system of Météo-France. They found that a digital filter weak constraint imposed on the low-resolution increments efficiently controlled the emergence of fast oscillations while maintaining a close fit to the observations. As the values required for input to the filter are already available, there is essentially no computational overhead in applying this procedure. The dynamical imbalance was significantly less in 4DVAR than in 3DVAR. Fuller details may be found in Gauthier and Thépaut (2001).

10. Conclusion

The very notion of eliminating what is called "noise" is open to debate. There is no doubt as to the presence of high frequency motions in the atmosphere, and some evidence suggests that they may have a function in the development of meso-scale systems. If the feedback from HF components to the meteorologically significant motion is found to be important in certain circumstances, the application of filtering may be injudicious. It is important to minimize spurious imbalances in the analysed fields, through improved modelling of the multivariate background error covariances, and thus reduce the size of the changes induced by the initialization process. Removal of gravity waves cannot be unequivocally justified; the problem becomes all the more acute as model resolution increases.

References

Antoniou, A, 1993: *Digital Filters: Analysis, Design and Applications*. 2nd Edn., McGraw-Hill, 689pp.

Bubnová, Radmila, Gwenaëlle Hello, Pierre Bénard and Jean-François Geleyn, 1995: Integration of the fully elastic equations cast in the hydrostatic pressure terrain-following coordinate in the framework of the ARPEGE/Aladin NWP system. *Mon. Wea. Rev.*, **123**, 515–535.

Gauthier, P. and J. N. Thépaut, 2001: Impact of the digital filter as a weak constraint in the pre-operational 4DVAR assimilation system of Météo-France. *Mon. Wea. Rev.*, **129**, 2089–2102.

Hamming, R.W., 1989: *Digital Filters*. Prentice-Hall International, 284pp.

Huang, Xiang-Yu and Peter Lynch, 1993: Diabatic digital filtering initialization: application to the HIRLAM model. *Mon. Wea. Rev.*, **121**, 589–603.

Huang, Xiang-Yu and Xiaohua Yang, 2002. A new implementation of digital filtering initialization schemes for HIRLAM [http://www.knmi.nl/hirlam/TechReports/TR53.pdf]. HIRLAM Technical Report No. 53

Lynch, Peter and Xiang-Yu Huang, 1992: Initialization of the HIRLAM model using a digital filter. *Mon. Weather Rev.*, **120**, 1019–1034.

Lynch, Peter, 1997: The Dolph-Chebyshev Window: A Simple Optimal Filter. *Mon. Wea. Rev.*, **125**, 655-660.

Lynch, Peter, Dominique Giard and Vladimir Ivanovici, 1997: Improving the efficiency of a digital filtering scheme. *Mon. Wea. Rev.*, **125**, 1976–1982.

Lynch, Peter and Ray McGrath, 2001: Boundary Filter Initialization of the HIRLAM Model. Pp. 31–44 in *Advances in Mathematical Modelling of Atmosphere and Ocean Dynamics*. Ed. P. F. Hodnett. Kluwer Acad. Publ., 298pp.

Lynch, Peter, Ray McGrath and Aidan McDonald, 1999: Digital Filter Initialization for HIRLAM. HIRLAM Tech. Rep. No. 42 [http://www.knmi.nl/hirlam/TechReports/TR42.ps.gz].

Machenhauer, B., 1977: On the dynamics of gravity oscillations in a shallow water model with applications to normal mode initialization. *Beitr. Atmos. Phys.*, **50**, 253–271.

Oppenheim, A.V. and R.W. Schafer, 1989: *Discrete-Time Signal Processing*. Prentice-Hall International, Inc., 879 pp.

Raymond, W. H., 1988: High-order low-pass implicit tangent filters for use in finite area calculations. *Mon. Wea. Rev.*, **116**, 2132–2141.

TREATING MODEL ERROR IN 3-D AND 4-D DATA ASSIMILATION

N.K. NICHOLS
Department of Mathematics, The University of Reading,
Box 220, Reading, RG6 6AX United Kingdom

1. Introduction

The aim of a data assimilation scheme is to use measured observations in combination with a dynamical system model to derive accurate estimates of the current and future states of the system. In operational schemes for atmosphere and ocean forecasting, the model equations are generally assumed to be a 'perfect' representation of the true dynamical system and are treated as strong constraints in the assimilation process. The model equations do not, in practice, represent the system behaviour exactly, however, and model errors arise due to lack of resolution and inaccuracies in physical parameters, boundary conditions and forcing terms. Errors also occur due to discrete approximations and random disturbances.

To account for model error, the system equations can be treated as weak constraints. Statistically the residual errors in the model are assumed to be unbiased white noise that is uncorrelated in time, and the 'best' linear unbiased estimates of the system states, given the available observations and their error covariances, is sought. This approach is not practicable, however, due to the excessive size of the optimisation problem. More significantly, the statistical assumptions are not satisfied in practice, since the model errors are expected to depend on the model states and to be *systematic* and *time-correlated*.

Recently the problem of accounting for model error in a cost-effective way has received more attention. Techniques for treating bias errors in the forecast using sequential and four-dimensional variational assimilation schemes (Dee and da Silva, 1998; Derber, 1989) and for treating time-correlated stochastic errors (Zupanska, 1997) have been investigated. A general formulation for the treatment of systematic time-varying model errors has been derived (Griffith and Nichols, 1996) and techniques based

R. Swinbank et al. (eds.), Data Assimilation for the Earth System, 127–135.

on this formulation have been explored (Griffith 1997; Martin *et al.*, 1999; Griffith *et al.*, 2000; Griffith and Nichols, 2000; Martin, 2001). These techniques have been applied successfully in practice to estimate systematic errors in an equatorial ocean model (Martin *et al.*, 2001; Bell *et al.*, 2001).

The framework for treating systematic, time-correlated model errors that we present here is based on the formulation of Griffith and Nichols (1996, 2000). Simple assumptions about the evolution of the error are made, enabling the systematic error to be estimated as part of the assimilation process. The model equations are augmented by the evolution equations for the error and standard data assimilation techniques can then be applied to the augmented state system.

In the next section, a general representation of model error for use in data assimilation is introduced. In Section 3 the technique of state augmentation for estimating the systematic time-varying components of the model error is described, and in Section 4 an application is presented. Conclusions are given in the final section.

2. Data Assimilation in Dynamical Systems with Model Errors

The evolution of the dynamical system, taking into account model errors, is described by the discrete nonlinear equations

$$\mathbf{x}_{k+1} = \mathbf{f}_k(\mathbf{x}_k, \mathbf{u}_k) + \boldsymbol{\epsilon}_k, \quad k = 0, \dots, N-1, \tag{1}$$

where $\mathbf{x}_k \in \mathbb{R}^n$ denotes the model states and $\mathbf{u}_k \in \mathbb{R}^{m_k}$ denotes the m_k inputs to the system at time t_k, $\mathbf{f}_k : \mathbb{R}^n \times \mathbb{R}^{m_k} \to \mathbb{R}^n$ is a nonlinear function describing the evolution of the states from time t_k to time t_{k+1} and $\boldsymbol{\epsilon}_k \in \mathbb{R}^n$ denotes model errors at time t_k. Prior estimates, or 'background estimates,' $\mathbf{x}_0^b$ of the initial states $\mathbf{x}_0$ are assumed to be known and the initial random error $(\mathbf{x}_0 - \mathbf{x}_0^b)$ is assumed to be Gaussian with covariance matrix $B_0 \in \mathbb{R}^{n \times n}$.

The observations are assumed to be related to the system states by the equations

$$\mathbf{y}_k = \mathbf{h}_k(\mathbf{x}_k) + \boldsymbol{\delta}_k, \quad k = 0, \dots, N-1, \tag{2}$$

where $\mathbf{y}_k \in \mathbb{R}^{p_k}$ is a vector of p_k observations at time t_k and $\mathbf{h}_k : \mathbb{R}^n \to \mathbb{R}^{p_k}$ is a nonlinear function that includes transformations and grid interpolations. The observational errors $\boldsymbol{\delta}_k \in \mathbb{R}^{p_k}$ are assumed to be unbiased, serially uncorrelated, Gaussian random vectors with covariance matrices $R_k \in \mathbb{R}^{p_k \times p_k}$.

Commonly the model errors $\boldsymbol{\epsilon}_k$ are also assumed to be stochastic variables that are unbiased and serially uncorrelated, with a known Gaussian distribution and co-variance matrices given by $Q_k \in \mathbb{R}^{n \times n}$. The observational errors, the model errors and the errors in the prior estimates are assumed to be uncorrelated.

For the 'optimal' analysis, we aim to find the best estimates $\mathbf{x}_k^a$ for the expected values of the true states $\mathbf{x}_k$, $k = 0, \ldots, N - 1$, given observations $\mathbf{y}_k$, $k = 0, \ldots, N - 1$, subject to the model equations (1) and prior estimates $\mathbf{x}_0^b$. Under the statistical assumptions made here, the optimal analysis is given by the maximum likelihood *a priori* Bayesian estimate of the system states, given the observations and the prior estimates of the initial states.

The 'optimal' assimilation problem reduces to minimising the square error between the model predictions and the observed system states, together with the square error in the model equations, all weighted by the inverses of the covariance matrices, over the assimilation interval. The model equations (1) are thus treated as weak constraints on the objective function. The initial states of the system and the model errors at every time step are the control parameters that must be determined. The data assimilation problem is defined mathematically as follows.

Problem 1 *Minimise, with respect to $\mathbf{x}_0$ and ϵ_k, $k = 0, \ldots, N - 1$, the objective function*

$$\mathcal{J} = \frac{1}{2}(\mathbf{x}_0 - \mathbf{x}_0^b)^T B_0^{-1}(\mathbf{x}_0 - \mathbf{x}_0^b) +$$

$$+ \frac{1}{2} \sum_{k=0}^{N-1} (\mathbf{h}_k(\mathbf{x}_k) - \mathbf{y}_k)^T R_k^{-1}(\mathbf{h}_k(\mathbf{x}_k) - \mathbf{y}_k) +$$

$$+ \frac{1}{2} \sum_{k=0}^{N-1} \epsilon_k^T Q_k^{-1} \epsilon_k, \tag{3}$$

subject to $\mathbf{x}_k$, $k = 1, \ldots, N - 1$, satisfying the system equations (1).

In order to find the 'optimal' analysis, either sequential schemes that use Kalman filtering, or four-dimensional schemes that use variational methods can be applied.

The Extended Kalman Filter (EKF) solves the data assimilation problem, Problem 1, sequentially (Kalman, 1961). The EKF propagates the analysis and the covariance matrices forward together in order to produce the optimal analysis at each time step, taking into account the current and all previous observations. For linear models, the solution obtained using the Kalman filter is the exact optimal and is equal to the solution of the minimisation problem at the end of the assimilation period. For nonlinear systems, approximate linearisations of the model and observation operators are introduced in the extended filter, and the optimality property is not retained. For nonlinear models, moreover, the EKF is highly sensitive to computational errors and may become unstable as a dynamical system (Bierman, 1977).

Variational techniques, in contrast, solve the optimal assimilation problem, Problem 1, for all the analysis states in the assimilation window simultaneously. A direct gradient iterative minimisation procedure is applied to the objective function (3), where the descent directions are determined from the associated adjoint equations. The full set of adjoint equations provides gradients of the objective function with respect to the initial states and with respect to all of the model errors at each time step. A forward solve of the model equations, followed by a reverse solve of the adjoint equations is needed to determine the gradients. Alternatively, the optimal assimilation problem can be solved by a dual variational approach in which the minimisation is performed in observation space.

For very large stochastic systems, such as weather and ocean systems, the variational and EKF four-dimensional data assimilation schemes are generally too expensive for operational use due to the enormous cost of estimating all of the model errors in the variational approach or, alternatively, propagating the error covariance matrices in the Kalman filter. Moreover, the model error is expected to depend on the model state and hence to be *systematic* and *correlated in time*. A more general form of the model error that includes both systematic and random elements is described in the next section.

3. Systematic Model Error and State Augmentation

In order to take into account systematic components in the model errors, we assume that the evolution of the errors is described by the equations

$$\epsilon_k = T_k(\mathbf{e}_k) + \mathbf{q}_k, \tag{4}$$

$$\mathbf{e}_{k+1} = \mathbf{g}_k(\mathbf{x}_k, \mathbf{e}_k), \tag{5}$$

where $\mathbf{q}_k \in \mathbb{R}^n$ are unbiased, serially uncorrelated, normally distributed random vectors with known covariance matrices and the vectors $\mathbf{e}_k \in \mathbb{R}^r$ represent time-varying systematic components of the model errors. The distribution of the systematic errors in the model equations is defined by the functions $T_k : \mathbb{R}^r \to \mathbb{R}^n$. The functions $\mathbf{g}_k : \mathbb{R}^n \times \mathbb{R}^r \to \mathbb{R}^r$, describing the systematic error dynamics, are to be specified.

In practice little is known about the form of the model errors and a simple form for the error evolution that reflects any available knowledge needs to be prescribed. Examples of simple forms of the error evolution include:

- *Constant bias error* : $\mathbf{e}_{k+1} = \mathbf{e}_k, \ T_k = I.$
This choice allows for a constant vector $\mathbf{e} \equiv \mathbf{e}_0$ of unknown parameters to

be found, which can be interpreted as statistical biases in the model errors. This form is expected to be appropriate for representing average errors in source terms or in boundary conditions.

- *Evolving error :* $\mathbf{e}_{k+1} = F_k \mathbf{e}_k, \quad T_k = I.$

Here $F_k \in \mathbb{R}^{n \times n}$ represents a simplified linear model of the state evolution. This choice is appropriate, for example, for representing discretisation error in models that approximate continuous dynamical processes by discrete-time systems.

- *Spectral form :* $\mathbf{e}_{k+1} = \mathbf{e}_k, \quad T_k = (I, \sin(k/N\tau)I, \cos(k/N\tau)I).$

In this case the constant vector $\mathbf{e} \equiv \mathbf{e}_0$ is partitioned into three component vectors, $\mathbf{e}^T = \left(\mathbf{e}_1^T, \mathbf{e}_2^T, \mathbf{e}_3^T\right)$, and τ is a constant determined by the timescale on which the model errors are expected to vary, for example, a diurnal timescale. This choice approximates the first order terms in a spectral expansion of the model error.

Other choices can be prescribed, including piecewise constant error and linearly growing error (see (Griffith, 1997)).

Together the system equations and the model error equations (1), (4) and (5) constitute an *augmented* state system model. The aim of the data assimilation problem for the augmented system is to estimate the expected values of the augmented states $(\mathbf{x}_k^T, \mathbf{e}_k^T)^T$ for $k = 0, \ldots, N-1$, that best fit the observations, subject to the augmented state equations. The solution delivers the maximum likelihood estimate of the augmented system states, given the covariances of the initial states, the observations and the random components of the model errors. Although this formulation takes into account the time evolution of the model errors, the data assimilation problem remains intractable for operational use. If, however, the stochastic elements of the error are ignored and the augmented system is treated as a 'perfect' deterministic model, then solving the augmented data assimilation problem becomes feasible. The aim of the data assimilation, in this case, is to estimate the systematic components of the model error along with the model states.

The 'perfect' augmented system equations are written

$$\mathbf{x}_{k+1} = \mathbf{f}_k(\mathbf{x}_k, \mathbf{u}_k) + T_k(\mathbf{e}_k), \tag{6}$$

$$\mathbf{e}_{k+1} = \mathbf{g}_k(\mathbf{x}_k, \mathbf{e}_k), \tag{7}$$

for $k = 0, \ldots, N-1$, where the observations are related to the model states by the equations

$$\mathbf{y}_k = \mathbf{h}_k(\mathbf{x}_k) + \boldsymbol{\delta}_k, \qquad k = 0, \ldots, N-1, \tag{8}$$

as previously. The covariance matrices R_k of the observational errors are known. It is assumed that prior estimates, or 'background estimates,' $\mathbf{x}_0^b$

and e_0^b of x_0 and e_0 are known and that the covariance matrix of the errors $((x_0 - x_0^b)^T, (e_0 - e_0^b)^T)^T$ is given by $W_0 \in \mathbb{R}^{(n+r) \times (n+r)}$. Note that the initial states and initial model errors may be cross-correlated. The observational errors and the errors in the prior estimates are not correlated.

The augmented data assimilation problem is to minimise the square errors between the model predictions and the observed system states, weighted by the inverse of the covariance matrices, over the assimilation interval. The problem is written

Problem 2 *Minimise, with respect to* $(x_0^T, e_0^T)^T$, *the objective function*

$$
\mathcal{J} = \frac{1}{2}((x_0 - x_0^b)^T, \ (e_0 - e_0^b)^T)W_0^{-1}((x_0 - x_0^b)^T, \ (e_0 - e_0^b)^T)^T
$$

$$
+ \frac{1}{2} \sum_{j=0}^{N-1} (h_j(x_j) - y_j)^T R_j^{-1} (h_j(x_j) - y_j), \tag{9}
$$

subject to the augmented system equations (6)–(7).

The initial values x_0 and e_0 of the model state and model error completely determine the response of the augmented system and are taken to be the control variables in the optimisation. The augmented system equations (6)–(7) are treated as strong constraints on the problem. Since the covariance matrix W_0 is nonsingular, the problem is well-posed and may be solved by any of the standard data assimilation schemes described in this volume.

The state augmentation technique for estimating systematic model error is illustrated in the next section by a simple example, using a sequential data assimilation scheme.

4. Example

The performance of the augmented data assimilation scheme is demonstrated for the nonlinear chaotic Lorenz equations (Miller *et al.*, 1994), used frequently as a simplified model of thermal convection. The system equations are

$$
\dot{x} = -\sigma(x - y), \tag{10}
$$
$$
\dot{y} = \rho x - y - xz, \tag{11}
$$
$$
\dot{z} = xy - \beta z, \tag{12}
$$

where $\sigma = 10$, $\rho = 28$, and $\beta = 8/3$. The system has three equilibrium points: an unstable saddle point at the origin and two unstable spiral points at the co-ordinates $(\pm\sqrt{\beta(\rho - 1)}, \pm\sqrt{\beta(\rho - 1)}, \rho - 1)$.

The system is discretised using a second order Runge-Kutta method with time steps of 0.01. Observations of the true solution to the discrete nonlinear system, starting with initial conditions $x_0 = -5.1551$, $y_0 = -8.4334$, $z_0 = 15.0667$, are taken every five time steps. Random Gaussian noise, with a standard deviation of 0.1 and zero mean, is added to the true data to generate the observational data used in the experiment.

The model containing systematic errors is obtained by adding a constant bias of 0.5 to each of the discrete system equations. The incorrect 'biased' model has a stable equilibrium point in the right half plane to which the background solution is attracted. The systematic model error is assumed to be constant and the model is augmented by the evolution equations for a constant bias, as defined in Section 3. The optimal interpolation sequential assimilation scheme is applied to the augmented system, with the error covariance matrix given by the averaged error statistics over the assimilation period. The analysis is performed on the interval $t \in [0, 12]$ and a forecast is then made on the interval $t \in [12, 20]$.

The true solution is shown in the x-z phase plane in Figure 1(a) and the dynamic response of the true state variable $x(t)$ is shown in Figure 1(b). In Figure 1(c) the assimilated and forecast solutions to the incorrect model are shown for the case without model error correction. The assimilation of the observations ensures that the results follow the true solution during the assimilation period, but in the forecast, errors appear immediately due to the bias errors in the model. The forecast is rapidly attracted to the incorrect stable equilibrium point. In Figure 1(d) the results of applying the assimilation to the augmented model are shown. In this case an accurate estimate of the bias error in the equations is obtained on the assimilation interval and is retained in the forecast period to give an improved prediction of the system states. The forecast remains accurate on the interval $[12, 17]$, but loses accuracy subsequently due to the chaotic nature of the dynamical system.

The correct averaged statistics are used here in the background covariance matrices for the augmented states. In practice the correct statistics may not be available. With sequential assimilation schemes, oscillations in the analysed bias error can then occur, giving a poorer forecast. The problem is treated by smoothing the analysed bias error over a moving time window during the assimilation. Details of this technique and additional results are described in Martin *et al.* (1999). With four-dimensional assimilation schemes, the analysed error is automatically smoothed over the assimilation window and this behaviour does not arise.

5. Conclusions

Techniques are described here for using sequential and four-dimensional variational data assimilation to estimate the time-correlated systematic components of model errors along with the dynamical model states of a system. A simple form for the evolution of the model error is assumed and an augmented system for both the model states and model errors is obtained. For different types of error, different forms for the model error evolution are appropriate. A modified objective function is minimised to determine the solution of the augmented system that best fits the available observations over the assimilation interval. This technique is effective and can lead to significantly improved forecasts.

Various applications of this approach to model error estimation, using both sequential and four-dimensional assimilation methods, are described in the literature (Griffith, 1997; Martin, 2001; Martin *et al.*, 1999; Griffith *et al.*, 2000; Griffith and Nichols, 2000; Martin *et al.*, 2001; Bell *et al.*, 2001).

References

Bell, M.J., Martin, M.J. and Nichols, N.K. (2001) *Assimilation of Data into an Ocean Model with Systematic Errors Near the Equator*, The Met Office, Ocean Applications Division, Tech. Note No. 27.

Bierman, G.L. (1977) *Factorization Methods for Discrete Sequential Estimation*, Mathematics in Science and Engineering, V. 128, Academic Press, New York.

Derber, J.C. (1989) A variational continuous assimilation technique, *Monthly Weather Review*, **117**, 2437–2446.

Dee, D.P. and da Silva, A.M. (1998) Data assimilation in the presence of forecast bias, *Quart. J. Roy. Met. Soc.*, **117**, 269–295.

Griffith, A.K. (1997) *Data Assimilation for Numerical Weather Prediction Using Control Theory*, The University of Reading, Department of Mathematics, PhD Thesis.

Griffith, A.K., Martin, M.J. and Nichols, N.K. (2000) Techniques for treating systematic model error in 3D and 4D data assimilation, in *Proceedings of the Third WMO Int. Symposium on Assimilation of Observations in Meteorology and Oceanography*, World Meteorological Organization, WWRP Report Series No. 2, WMO/TD - No. 986, pp. 9–12.

Griffith, A.K. and Nichols, N.K. (1996) Accounting for model error in data assimilation using adjoint methods, in M. Berz, C. Bischof, G. Corliss and A. Greiwank (eds.), *Computational Differentiation: Techniques, Applications and Tools*, SIAM, Philadelphia, pp. 195–204.

Griffith, A.K. and Nichols, N.K. (2000) Adjoint techniques in data assimilation for estimating model error, *Journal of Flow, Turbulence and Combustion*, **65**, 469–488.

Kalman, R.E. (1961) A new approach to linear filtering and prediction problems, *Transactions of the ASME, Series D*, **83**, 35–44.

Martin, M.J. (2001) *Data Assimilation in Ocean Circulation Models with Systematic Errors*, The University of Reading, Department of Mathematics, PhD Thesis.

Martin, M.J., Bell, M.J. and Nichols, N.K. (2001) Estimation of systematic error in an equatorial ocean model using data assimilation, in M.J. Baines (ed.), *Numerical Methods for Fluid Dynamics VII*, ICFD, Oxford, pp. 423–430.

Martin, M.J., Nichols, N.K. and Bell, M.J. (1999) *Treatment of Systematic Errors in Sequential Data Assimilation*, The Met Office, Ocean Applications Division, Tech. Note No. 21.

Miller, R.N., Ghil, M. and Gauthiez, F. (1994) Advanced data assimilation in strongly nonlinear dynamical systems, *J. Atmos. Sc.*, **51**, 1037–1056.

Zupanski, D. (1997) A general weak constraint applicable to operational 4DVAR data assimilation systems, *Monthly Weather Review*, **123**, 1112–1127.

(a) True solution (x-z phase plane)

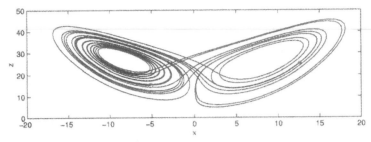

(b) True solution (state variable x(t))

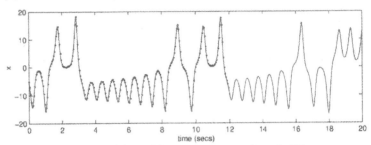

(c) Assimilation without bias error correction (x(t))

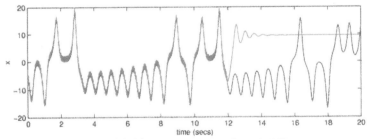

(d) Assimilation with bias error correction (x(t))

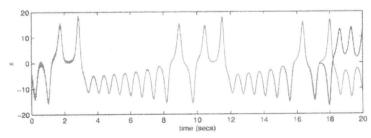

Figure 1. (a)-(b) True solution to the chaotic Lorenz equations. (c)-(d) Assimilated solution on the interval [0; 12], followed by a forecast on the interval [12; 20]. Without bias correction the forecast departs rapidly from the true behaviour of the system, but with the error estimation technique, good accuracy in the forecast is retained for a significant period.

OBSERVING THE ATMOSPHERE

R. SWINBANK
NWP Division, Met Office
Bracknell, Berkshire, UK

1. Introduction: the observing system

The aim of this chapter is to give an overview of the techniques used to observe the atmosphere, and the physical principles involved. In particular, we attempt to highlight characteristics of different observation types that need to be taken into account when using them in an atmospheric data assimilation system.

In the next two sections, we survey the main types of observations of the atmosphere, with emphasis on those regularly used in operational Numerical Weather Prediction (NWP) systems. We have divided up the different types of observations into two broad categories: *in situ* observations and remotely sensed observations. Different observation types have different characteristics; broadly speaking, remote sensing data need to be treated rather differently from conventional in situ data, taking into account the indirect nature of remote sensing measurements. In section 4 we highlight some considerations about how well the different types of observations are suited for data assimilation. Finally we include a brief overview of some new measurement techniques which are likely to grow in importance over the coming years.

2. In situ observations

We sometimes refer to "conventional observations" as those which were in general use before the era of satellite data. For the most part, they are *in situ* measurements of meteorological parameters such as temperature and pressure (for a general overview of meteorological observations, see Meteorological Office, 1982). For in situ data it is generally straightforward to write an observation operator for data assimilation schemes, since essentially all that is required is to interpolate appropriate model parameters to the observation locations. (An observation operator is an operator that derives synthetic observation values from a model state.)

2.1. SURFACE OBSERVATIONS

Measurements of temperature, wind and humidity at (or close to) the surface have been made for several centuries. Stevenson screens are used to house thermometers to measure temperature (at a standard height of 2m), a wet bulb thermometer to determine

R. Swinbank et al. (eds.), Data Assimilation for the Earth System, 137–148.

138

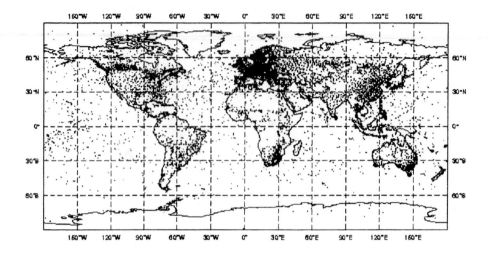

Figure 1. Typical distribution of surface observations. This is the set of observations used for the 12 UTC global forecast at the Met Office on 8th August 2002. The total number of observations was 9554, including manual synoptic observations from land and sea, plus buoy data.

dewpoint depression (hence humidity), and a barometer for pressure. Wind velocity is measured at a standard height (10m) with an anemometer. To aid comparison of pressure measurements from different stations they are normally adjusted to mean sea level, which can cause problems in mountainous areas. At the same time as reading the instruments, the observer makes visual observation of clouds, visibility, and current weather type. While all these data are very useful for forecasters, it is still hard to use all parts of a surface observation report in a data assimilation system designed for NWP.

Surface observations are available over much of the densely populated regions of the world, particularly in the northern hemisphere, although there are extensive data voids over regions such as Africa (Fig. 1). Similar types of observations are made from many ships, which helps fill the gaps over the parts of the oceans that are well covered by shipping routes. In recent years many drifting (and moored) buoys have been deployed to help fill data voids.

2.2. RADIOSONDES

First attempts at making observations of the upper atmosphere (in this context, the free troposphere) were made during the second half of the nineteenth century. Labitzke and van Loon (1999) give some fascinating accounts of early exploration of upper levels of the atmosphere, with particular emphasis on discoveries related to the stratosphere. Radiosondes came to be a crucial part of the Global Observing System following the International Geophysical Year, or IGY (1957).

Radiosondes are generally launched twice a day (at 0 and 12 UTC) from radiosonde stations spread across the world. As in the case of surface data, the stations are concentrated on the land masses of the northern hemisphere (Fig. 2). Each radiosonde consists of a balloon carrying an instrument package, from which measurements are

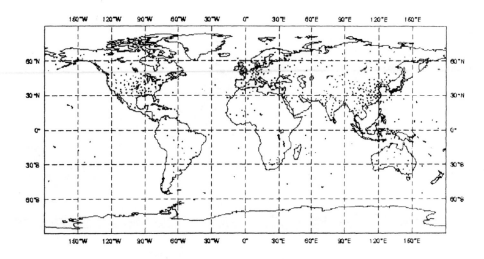

Figure 2. Distribution of upper air observations for the same case as Fig. 1. The total number of observations was 1293, mostly radiosonde and pilot balloon ascents, but including some wind profiler data

relayed to the ground station. The instrument package makes *in situ* measurements of pressure, temperature and humidity. Radar, or more recently, GPS navigation, is used to track the balloon and so ascertain the wind at the height of the balloon. In a radiosonde sounding, weather elements are reported at standard pressure levels. The reports also include "special levels" to allow details of the measurement profile to be reconstructed between the standard levels.

Radiosondes are a crucial part of the observation network. They are still heavily used by forecasters, particularly in developing countries. They make a major contribution to the NWP scores (Bouttier and Kelly, 2001), primarily because it is difficult to use satellite data over land. Radiosondes are also essential for the calibration of satellite data.

Since there are many fewer radiosonde observations than surface data, the data voids are even more severe. Some radiosonde ascents are also made from special weather ships, but, because of their high cost, they are being replaced to some degree by ASAP (Automated Shipboard Aerological Programme) balloons that can be automatically launched from commercial ships. Radiosondes are also complemented by pilot balloons, which are launched simply to measure wind profiles, without carrying an instrument package.

Dropsonde observations are similar to radiosondes, except that the instrument packages are dropped from aircraft rather than flown from a balloon. These are often used in experimental campaigns, rather than in routine operations. They also also employed when targeting observations (see section 4).

140

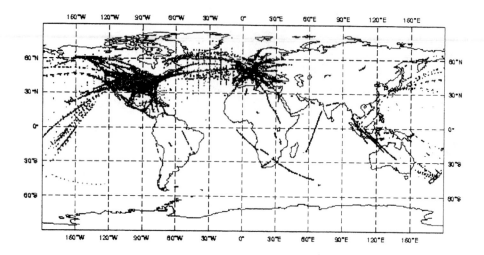

Figure 3. Distribution of aircraft observations for the same case as Fig. 1. The total number of observations was 12085. Most of the data are automatic AMDAR measurements, but some manual AIREPs are also included.

2.3. AIRCRAFT

Many commercial aircraft make *in situ* measurements of temperatures and winds during their flights. (The wind measurements also need input from navigation system, since the wind velocity is the difference between the aircraft ground velocity and its air speed). Conventionally, aircraft observations are reported by the pilot at particular locations along aircraft routes. In recent years, the most important developments have been in the automatic dissemination of observations, for example in the AMDAR (Aircraft Meteorological Data and Reporting) programme. This has led to far more data becoming available (Fig. 3), both at altitude and during takeoff and landing.

3. Remote-sensing observations

Satellite observations are perhaps the most important general category of observations used by a modern operational NWP system. Although we often talk of satellites measuring temperatures, winds or constituents, satellites of course only ever measure radiance. The key to using satellite observations is the ingenious ways in which those radiance measurements can be used to reveal information about meteorological parameters.

While most people are familiar with satellite imagery or "satellite pictures" (viewed at either visible or infrared wavelengths), there are many other useful ways in which satellite measurements can be used for weather forecasting. Houghton et al (1984) gives a good general introduction to the use of satellite observations. A recent ECMWF seminar report (ECMWF, 2001) includes a range of papers describing methods to exploit observations from recent, and planned, satellite instruments.

Satellites fall into two broad categories. Geostationary satellites are in a very high orbit (about 36,000 km above the earth's surface), where they orbit at the same rate as the earth's rotation, so that they continually view the same part of the earth's surface. The are sometimes referred to as GEO (Geostationary Earth Orbit) satellites. By contrast LEO (Low Earth Orbit) satellites are in orbits of perhaps 800 km altitude. Meteorological satellites are often placed in a sun-synchronous (LEO) orbit, so that they always pass over a particular latitude at the same local time (or two local times, counting both the ascending and descending node of the orbit).

3.1. CLOUD TRACK WINDS

Cloud track winds are often also referred to as atmospheric motion vectors. The method was initially developed to measure winds by tracking cloud movement between a pair of images taken from a geostationary satellite. More generally, winds can be measured by tracking features in any type of frequent satellite image, for example, water vapour images. Although the speeds can be very accurate, the height assignment of this type of observation can be problematic, for example because the satellite sees cloud tops while cloud motion is more closely related to winds at cloud base. Another problem is that cloud may not be moving with the wind (for example in the case of orographic clouds), although such cases can normally be detected and removed automatically.

3.2. NADIR SOUNDERS

Nadir sounders are those that measure thermal emissions from a field of view below the satellite, although they may be scanned away from the nadir. They are generally deployed on polar LEO satellites (although sounders are also starting to be deployed on GEO satellites). Since the late 1970's these data have become a key component of the global observing system. In particular, they have opened the door to understanding the stratosphere - which is largely beyond the reach of conventional observations.

Nadir sounders measure radiation emitted by the atmosphere at wavelengths where the atmosphere is optically thick (i.e. the atmosphere is only partially transparent), so measured radiation is coming from some layer within the atmosphere. The brightness of the emission at that particular wavelength gives us information about the atmospheric temperature at the particular layer at which it was emitted. For example at a wavelength where the atmosphere is quite opaque, the satellite will receive radiation emanating high up in the atmosphere, but at another wavelength where the atmosphere is more transparent the satellite will receive radiation from lower altitudes. By taking a set of measurements at different wavelengths, it is possible to derive temperature profiles. These nadir soundings suffer from poor vertical resolution (typically 5-10 km), but, since they are viewing the atmosphere from above, high horizontal resolutions can be achieved. Figure 4 shows a typical distribution of soundings.

Meteorological satellites, as typified by the NOAA polar orbiters, carry sounders that measure radiance in both the infrared (IR) and microwave (eg. HIRS, High-resolution Infrared Radiation Sounder, and AMSU, Advanced Microwave Sounding Unit, collectively referred to as ATOVS, Advanced TIROS Operational Vertical Sounder). While the IR sounders have generally better resolution, they are more

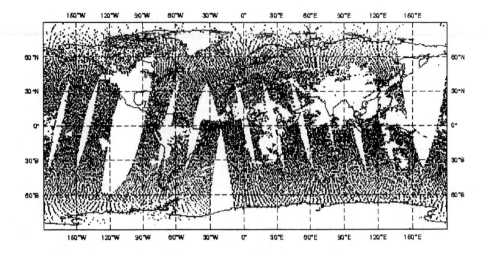

Figure 4. Distribution of satellite soundings for the same case as Fig. 1. The total number of observations was 14883. The measurements are from the polar orbiters NOAA-15 and NOAA-16 (after thinning, prior to assimilation).

susceptible to cloud contamination, rendering tropospheric IR soundings in cloudy areas rather inaccurate. This was a key reason for the development of AMSU, which is a major improvement over its predecessor, MSU (Microwave Sounding Unit).

For temperature soundings, the sounders use wavelengths affected by well-mixed gases (for example some of the AMSU sensors, designated AMSU-A, measure wavelengths affected primarily by oxygen). This means that temperature profiles can be derived as a function of pressure, rather than just the mass of oxygen. The AMSU-B sensors are tuned to wavelengths dominated by absorption by water vapour, which is not well mixed. In combination with information from AMSU-A, it is possible to derive profiles of water vapour concentration as a function of pressure.

It took a number of years before good use was made of nadir sounding data for operational weather forecasting. Because of their low vertical resolution, poor results were obtained by retrieving temperature profiles consistent with the measured radiances, and then simply assimilating those profiles. To make optimum use of the soundings, it proved necessary to use observation operators that incorporate a forward model to simulate radiances from model temperature profiles.

Further discussion of nadir soundings, including the use of a new generation of high resolution infrared spectrometers, will be presented in later chapters. Rodgers (2001) gives a more detailed account of the inversion of atmospheric sounding measurements.

3.3. LIMB SOUNDERS

Although limb-viewing geometry is not currently used for any operational meteorological satellites, they are important for stratospheric soundings. Limb-viewing instruments look at the atmospheric limb (i.e. the atmosphere above the horizon, as

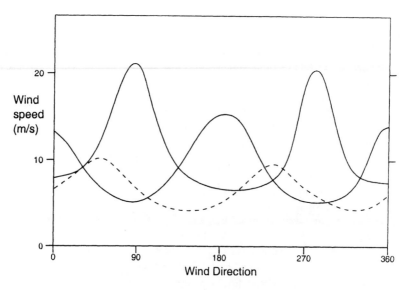

Figure 5. The signal returned from each scatterometer radar beam is consistent with a range of possible wind speeds and directions, as shown by each curve in this schematic diagram. If a scatterometer uses two radar beams (solid lines), there are typically four possible wind velocities, where the curves intersect. If a third beam is added (dashed line), these ambiguities can be almost entirely eliminated.

viewed from the satellite). The instruments measure radiation emitted from close to the tangent point. The instruments need to measure at wavelengths where the atmosphere is optically thin (otherwise the emission is coming from part of the atmosphere closer to the satellite than the tangent point, which complicates interpretation of the measurements). For this reason, and because it is not possible to measure below cloud level, limb sounders are essentially limited to the stratosphere. The chapter *Research Satellites* by W. Lahoz gives more details of a range of research satellite instruments which use limb viewing geometry.

In contrast with nadir sounders, limb sounders are characterised by good vertical resolution (typically 2 km) and poor horizontal resolution. The poor horizontal resolution is due to the sounder viewing a path length of roughly 300 km of atmosphere close to the tangent point. Geometrically, limb sounders are more closely matched (than nadir sounders) to typical atmospheric model resolutions. So, an approximate observation operator, assuming a simple profile at the nominal sounding location is probably satisfactory for low horizontal resolution stratospheric models. However, observation operators for limb sounders that take proper account of variations along the instruments' line of sight are rather complex.

One variant of the limb sounder is the limb occultation instrument. Typically, an instrument would view the Sun through the earth's limb as it rises (or sets) as viewed from the satellite. These can give very precise and high-resolution measurements, although the horizontal resolution along the line of sight is still very poor. A drawback of the solar occultation technique is that very few observations are obtained – only two per orbit. In addition to, or instead of, solar occultation the instruments can use lunar or stellar occultation, which increases the number of measurements.

3.4. SCATTEROMETERS

In contrast to the previously discussed, passive, satellite instruments, scatterometers are active instruments designed to measure winds close to the ocean surface. Microwaves, typically of wavelength of order 1 cm, are beamed down from the satellite. Some of the signal is reflected back to the satellite by Bragg reflection (this is constructive interference of microwaves scattered from ocean waves of wavelength comparable to the microwaves). The strength of the reflection is a function of both the amplitude of the ocean waves and the direction between the microwave beam and the wave orientation. In turn, the ocean waves are dependent on the wind near the ocean surface. One can determine how a particular strength of reflection can fit a range of possible wind speeds and directions, as shown in Figure 5.

By using twin beams to direct microwaves at the same patch of ocean, from two different points in the satellite orbit, the wind speed and direction can be narrowed down to four possibilities (in most cases). Some scatterometers (such as those on ERS-1 and ERS-2, European Remote Sensing Satellites 1 and 2) include a third beam, allowing the ambiguities to be almost entirely eliminated. Within the context of a data assimilation system, it is possible to make good use of data with these directional ambiguities; the recent QuickScat instrument has proved useful at many NWP centres.

4. How well are observations targeted at data assimilation requirements?

4.1. SYNOPTIC VERSUS ASYNOPTIC DATA

Traditionally, observations have been made every 6 or 12 hours for synoptic analyses and forecasts. This allowed synoptic charts to be readily constructed at the main synoptic hours of 00 and 12 UTC, or the intermediate synoptic times of 06 and 18 UTC. It follows that these synoptic observations are also eminently suited to a conventional six-hour analysis-forecast cycle, run at those synoptic times.

However modern techniques often involve asynoptic data from satellites, or aircraft, where observations can only be made when the instruments are in the right location. In addition, many data are being produced with high time resolution for short-range forecasting or "nowcasting". Data assimilation systems need to be designed to use these data to the best advantage. The improved use of such asynoptic data is probably the main reason why 4D-VAR (4 dimensional variational) data assimilation systems give better results than 3D-VAR (3D variational). Not only is a 4D-VAR system able to use data at the proper observation time, but it can extract information from a time series of observations (e.g. Järvinen et al, 1999).

4.2. INSTRUMENTAL ERRORS AND ERRORS OF REPRESENTATIVENESS

Although many meteorological measurements are very accurate, we cannot, generally speaking, use the instrument errors directly in the specification of a data assimilation system. The main reason is that observations (especially in situ data) are point values, and may not be representative of the actual state of the atmosphere on the scale of the assimilation model. We often refer to this discrepancy as an error of representativeness.

The observation error covariance $\mathbf{R}$ is often written as a combination of instrument error and error of representativeness ($\mathbf{R=E+F}$). The error of representativeness can also be considered to be the error covariance due to errors in the observation operator $\mathbf{H}$.

In many cases errors of representativeness are dominant, so they cannot be ignored in a practical implementation of a data assimilation scheme. It is possible to assess the representativeness error objectively in some cases (e.g. Lorenc et al, 1992), but in other cases the allowance for representivity error tends to be rather pragmatic. Furthermore, data assimilation systems usually make the assumption that the observation error covariance matrix $\mathbf{R}$ is diagonal. This approximation is generally compensated by tuning the (diagonal) error variances, effectively adjusting the errors of representativeness.

4.3. MAKING THE MOST OF THE OBSERVATIONS

We have already noted that we don't fully exploit all the information in conventional observations. Unused elements, such as reports of cloud type or rainfall, might contain enough information to help a forecaster adjust a computer analysis (although that is not without pitfalls). Satellite imagery also has a lot of information about cloud distribution and types. While the use of cloud data is not straightforward, the presence (or absence) of cloud allows one to draw some inferences about humidity (or liquid water content) and vertical velocity, which should be useful information for a data assimilation scheme. The use of cloud information is currently the subject of active research at the moment (e.g. Chevallier et al, 2002).

Another area of current research is the exploitation of rainfall observations. Some of these data are beginning to be used in data assimilation systems, in mesoscale models for short range forecast models, as well as global models. For example, Hou et al (2001) studied the impact of assimilating rainfall and total precipitable water information from the TRMM (Tropical Rainfall Measuring Mission) and SSM/I (Special Sensor Microwave / Imager). They showed that these data not only improve the hydrological cycle but also cloud and radiation, as well as large-scale motion in the tropics.

4.4. TARGETED OBSERVATIONS

Generally speaking, the locations of observations are fixed by where weather stations happen to be, or by the orbit of weather satellites. As we have already seen, there are some regions that are well covered by observations, and others that are only sparsely covered. In many cases the observations are not located where data assimilation systems can make best use of them.

It is possible to identify in advance, regions of the atmosphere where forecasts are particularly sensitive to initial conditions. If those areas are not well covered by routine observations, it would be advantageous to get more accurate analyses by making additional observations. Preliminary tests were carried out in the recent FASTEX (Fronts and Atlantic Storm Track Experiment) and NORPEX (North Pacific

Experiment) campaigns in the North Atlantic and North Pacific, respectively. Additional, targeted, observations were made by dropping dropsondes from aircraft which flew through sensitive regions of the atmosphere. Pu and Kalnay (1999) identified sensitive regions of the atmosphere using adjoint and quasi-linear inverse models, while Szunyogh et al (1999) used an ensemble transform technique. Targeted observations are now regularly used in the US Winter Storms Reconnaisance programme, carried out over the Pacific.

5.   Some novel measurement techniques

In sections 3 and 4 we have outlined the main types of observations used for operational NWP. In this section, we will briefly mention some other techniques that might become important in the future.

The time delay of signals sent from GPS (Global Positioning System) satellites to ground stations can be measured extremely accurately. Since the refractive index of air varies with density and water vapour, this delay gives useful information about the integrated water vapour and temperature along the slant path between the ground station and the GPS satellites. GPS receivers can also be flown on LEO satellites. As the GPS satellites rise or set, as viewed from the LEO satellite, temperature profiles can be measured to accuracies of around 1 K at 1 km vertical resolution - superior to most nadir sounding data. Initial experiments with the GPS/MET instrument showed good results (e.g. Kursinski, 1997), and a few other GPS receivers have since been deployed. The technique will probably come into its own once a small network of six receivers is deployed in COSMIC (Constellation Observing System for Meteorology, Ionosphere and Climate; Anthes et al, 2000). Compared to other satellite-based techniques this is relatively cheap, since it uses the pre-existing network of navigational satellites.

To supplement, or in some cases replace, radiosonde wind measurements, wind profilers are being deployed to measure wind profiles in the troposphere. Wind profilers are upward looking, highly sensitive, Doppler radars that are specifically designed to measure vertical profiles of wind speed and direction above the profiler site. Lidars are another technology that can be used to measure wind profiles from the ground.

For a number of years there have been proposals for satellite-borne lidars which would measure Doppler shift along the lidar line of sight; this wind information could be potentially very beneficial for forecasting. However, these instruments are expensive, heavy and power hungry; studies are still being carried out to determine the likely benefit of flying this type of instrument (e.g. Masutani et al, 2002). Recent advances in laser technology have improved the feasibility of this technique, and it is planned to launch a Doppler wind lidar on the ESA Atmospheric Dynamics Mission (ADM-Aeolus) in about 2008.

A cheaper option is to use passive techniques to measure winds, although this is not feasible in the troposphere. The HRDI (High-resolution Doppler Imager, Hays et al, 1993) and WINDII (Wind Imaging Interferometer, Shepherd et al, 1993) instruments on the Upper Atmosphere Research Satellite measured winds directly, using Doppler shifts of particular spectral lines in conjunction with limb-viewing geometry. Both

Table 1. - Summary of the typical characteristics of two broad categories of observations. While most observation types follow this pattern, there are others that do not (e.g. aircraft observations).

IN SITU	REMOTE SENSING
Conventional	**Satellite**
"Point" measurements	**Average measurements**
Simple interpolation	More complex observation operator
Synoptic	**Asynoptic**
Suits Analysis-Forecast Cycle	Suits continuous assimilation process

instruments measured winds in the mesosphere, and revealed a lot about mesospheric dynamics. HRDI also measured stratospheric winds, but not with sufficient accuracy to improve on current knowledge of winds in the stratosphere. A proposed instrument, known as SWIFT (Stratospheric Wind Interferometer for Transport Studies), is expected to give much better stratospheric wind measurements.

There are proposals to fly a constellation of balloons in the lower stratosphere (e.g. GAINS Global Air Ocean In-situ System). Winds can be derived from the balloon trajectories, and temperatures measured in situ. As well as giving useful observations of the lower stratosphere, the balloons could also be able to deploy dropsondes, targeting regions of particular interest.

6. Conclusions

This article has described many of the observations currently used for operational numerical weather prediction, and some of the physical principles upon which those measurements are based. We have highlighted particular features of different types of observations that need to be taken into account when using them in data assimilation systems. In particular, we have contrasted the general characteristics of conventional, in situ, data (reviewed in section 3) with remote sensing data (reviewed in section 4). Table 1 gives a summary of the main features of these classes of data.

Further information about some types of observations is given in subsequent chapters of this book. In particular, *Research satellites* by W. Lahoz describes some of the instruments deployed on research satellites and *Assimilation of remote sensing observations in Numerical Weather Prediction* by J-N Thépaut gives a detailed account of how satellite sounding data are used in modern NWP systems.

Acknowledgements: I would like to thank colleagues in the Met Office for help in preparing the ASI lecture, and this article - in particular Mike Keil and John Eyre. Jean-Nöel Thepaut and Pierre Gauthier also provided helpful comments on an early draft of this chapter

148

References

Anthes, R.A., C. Rocken and Y.H. Kuo 2000: Applications of COSMIC to meteorology and climate, *Terr. Atmos. Ocean. Sci.*, **11**, 115-156.

Bouttier, F. and G. Kelly, 2001: Observing-system experiments in the ECMWF 4D-Var data assimilation system, *Quart. J. R. Meteor. Soc.*, **127**, 1469-1488

Chevallier, F. P. Bauer, J.F. Mahfouf and J.-J. Morcrette, 2002: Variational retrieval of cloud profiles from ATOVS observations, *Quart. J. R. Meteor. Soc.*, **128**, 2511-2525

ECMWF, 2001: *Exploitation of the New Generation of Satellite Instruments for Numerical Weather Prediction*, Seminar Proceedings, 4-8 Sept 2000. Available from The Library, ECMWF, Shinfield Park, Reading RG2 9AX, UK, 287pp.

Hou, A.Y., S.Q. Zhang, A.M. da Silva, W.S. Olson, C.D. Kummerow, J. Simpson, 2001: Improving global analysis and short-range forecast using rainfall and moisture observations derived from TRMM and SSM/I passive microwave sensors, *Bull. Amer. Meteor. Soc.*, **82**, 659-679.

Houghton, J.T, F.W. Taylor and C.D. Rodgers, 1984: *Remote Sounding of Atmospheres*, Cambridge University Press, Cambridge, 343pp.

Hays, P.B., V.J. Arbreu, M.E. Dobbs, D.A. Gell, H.J. Grassl and W.R. Skinner, 1993: The High-Resolution Doppler Imager on the Upper Atmosphere Research Satellite'. *J. Geophys. Res.*, **98**, 10713-10723.

Järvinen, H., E. Andersson and F. Bouttier, 1999: Variational assimilation of time sequences of surface observations with serially correlated errors, *Tellus*, A, 51, 469-488.

Kursinski E.R., G.A. Hajj, J.T. Schofield, R.P. Linfield and K.R. Hardy, 1997: Observing Earth's atmosphere with radio occultation measurements using the Global Positioning System, *J. Geophys. Res.*, **102**, 23429-23465.

Labitzke, K.G and van Loon, H, 1999: *The Stratosphere Phenomena, History and Relevance*, Springer-Verlag, 179pp.

Lorenc, A.C., R.J. Graham, I. Dharssi, B. Macpherson, N.B. Ingleby and R.W. Lunnon,1992: Preparation for the use of Doppler wind lidar information in meteorological data assimilation systems, *European Space Agency Contract Report, S-div. Sci Paper No. 3*, Met Office.

Masutani, M. et al, 2002: Impact Assessment of a Doppler Wind Lidar for OSSE/NPOESS, Preprints of 15th Conference on Numerical Weather Prediction, American Meteorological Society.

Meteorological Office, 1982: *The Observer's Handbook*, HMSO, London

Pu, Z.X. and E. Kalnay, 1999: Targeting observations with the quasi-inverse linear and adjoint NCEP global models: Performance during FASTEX, *Quart. J. R. Meteor. Soc.*, **125**, 3329-3337.

Rodgers, C.D., 2001: *Inverse Methods in Atmospheric Sounding: Theory and Practice*, World Scientific, Singapore, 238pp.

Shepherd, G.G., et al, 1993: WINDII, The Wind Imaging Interferometer on the Upper-Atmosphere Research Satellite, *J. Geophys. Res.*, **98**, 10725-10750.

Szunyogh I., Z. Toth, K.A. Emanuel, C.H. Bishop, C. Snyder, R.E. Morss, J. Woolen and T. Marchok, 1999: Ensemble-based targeting experiments during FASTEX: The effect of dropsonde data from the Lear jet, *Quart. J. R. Meteor. Soc.*, **125**, 3189-3217

ATMOSPHERIC MODELLING

W. A. LAHOZ
DARC, Department of Meteorology, University of Reading
Reading RG6 6BB, UK

1. Introduction

Numerical models play a key role in data assimilation for the Earth System, since they are the means by which information from observations is organised and summarised. To do the best possible job, data assimilation systems are built on state-of-the-art models that embody our understanding of how the Earth System evolves, *i.e.*, the physical laws governing its behaviour.

It is well known from the past record that climate change has taken place on a wide range of time-scales, the best known being the alternation of glacial and interglacial periods at intervals of about 100,000 years over the last half million years. This climate variation has a large impact on human activities and the human economy. That climate change can be anthropogenic is an important current concern. Another important concern is the weather, which is change in the Earth System at very short time-scales (typically of order hours to days). Understanding and responding to these changes in the Earth System requires prediction tools. The socio-economic impact of, *e.g.*, weather extremes and climate change demand this. Modelling tools are used to study the Earth System and make predictions on a wide range of time-scales, from those associated with the weather to those associated with climate change.

2. Earth System Models

Modelling change in the Earth System requires consideration of all components of the Earth System and the feedbacks between them. These components are: (a) atmosphere, (b) ocean, (c) cryosphere, (d) land, and (e) biosphere. Depending on the situation (*e.g.* short-term weather forecasting), it can be appropriate to just consider in detail one component, *e.g.*, the atmosphere. This chapter concerns atmospheric modelling.

One must distinguish between "natural" and "anthropogenic" variability. Natural variability occurs at a variety of time- and space-scales, and can be purely internal (due to complex interactions between individual components of the Earth system, such as the atmosphere and ocean) or externally driven by changes in solar activity or the volcanic aerosol loading of the atmosphere.

R. Swinbank et al. (eds.), Data Assimilation for the Earth System, 149–166.

150

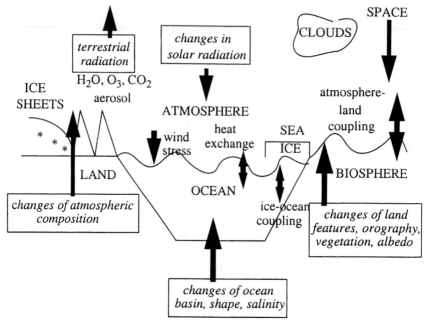

Figure 1. Schematic of the Earth System

Modelling tools are designed to provide a realistic representation of the Earth System (Figure 1). They are of varying complexity and ideally should be complete (inclusion of all significant components of the Earth System) and consistent (correct representation of the interrelationships between these components). Limitations in achieving these goals chiefly arise from (a) an incomplete understanding of the behaviour of these components, and (b) computing constraints. These models are used to test hypotheses, elucidate cause and effect, make hindcasts, make predictions, and monitor the environment. An important requirement is that results should be model and/or algorithm independent.

To set up a model integration one needs initial conditions (*e.g.* from analyses or a model) and boundary conditions (*e.g.* sea-surface-temperatures -- SSTs, or other forcings). Because the initial conditions may not be entirely appropriate for the integration of interest due to, *e.g.*, imbalances in the initial conditions, or lack of equilibrium in the model conditions, a "spin-up" period may be required before model results become useful.

Our knowledge of the Earth System ultimately comes from observations. Although observations have uncertainties and biases, they are the "truth" against which models (as well as theories) must be confronted and evaluated. Examples of observing platforms include ground-based instruments, balloons, aircraft and remote sounding satellites.

3. Atmospheric Models

An atmospheric General Circulation Model (AGCM) is the most complex model in the hierarchy of atmospheric models. It typically has the following components: (a) dynamics, (b) radiation, (c) parametrizations, and (d) chemistry. Due to the complexity of AGCMs it is often difficult to establish cause and effect of phenomena; furthermore, it is often expensive to run the computer experiments required for the comprehensive analysis of phenomena. To remedy these shortcomings, simpler models (often simplified GCMs) are used. Despite the advantages of ease of understanding and of relatively inexpensive computer costs, these simpler models have a significant shortcoming compared to GCMs, *viz.*, that they are less physically realistic and less complete due to the omission of components of the Earth system.

Examples of simpler physically-based models include, in increasing order of complexity: energy balance models (which determine the effective radiative temperature of the Earth climate system by assuming a balance between the incoming solar radiation and the outgoing long-wave radiation), radiative-convective models (which determine the vertical distribution of globally-averaged temperatures for the atmosphere and the underlying surface), zonally-averaged models (which are 2-dimensional models capable of simulating vertical and meridional variations in surface and atmospheric properties averaged around latitude circles), and mechanistic models (*e.g.* GCMs which exclude the troposphere). In this chapter, we focus on AGCMs and their components.

3.1 DYNAMICS

The dynamics component of the AGCM is concerned with solving the Navier-Stokes equations (or the primitive equations, which are an approximation to the Navier-Stokes equations, see Andrews *et al.* 1987) on a grid. Coordinates must be chosen for the horizontal and vertical dimensions, as well as for the representation of the flow.

Because the Earth is spherical, it is common to use the latitude-longitude horizontal grid based on spherical coordinates. The primitive equations are often solved in a coordinate system in which geometric height is replaced by hydrostatic pressure. To account for the Earth's orography, either sigma coordinates (pressure normalised by surface pressure), or hybrid coordinates (a mixture of sigma and pure pressure coordinates) are used. A number of specialised models use potential temperature as the vertical coordinate. Because under certain conditions, 2-dimensional flow remains on a potential temperature surface, this vertical coordinate is useful for studying tracer transport.

There are two basic representations of the flow: (a) Eulerian, (b) Lagrangian. In the former the grid is fixed with respect to the flow, in the latter the grid moves with the flow. As the Lagrangian method used in modelling involves periodic interpolation to a grid, it is termed semi-Lagrangian. Semi-Lagrangian schemes are of considerable interest because they can be more efficient than competing Eulerian schemes as they can use a longer time-step (see, *e.g.*, Staniforth and Côte 1991).

The domain over which the primitive equations is solved is finite. In the case of atmospheric modelling this requires a bottom and top boundary. The bottom boundary of an AGCM is the Earth's surface. The top boundary has to be placed at an arbitrary location. Currently, many AGCMs have a top boundary located between 0.1 hPa (about 65 km) and 0.01 hPa (about 80 km). There are a number of technical issues concerning the treatment of boundary conditions located within the body of a fluid. For example, the boundary conditions should allow outward-travelling disturbances to pass through the boundary without generating spurious reflections that propagate back toward the interior. Details of these technical issues can be found in Durran (1999).

There are two conceptually different methods to represent continuous functions on digital computers: (a) as a finite set of grid-point values (*e.g.* finite differences), (b) as a finite set of series-expansion functions (*e.g.* spectral methods). In this chapter we discuss finite differences and spectral methods. Durran (1999) discusses finite element and finite volume methods.

The computation of space and time derivatives in a numerical scheme involving grid-point values presents different sets of practical problems. After the nth step of the numerical integration, the numerical solution will be known at every point on the spatial grid. Thus, it is easy to construct high-order centred approximations to space derivatives. In contrast, storage considerations dictate that the numerical solution be retained at as few time levels as possible, and the only time levels available are those from previous iterations. Thus, higher-order finite-difference approximations to time derivatives tend to be one-sided. The storage requirements of AGCMs mean that models tend to use 2- or 3-time level schemes. Durran (1999) discusses the general family of 3-time-level schemes, including the leapfrog and Adams-Bashforth schemes. Currently, AGCMs tend to use 2-time level schemes because: (a) provided they converge to a physical solution they do not have an unphysical "computational mode", and (b) they are more efficient. Examples of 2-time level schemes which do not have a computational mode include the forward explicit and the backward implicit schemes. The former is unstable, whereas the latter is stable: a difference scheme is stable if its solutions remain bounded for all small time steps. Haltiner and Williams (1980) provides a more rigorous definition for stability, as well as for other concepts concerning the properties of numerical schemes.

Equation 1 gives examples of 2-, 3-level schemes applied to a simple linear system, $\frac{\partial h}{\partial t} = F$, where r is the rth time-step, Δt is time-step:

$$h^{r+1} - h^{r-1} = 2\Delta t F^r$$

Leapfrog   (1.1)

$$h^{r+1} - h^r = \Delta t(3/2\ F^r - 1/2\ F^{r-1})$$

Adams-Bashworth   (1.2)

$$h^{r+1} - h^r = \Delta t F^r$$

Forward explicit   (1.3)

$$h^{r+1} - h^r = \Delta t F^{r+1}$$

Backward implicit   (1.4)

When computing horizontal derivatives using finite differences, much of the data comes from adjacent points, which suggests it would be more efficient if the variables are staggered in the horizontal. The Arakawa B and C horizontal grids (Figure 2) have the most desirable properties for AGCMs; for example, they are best at adjusting the initial conditions to a balanced state (Arakawa and Lamb 1977). Similar considerations hold for computing vertical derivatives. Examples of staggered vertical grids include the Lorenz and Charney-Phillips grids (Figure 3). Durran (1999) discusses the energy-conservation properties of vertical differencing schemes.

The equations associated with atmospheric modelling permit both relatively slow meteorological modes and relatively fast inertial gravity waves. The latter are important for the "geostrophic adjustment" process whereby real and spurious imbalances between the wind and mass fields created by, *e.g.*, observation errors, mountains and point heat sources, are eliminated. The inertial gravity waves impose a stringent computational stability condition on the timestep Δt, and require a much smaller Δt than that needed to maintain a high degree of numerical accuracy for the meteorological modes with respect to the time truncation error. This can be dealt with in AGCMs by the use of a semi-implicit time-stepping scheme in which the fast gravity waves are treated implicitly and the remainder explicitly.

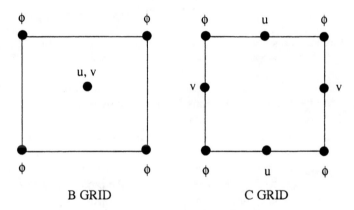

Figure 2. Arakawa B and C grids. u, v are wind components, ϕ are temperature, geopotential height.

Numerical schemes can have exponential amplification of the solution of the difference equation (this is known as a computational instability). A necessary condition for computational stability is the Courant-Friedrichs-Levy (CFL) criterion: $|(c\Delta t)/(\Delta x)| \leq 1$ where c is a characteristic speed of the system (*e.g.* zonal wind), and Δt and Δx are the time and space discretization steps, respectively. The CFL criterion can be interpreted as requiring that the numerical dependence of a finite-difference scheme include the domain of dependence of the associated partial differential equation. Numerical schemes can also generate unphysical small-scale, high frequency waves ("noise") which contaminate the desired physical solution. A common method of removing this noise is to introduce an explicit artificial diffusion term which is scale-selective, *i.e.*, can discriminate between different space-scales.

154

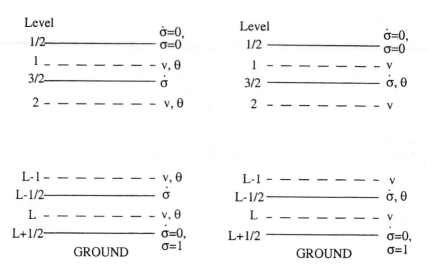

LORENZ GRID   CHARNEY-PHILLIPS GRID

Figure 3. Lorenz and Charney-Phillips grids. v, θ are horizontal wind components and potential temperature. σ is the vertical coordinate (1 at the model lower boundary, 0 at the model upper boundary); $\dot{\sigma}$ is rate of change of vertical coordinate. Solid lines are model full-levels, dashed lines are model half-levels.

The characteristic that distinguishes the spectral method from other series-expansion methods is that the expansion functions form an orthogonal set. Spherical harmonics (involving the product of a Fourier mode and a Legendre function) are often used in AGCMs. Let λ be the longitude, θ the latitude and define μ=sinθ. If ψ is a smooth function of λ and μ, it can be represented by a convergent expansion of an infinite sum of appropriately weighted spherical harmonic functions, where each spherical harmonic function $Y_{m,n}(\lambda, \mu) = P_{m,n}(\mu)\exp(im\lambda)$ is the product of a Fourier mode in λ and an associated Legendre function in μ. n is summed from $|m|$ to ∞; m is summed from −∞ to ∞. Durran (1999) discusses the properties of spherical harmonics and Legendre functions.

In all practical applications the infinite series associated with the spherical harmonics must be truncated at a particular wavenumber by replacing the infinite limits of the sum over m by m=-M and M, and the upper (infinite) limit of the sum over n by N(m). The "triangular truncation", in which N(m)=M, is unique among the possible truncations because it is the only one that provides uniform spatial resolution over the entire surface of the sphere.

Despite its elegant properties, the triangular truncation may not be optimal in cases where the typical scale of the approximated field has a systematic variation over the surface of the sphere. An example is the geopotential height field, which has much weaker anomalies in the tropics than at middle latitudes. Several alternative truncations have, therefore, been used in low-resolution (M<30) AGCMs. The most common of these is the "rhomboidal truncation", in which $N(m) = |m| + M$ in the truncated version of the spherical harmonics. The set of indices (m, n) retained in triangular and rhomboidal truncations are shown in Figure 4.

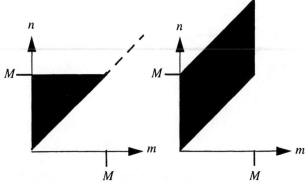

Figure 4. Triangular truncation (left) and rhomboidal truncation (right). Shading indicates the portion of the $m \geq 0$ half-plane in which the m and n indices are retained in an Mth-order truncation.

An advantage of using spherical harmonics is that derivatives such as the Laplacian are easily computed using the known properties of the expansion functions. Another advantage is that explicit finite-difference methods applied in a grid-point model can require very small time-steps to maintain stability if the grid points are distributed in the sphere in a uniform latitude-longitude grid. This time-step restriction arises because of the convergence of the meridians near the pole and the CFL criterion (the "pole" problem). This difficulty can be avoided by the use of spherical harmonics. In grid-point models it is often treated by Fourier filtering, or by using a rotated pole.

A disadvantage of spherical harmonics (as well as general spectral methods) with respect to grid-point methods is the occurrence of unphysical Gibbs oscillations. Another disadvantage is the lack of highly efficient algorithms to transform the spectral formulation to grid-point space (typically a latitude-longitude grid). A practical complication when using spherical harmonics in an AGCM is that the calculation of the model physics (*e.g.* radiation) and of non-linear quantities are performed in grid-point space, with the result that in these cases the model's algorithm must switch frequently from a spectral formulation to a grid-point formulation and back.

One problem arising with spherical harmonics is "aliasing" (see Figure 5). Aliasing error arises because the product of two or more truncated spherical harmonics contains high-order Fourier modes and high-order Legendre functions that are not present in the original truncation. To avoid aliasing error at truncation to wavenumber M, the computation of the coefficients in the spherical harmonic expansion requires the evaluation of polynomials of degree 3M for a triangular truncation and of degree 5M for a rhomboidal truncation (Durran 1999).

Two extra considerations which arise in using spherical harmonics for practical applications are: (a) the evaluation of derivatives with respect to the meridional coordinate, and (b) the representation of the vector velocity field. In (a), binary products are computed in grid-point space and the result transformed in spectral space, where the properties of the expansions are used to evaluate the derivatives within the truncation error of the spectral approximation. In (b), the zonal and meridional velocity components (u and v, respectively) are not conveniently approximated by spherical harmonics because artificial discontinuities are present at the poles unless the wind speed at the pole is zero. This difficulty is commonly avoided by replacing the prognostic equations for u and v by prognostic equations for the vorticity and divergence and by rewriting all remaining expressions involving u and v in terms of the transformed velocities $U=u\cos\theta$ and $V=v\cos\theta$ (θ latitude). Both U and V are zero at the poles and free of discontinuities (see Durran, 1999 for details).

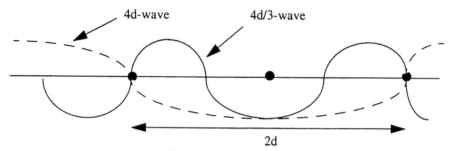

Figure 5. A 4d/3 wave is mis-represented as a 4d wave.

Ideally, the numerical techniques used in AGCMs should have the following attributes: (a) consistency, *i.e.*, the scheme approaches the corresponding differential equation as the space and time increments in the scheme approach zero, (b) accuracy (generally, the aim is to have second-order accuracy, *i.e.*, the space truncation error is $(\Delta x)^2$), (c) stability, (d) convergence to the physical solution, (e) efficiency, (f) conservation of physical quantities such as mass, energy, and momentum, and (g) monotonicity, *i.e.*, the absence of unphysical overshoots and undershoots in, *e.g.*, tracer fields or humidity. Some of these attributes are linked. For example, there is a theorem by Lax which states that a linear, stable, consistent scheme must converge. These requirements are discussed in more detail in Haltiner and Williams (1980).

The above attributes are ideal and the numerical schemes implemented often have conflicting requirements. Examples include the following. (1) Semi-Lagrangian schemes are efficient and stable for large Δt, but are not fully conservative. (2) The semi-implicit treatment of gravity waves is stable for large Δt, but is not accurate. This is not generally a problem if the gravity waves are physically unimportant, but can cause problems such as spurious orographic resonances. (3) Spectral methods give a high order of spatial accuracy and convergence but are associated with unphysical Gibbs oscillations. (4) Total variation diminishing (TVD) schemes used for advection schemes are monotonic but reduce to first order accuracy near extrema. Within the TVD approach, "flux-limiters" are used to preserve monotonicity in the scheme and avoid spurious ripples. Examples of flux-limiters used in AGCMs include the "superbee" and van Leer limiters. (5) A linear monotonicity preserving method is at most first order accurate (this is Godunov's theorem). Therefore, monotonic schemes of higher order have to be non-linear even for linear problems (*e.g.* for the advection of tracers).

Two state-of-the-art AGCMs are the Met Office Unified Model (UM), and the model of the European Centre for Medium-range Weather Forecasts (ECMWF). The UM is a grid-point model used for Numerical Weather Prediction and climate change studies (Cullen 1993). The UM formulation is being changed to semi-Lagrangian dynamics, and Arakawa C and Charney-Phillips grids. The ECMWF model (Simmons *et al.* 1999) is a spectral model used for seasonal forecasting and producing re-analyses of the atmosphere (*e.g.* the 40-year ECMWF re-analyses, ERA-40).

3.2 RADIATION

In the atmosphere, the circulation of air masses depends strongly on the magnitude and distribution of the net diabatic heating rate. In the troposphere, the net diabatic heating rate is dominated by the imbalance between two large terms, *viz.*, transfer of heat from the surface and thermal emission of radiation to space. Latent heat is a major component of the flux from the surface to the atmosphere, and clouds play a major role in the emission of radiation to space.

In the stratosphere and mesosphere net heating depends on the imbalance between local solar absorption of ultra-violet (UV) radiation and infrared (IR) radiative loss (Figure 6). In this region, O_3 is the dominant absorber (and contributor to the heating of the atmosphere) and CO_2 is the dominant emitter (and contributor to the cooling of the atmosphere). IR emission by O_3 and H_2O and solar absorption by H_2O, O_2, CO_2 and NO_2 play secondary roles. The distribution of the radiative sources and sinks due to these gases controls strongly the large-scale seasonally varying mean temperature and zonal wind fields of the middle atmosphere. IR emission also provides an important mechanism for damping dynamically forced temperature variations.

Solar radiation is termed short-wave (SW) radiation and thermal radiation (typically IR radiation) long-wave (LW) radiation. SW radiation includes wavenumbers 1,000-50,000 cm^{-1} (wavelengths 0.2-10 μm). LW radiation includes wavenumbers less than 3,000 cm^{-1} (wavelengths longer than 3 μm). Whereas LW radiation contributes to heating and cooling of the Earth's atmosphere, SW radiation produces only heating because the Earth's atmosphere does not emit at these wavelengths. Clouds tend to have a net heating effect in the LW (as they emit in the LW) and a net cooling effect in the SW (as they reflect or scatter radiation in the SW). The net radiative effect of different cloud types depends on their properties and altitude and, in general, is not fully understood.

The radiation code in AGCMs solves the equations of radiative transfer and converts the model input parameters into LW/SW heating rates which then modify the temperature distribution. The algorithms use diffuse (*i.e.*, it has contributions from all directions) irradiances. The irradiance (flux) is the integral of the component of intensity (radiance) normal to a surface over the half-space of solid angle in the positive direction (upward irradiance) or the opposite half-space of solid angle in the negative direction (downward irradiance). The net flux in the positive direction is the difference between the upward and downward fluxes. The heating rate is computed by taking the divergence of the net flux (*i.e.*, rate of change of the next flux with respect to pressure or optical depth) integrated through an atmospheric column with appropriate boundary flux conditions.

158

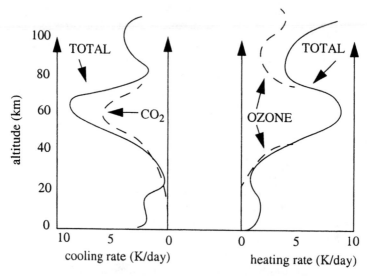

Figure 6. Schematic of long-wave cooling and short-wave heating

This integration takes account of the optical depth of the atmospheric column and the source terms. In Local Thermal Equilibrium (LTE), and in the absence of scattering, Kirchhoff's law applies and the source term is given by the Planck function. The conditions for LTE are satisfied when energy transitions are dominated by molecular collisions, which is the case for the most important radiatively active gases in the LW at pressures greater than 0.1 hPa (at altitudes below about 65 km). At higher altitudes LTE breaks down because the interval between collisions is no longer short compared to the lifetime of excited states associated with absorption and emission, and Kirchhoff's law is no longer valid.

The algorithms that calculate LW heating rates divide the spectrum into bands which include the contribution from the important gases: CO_2 (15 μm band), O_3 (9.6 and 14 μm bands) and H_2O (6.3 μm band). Most of the LW energy emitted by the Earth's surface is absorbed in the troposphere. Only in the atmospheric window 8-12 μm is absorption weak enough for much of the LW radiation to pass freely through the atmosphere. At these wavelengths, the 9.6 μm O_3 band is the only strong absorber, and most of that absorption takes place in the stratosphere where O_3 concentrations are large.

The most important property of the SW spectrum is the solar constant, which represents the flux of radiant energy integrated over wavelength reaching the top of the atmosphere. This constant is weakly variable, having a value of about 1370 Wm^{-2}. Solar insolation must be taken into account via the solar zenith angle (SZA). The algorithms that calculate SW heating rates divide the SW spectrum into bands including most important gases: H_2O, CO_2, O_3, N_2O, CH_4 and O_2.

In principle, the absorption characteristics of the band determine the corresponding transmission function (a value of 1 indicates no absorption; a value of 0 indicates full absorption). In practice, absorption bands are so complex that direct calculation is impractical. Instead, the transmission function is evaluated using band models that capture the main features of the absorption spectrum in terms of spectroscopic properties such as mean line strength. The LW/SW radiation algorithms used by AGCMs can be very different (Phillips 1994). There are three main methods commonly used in AGCMs to parametrize gaseous absorption in a spectral band: (a) look-up tables, (b) analytical band models (ABM), (c) the k-distribution approach, which considers the contribution from absorption coefficients in terms of a cumulative distribution probability (absorption coefficients k less than a fixed amount).

In the Met Office AGCM the k-distribution approach is used for the LW and SW. The perceived advantages of the k-distribution approach are (a) scattering can be treated more accurately, and (b) the total computation amount is linearly proportional to the number of layers, rather than the quadratic dependence of the look-up tables and the ABMs.

Three methods for generating the coefficients for the k-distribution can be identified (k is absorption coefficient). (1) The exponential-sum fitting of transmissions technique, in which the observed or predicted transmission of each gas in each band and at various absorber masses, is used to fit a sum of exponentials, with each exponent equal to a unique coefficient k times the absorber mass. (2) Probability distributions of k values from ABMs have been used from which the coefficients are expressible in terms of ABM parameters. This method, however, requires too much computation for studies which use AGCMs. (3) k-distributions can be created from spectroscopic databases, and with suitable averaging of k values the coefficients can be found. All the methods assume correlation in wavenumber space of the k values at different pressures, so the integral over wavenumber to find the total broad-band transmission can be replaced by an integral over the cumulative probability distribution of k values. This procedure has the advantage that the cumulative probability distribution is monotonic, whereas the dependence of the absorption coefficient with wavenumber is not. Thus, this integral is relatively straightforward to compute. The line-transmission data are derived typically from established molecular databases such as HITRAN92.

Calculation of realistic cloud properties is an important task for the radiation algorithm. Much effort has been expended over recent years to develop numerically fast yet accurate algorithms. In the SW both water and ice clouds must be considered. "Thick" and "thin averaging" approximations are used (these are exact in the limit of optically thick and thin clouds, respectively). High clouds exert the largest influence of any cloud type in the outgoing LW radiation. Cirrus clouds pose a particular problem as they are difficult to observe and model. In the LW it is usually assumed that absorption dominates scattering, and calculations of the influence of clouds on LW fluxes are often done with emissivity schemes. However, there is evidence that scattering may sometimes have an effect on broad-band fluxes and heating rates.

Atmospheric aerosols play an important role in the Earth's radiation budget. They scatter and absorb radiation in the SW, causing an increase in the planetary albedo and a reduction in the amount of radiation reaching the surface. The impact on the LW is generally smaller, although it can be significant. Anthropogenic aerosols are believed to reduce the magnitude of global warming, resulting from increased greenhouse gases (Houghton et al. 1995).

Aerosols are only recently being included in AGCMs. Independent evidence that aerosols need to be included comes from comparisons with data from the ERBE and

ScaRaB Earth radiation budget experiments. For example, the clear-sky planetary albedo in the Met Office UM was lower than that from ERBE and ScaRaB and the distribution of the deficit suggested that the lack of aerosols made a substantial contribution. The inclusion of aerosols in the Met Office UM in a formulation where spatially averaged quantities are used (rather than the full temporal and spatial variation of the aerosol optical properties), has the potential to alter the surface and the top of the atmosphere radiation budgets in AGCMs, and to improve the agreement of these quantities with measurements significantly.

3.3 PARAMETRIZATIONS

Many physical processes, *e.g.*, clouds, have much smaller space-scales than can be resolved and modelled explicitly by AGCMs. The solution is to parametrize the average effects of sub-grid-scale phenomena by making use of physically based relationships between the small- and large-scale variables. Requirements of parametrizations are: (a) they must be physically and dimensionally correct, (b) they must be invariant under coordinate transformations, (c) they must be consistent with the budgets and constraints (*e.g.* they must not produce negative humidities). As computing power increases, and the resolution of AGCMs increases, some parametrizations may become redundant.

3.3.1 Clouds
The recognition that small changes in cloud forcing may have a big impact on climate change, and the desire for improved precipitation forecasts in Numerical Weather Prediction (NWP) models, have led to the use of complex stratiform clouds parametrizations in AGCMs. In these schemes cloud cover and condensed water are predicted, allowing cloud radiative properties to be calculated interactively rather than being prescribed. Predictive schemes are termed "prognostic"; prescriptive schemes are termed "diagnostic". However, even with these advances the representation of clouds in AGCMs still remains an area of great uncertainty.

One area of uncertainty in current AGCM cloud parametrizations is the estimation of precipitation from clouds in the temperature range in which ice and water are observed to exist within the same cloud layer (*i.e.*, "mixed-phase clouds"). Studies suggest that sensitivity of modelled climate to SST changes is highly dependent upon the specification of the temperature at which cloud condensate is treated as ice. Given the uncertainty in estimates of climate change from climate models due to the specification of ice precipitation, it is important that more observationally-based parametrizations of mixed-phase clouds are developed.

3.3.2 Planetary Boundary Layer
The Planetary Boundary Layer (PBL) extends about 2 km above the Earth's surface. This is the lowest part of the troposphere where the direct influence from the surface is felt through turbulent exchange with the surface. Boundary layer schemes may be divided into two classes according to the vertical resolution of the model close to the ground. If the vertical resolution is such that, at most, one level lies within the boundary layer, "bulk" parametrizations must be used to represent the boundary layer. Alternatively, the

boundary layer may be explicitly (albeit crudely) resolved by locating several levels within the lowest 2km of the model atmosphere.

In the simplest bulk approach, surface fluxes are calculated from a basic surface drag law using either the wind at the lowest model level, or a wind extrapolated from more than one level. The drag coefficients and turning of the wind through the boundary layer may depend on the underlying surface and the stability of the lowest model layer. Turbulent fluxes either vanish in the free atmosphere or are treated by simple eddy diffusivities.

In explicit boundary-layer models the lowest level is chosen to be within a few tens of metres from the ground, and surface fluxes are determined using either a logarithmic wind profile or, more generally, Monin-Obukhov similarity theory. Above the lowest level, turbulent vertical fluxes are represented as the product of eddy diffusivities and vertical gradients of the explicitly resolved fields. This diffusivity is often computed from the local wind shear and a mixing length which decreases linearly with height from the lowest model level to vanish just above the top of the PBL. The diffusivity also depends on the local stability, measured by the Richardson number -- some schemes use a mixing length which does not vanish in the free atmosphere. So-called "higher-order closure" schemes in which diffusivities are related to an additional predicted variable, the local turbulent energy, are also used.

3.3.3 Moist Convection

Convective parametrization schemes aim to represent the "apparent convective source" and "apparent moisture sink" due to convection, in terms of the large-scale atmospheric variables predicted by the model. In doing so, they must: (1) predict the vertical distribution of latent heating and transport properties due to convection via simple cloud models, (2) predict the overall magnitude of the energy release from the convection. Further requirements include: (a) the resultant thermodynamic structure must be realistic, (b) the schemes must be able to represent the mean distribution of precipitation accurately, (c) the schemes must have a realistic representation of atmospheric variability.

Currently, there are four types of convective parametrization schemes: (a) moist convective adjustment schemes, (b) Kuo-type schemes, (c) Betts-Miller adjustment schemes, (d) mass flux convection schemes. With increasing computing power, mass flux convection schemes are becoming more popular and are used in many AGCMs. Furthermore, they provide a physical understanding of how local convection affects the large-scale atmosphere.

Application of the mass flux theory in AGCMs is a two-stage process: (1) a cloud model must be used to estimate the vertical distribution of the cloud quantities, (2) the magnitude of the mass flux at the cloud base must be determined (the "closure" problem). Two approaches have been suggested for using cloud models to estimate cloud quantities: (1) spectral cloud ensembles, (2) the bulk cloud model, where a single cloud model is used. The latter is often used in AGCMs as it is cheaper.

In recent years several mass flux schemes have been updated to incorporate the effect of convective-scale downdraughts (which are an important component of many deep convective systems). They are represented by inverted entraining plumes, adiabatic warming through the descent of air being compensated by cooling due to the evaporation of precipitation.

The production of precipitation within clouds is governed by complex microphysical processes. These are poorly represented in most convection schemes. Although the latent heat of freezing is included, little distinction is made between the formation of rain and ice precipitation.

3.3.4 Large-scale Precipitation

This is computed after other dynamical and physical processes which change the temperature and water content and, in general, involves the condensation (with associated latent heat release), of sufficient water vapour to keep the relative humidity below a threshold value (80-100%). Values lower than 100% represent the small-scale nature of precipitation, which may occur at points within a grid square, even if the grid square is not saturated. Too low a threshold in the boundary layer may, however, lead to unrealistic amounts of precipitation from the lowest level above a wet surface.

In the simplest schemes, all condensed moisture falls instantly to the ground, but the evaporation of precipitation may also be taken into account (as can also happen in the convective case). A range of further refinements may also be adopted, including an explicit representation of liquid water.

3.3.5 Gravity Wave Drag

Gravity waves (GWs) have relatively small space-scales which are not resolved at the typical grid-scale of AGCMs. GWs transfer mean horizontal momentum from the ground to levels in the atmosphere or from one atmospheric layer to another. Flow over topography can generate stationary GWs that break non-linearly in the troposphere and lower stratosphere. These GWs transfer momentum from the Earth's surface to the breaking region, and this is thought to act as a significant drag on the eastward mean winds in the mid-latitude troposphere. Other processes (*e.g.* convection, jet stream instabilities) can produce GWs with non-zero horizontal phase speeds which transfer mean momentum between the troposphere and the stratosphere/mesosphere. AGCMs are known to produce unrealistic simulations of the extra-tropical stratospheric/mesospheric circulation unless some account is taken of the effects of these GWs. In these cases, AGCMs have large cold biases ("cold pole" problem) and unrealistically strong polar night jets. If credible climate simulations (and predictions of climate response to anthropogenic change) are to be obtained, a physically justifiable parametrization of the momentum transport due to unresolvable GWs needs to be formulated.

Over the last few years, a number of AGCMs have managed to produce reasonably realistic simulations of the quasi-biennial oscillation (QBO) by incorporating GWD parametrizations (see, *e.g.*, Scaife *et al.* 2000). However, despite these relative successes, the parametrization of gravity waves is still one of the most important challenges in dynamical meteorology. One particular problem is the lack of global, long-term GW observations to evaluate GWD parametrizations. Hamilton (1996) discusses this issue.

There are three main GW drag schemes now being tested: the Lindzen, Hines and Fritts and Lu schemes. Each scheme is essentially 1-dimensional and explicitly considers statistically-steady conditions. Each makes assumptions concerning the spectrum of upward-propagating GWs at the lowest level considered (typically near the tropopause) and then uses a simplified treatment of the dynamics to compute the propagation and dissipation of the GWs through the middle atmosphere (and hence the mean-flow modifications induced by the waves). A number of AGCMs use the crude Rayleigh Friction (RF) parametrization in lieu of elements of the GWD parametrization. Typically, in the RF parametrization the winds in the upper stratosphere and mesosphere are damped towards zero using a damping coefficient that depends on model level.

3.4 CHEMISTRY

As climate models become more realistic, the need to include atmospheric chemistry is increasingly apparent. For example, of the main radiative species, O_3 is controlled by both dynamical and photochemical processes with a wide-range of time-scales in the atmosphere. There are also many chemical feedbacks on climate, *e.g.*, chlorofluorocarbons, which have both a greenhouse warming effect and deplete stratospheric O_3.

Many AGCMs are now starting to include photochemistry (see also chapter *Introduction to Photochemical Modelling*). Typical stratospheric photochemistry schemes include several photochemically active tracers (some treated as families, *e.g.*, total odd oxygen). In general, the species are assumed to be in the gas phase, except for those which are in the solid phase in Polar Stratospheric Clouds (PSCs; *e.g.* H_2O, HNO_3). A further scheme can be included to represent heterogeneous chemistry taking place on PSCs. In addition to the transported species, a number of species are derived from the reservoirs by photostationary state and conservation assumptions. Photolytic reactions are included and their rates commonly determined from the altitude, O_3 column, and SZA using a look-up table. Tropospheric chemistry schemes are not as advanced as stratospheric schemes mainly due to the complexity of tropospheric chemistry and the need to include effects such as deposition.

The 1999 Ozone assessment (WMO 1999) provides a table of 3-d chemistry climate models. Multi-year integrations with AGCMs with comprehensive chemistry have recently been run with the Hamburg model (Dameris *et al.* 1998) and the Met Office chemistry-climate model (Austin 2002).

Other models such as the Goddard Institute for Space Studies (GISS) climate model (*e.g.* Rind *et al.* 1998) have been run with a simple representation of stratospheric chemistry. However, because the chemical and dynamical transport time-scales for O_3 are roughly comparable in the stratosphere, using such simplified chemistry has limitations.

4. Ensembles

An increasingly important aspect of the statistical approach is the idea of multiple realizations of the Earth System. In this approach, several members of the ensemble are constructed so as to capture a significant fraction of the model's variability. This estimates a probability distribution function (pdf) for the model variables. The ensemble mean is a measure of the climatology, whereas the ensemble spread is a measure of the variability. The advent of increased computer power makes this approach feasible, and it is rapidly becoming the norm in climate prediction (this approach is also becoming the norm in

Numerical Weather Prediction). With ensembles one speaks of predictability (represented by a pdf), rather than of deterministic prediction.

When setting up an ensemble of climate runs one must decide: (1) how many members it should have, and (2) how are the members to be initialised. (1) is largely determined by computing cost constraints, which means that ensemble size is typically less than 10 (note that studies with the ECMWF ensemble system suggest that 30 members are enough to capture atmospheric variability). (2) is commonly accomplished by initialising each ensemble member from climate states one day apart, everything else being the same, although more sophisticated approaches can be taken (Toth and Kalnay 1993). In this way, the experimenter can capture the uncertainties associated with the initial state, which as demonstrated by Lorenz (1963) can lead to chaotic behaviour.

Atmospheric conditions which are highly predictable will tend to be associated with low spread in the ensemble, whereas conditions which are highly unpredictable will tend to be associated with high spread in the ensemble. This is related to the concept that repeatability in atmospheric conditions is associated with high predictive skill. The extent to which the estimates of the prediction are reliable will depend on ensemble size.

A key feature of the ensemble approach is that the squared error of the ensemble forecast (*i.e.*, the ensemble mean) is smaller than the mean squared error of all the individual forecasts. However, in the case of weather extrema, the ensemble mean is of less interest than the wings of the pdf, represented by the ensemble members which are "outliers".

5. Predictability

What can we predict? How far into the future can we predict? These questions presuppose: (1) there exist modelling tools which are useful for prediction, (2) there exist observations (not error free) which are the "truth" and enable the evaluation of predictive skill, (3) there are limitations to our predictive capability. Limitations to predictive capability arise from: (a) shortcomings in our understanding (are equations missing?), (b) non-linearity (is there chaotic behaviour?), (c) lack of consistency (are feedbacks missing?), (d) lack of completeness (are significant components missing?), (e) uncertainties in the initial conditions and external forcings.

Useful weather forecasts can be made for periods up to 10 days. In the stratosphere, AGCMs tend to have higher forecast skill than in the troposphere. This is because flow regimes in the stratosphere are dominated by lower wavenumbers than in the troposphere. This illustrates the impact of space-scales on the predictive skill of the model. The different time-scales in weather forecasts (short) and climate simulations (long) have an impact on the predictive skill of the model. Longer time-scales mean that the internal component of the natural variability of the model tends to have a larger influence on model variability in relation to the influence from the initial and boundary conditions. Thus, the deterministic approach often used in weather forecasts is generally not very useful for climate prediction and a statistical approach is required. In this approach, the statistics of the Earth System are of interest rather than the daily evolution of the weather. These statistics include measures of variability (variance) and mean (climatology) and are taken over many weather systems and for several months, years or more.

To test hypotheses of climate change forced by natural and/or anthropogenic forcings, a control scenario is needed (the natural variability of the model). The pdfs of the control and the perturbed scenarios can then be compared under, *e.g.*, the null hypothesis that the means are equal given the standard deviations. Although this is a simple example, it illustrates the philosophy behind the approach (often involving sophisticated techniques) used to detect and/or attribute anthropogenic climate change.

Typically, the natural variability of a climate model is estimated by performing long (*e.g.* multi-century) control runs with constant forcing parameters (natural and/or anthropogenic). Such simulations provide estimates of the internally generated natural climate variability on a range of space- and time-scales. Typically, evaluation of the estimates of model natural variability is done by comparing the patterns in model variability on a range of spatial and time-scales against those in observations of, *e.g.*, surface temperature.

A common overall approach has emerged to the detection of anthropogenic climate change. A detection statistic is defined and evaluated in an observational dataset. This might be: (1) a global mean quantity, (2) a model versus observation pattern correlation, (3) the observed trend in pattern correlation, (4) a form of optimized fingerprint. The same detection statistic is then evaluated treating sections of a control run of a climate model (in which the forcing is constant) as "pseudo-observations" to provide an estimate of the distribution of that statistic under the null-hypothesis of no anthropogenic change. If the observed value of the chosen statistic lies outside pre-assigned confidence level (typically 95% or 99%), then detection is claimed with a 0.05 (at the 95% confidence level) or 0.01 (at the 99% confidence level) probability of a false positive result (a "type-1" error). This approach to quantifying the risk of error requires that the model's simulation of internal climate variability (the natural variability) be realistic.

Typically, one wishes to discriminate between "signal" and "noise" when analysing time series of the climate record (*e.g.* temperatures). While the meaning of signal and noise varies with context, there will always be a non-zero probability of incorrectly identifying noise as a deterministic trend or oscillation given limited data. The acceptable probability of such a "type-1" error must be specified, being the "nominal level" of any statistical test. Allen and Smith (1996) discuss the use of sophisticated statistical techniques to analyse records of climate data. In particular, they discuss how to extract physically-meaningful oscillations and how to construct hypotheses for testing against "white noise" or "red-noise".

A common approach when evaluating climate change is to assume that the observations (y) may be represented as the sum of simulated responses or signals (x_i, modified by an amplitude β_i) and internal climate variability (u, assumed to be normally distributed). The amplitude β_i represents the amount by which the ith signal has to be scaled to give the best fit to the observations. The optimal fingerprinting algorithm (a form of multi-variate regression) is then used to estimate the amplitudes β_i and the uncertainty ranges. See Tett *et al.* (1999) for further details.

To test the robustness of the model results a number of approaches are taken. One approach is to use data from different models to explore the sensitivity of the results to model-dependent uncertainties in the definition of an anthropogenic signal. These uncertainties arise from model differences in, *e.g.*, the physical parametrizations and the model resolution. Santer *et al.* (1996) follows this approach. Another approach is to carry out sensitivity studies in which the analysis procedure is changed. Tett *et al.* (1999) follows this approach. In both approaches, independence of the results to the model used or the algorithm used is an indication of robustness.

References

Allen, M.R., and L.A. Smith, 1996: Monte Carlo SSA: Detecting irregular oscillations in the presence of colored noise. *J. Clim.,* **9**, 3373-3404.

Andrews, D.G., J. R. Holton, and C.R. Leovy, 1987: *Middle Atmosphere Dynamics.* Academic Press, London.

Arakawa, A., and V.R. Lamb, 1977: Computational design of the basic dynamical processes of the UCLA general circulation model. *Methods in Comput. Phys.,* **17**, 173-265.

Austin, J., 2002: A three-dimensional coupled chemistry-climate model simulation of past stratospheric trends. *J. Atmos. Sci.,* **59**, 218-232.

Cullen, M.J.P., 1993: The unified forecast/climate model. *Meteorol. Mag.,* **122**, 81-94.

Dameris, M., V. Grewe, R. Hein, C. Schnadt, C. Brühl, and B. Steil, 1998: Assessment of the future development of the ozone layer. *Geophys. Res. Lett.,* **25**, 3579-3582.

Durran, D.R., 1999: *Numerical Methods for Wave Equations in Geophysical Fluid Dynamics.* Springer-Verlag, New York.

Haltiner, G.J., and T.R. Williams, 1980: *Numerical Prediction and Dynamic Meteorology.* John Wiley & Sons, New York.

Hamilton, K. (Ed)., 1996: *Gravity wave processes and their parametrization in global climate models.* Springer-Verlag.

Houghton, J.T., L.G. Meira Filho, J. Bruce, Hoesung Lee, B.A. Callander, E. Haites, N. Harris, and K. Maskell K. (Eds.), 1995: *Climate change 1994: Radiative forcing of climate change.* Cambridge University Press, Cambridge.

Lorenz, E.N., 1963: Deterministic nonperiodic flow. *J. Atmos. Sci.,* **20**, 130-141.

Phillips, T.J., 1994: A Summary Documentation of the AMIP Models. PCMDI Report No 118.

Rind, D., D. Shindell, P. Lonergan, and N.K.Balachandran, 1998: Climate change and the middle atmosphere: Part III, The doubled CO_2 climate revisited. *J. Clim.,* **11**, 876-894.

Santer, B.D., K.E. Taylor, T.M.L. Wigley, T.C. Johns, P.D. Jones, D.J. Karoly, J.F.B. Mitchell, A.H. Oort, J.E. Penner, V. Ramaswamy, M.D. Schwarzkopf, R.J. Stouffer, and S.F.B. Tett, 1996: A search for human influences on the thermal structure of the atmosphere. *Nature,* **382**, 39-46.

Scaife, A.A., N. Butchart, C.D. Warner, D. Stainforth, W. Norton, and J. Austin, 2000: Realistic quasi-biennial oscillations in a simulation of the global climate. *Geophys. Res. Lett.,* **27**, 3481-3484.

Simmons, A.J., A. Untch, C. Jakob, P. Kållberg, and P. Undén, 1999: Stratospheric water vapour and tropical tropopause temperatures in ECMWF analyses and multi-year simulations. *Q. J. R. Meteorol. Soc.,* **125**, 353-386.

Staniforth, A.N., and J. Côte, 1991: Semi-Lagrangian integration schemes for atmospheric models -- A review. *Mon. Weather Rev.,* **119**, 2206-2223.

Tett, S.F.B., P.A. Stott, M.R. Allen, W.J. Ingram, and J.F.B. Mitchell, 1999: Causes of twentieth-century temperature changes near the Earth's surface. *Nature,* **399**, 569-572.

Toth, Z., and E. Kalnay, 1993: Ensemble forecasting at NMC: the generation of perturbations. *Bull. Amer. Meteorol. Soc.,* **74**, 2317-2330.

WMO, 1999: Scientific Assessment of Ozone Depletion: 1998. WMO, Global Ozone Research and Monitoring Project Report No 44.

OPERATIONAL IMPLEMENTATION OF VARIATIONAL DATA ASSIMILATION

PIERRE GAUTHIER
Meteorological Service of Canada
Dorval, Québec, Canada

1. Introduction

Over the last few years, the variational form of statistical estimation has been implemented at many operational centres. The motivation originated from the difficulties associated with the assimilation of satellite data such as TOVS (TIROS-N Operational Vertical Sounders) radiances. Lorenc (1986) showed that the statistical estimation problem could be cast in a variational form (3D-Var) which is just a different way of solving the problem that the so-called *optimal interpolation* attempts to solve directly. Eyre (1989) showed, in a 1D-Var context, that a variational formulation leads to a more natural framework for the direct assimilation of radiances instead of *retrieved* temperature and humidity profiles. This is also true for any indirect measurement of the state of the atmosphere. Talagrand and Courtier (1987) showed that the use of the adjoint of a numerical model makes it possible to determine the initial conditions leading to a forecast that would best fit data available over a finite time interval. These two formulations can be combined to yield what is now called the 4D variational formulation of the statistical estimation problem or *4D-Var*.

The first implementation of 3D-Var was done at NCEP (Parrish and Derber, 1992) and later on at ECMWF (Courtier *et al.*, 1998). Other centres like the Canadian Meteorological Centre (Gauthier *et al.*, 1999) and the Met Office (Lorenc *et al.*, 2000) also implemented operationally a 3D-Var scheme. Courtier (1997) noted that there exists a dual formulation of 3D-Var on which is based the assimilation of NASA's Data Assimilation Office (Cohn *et al.*, 1998) and of the US Naval Research Laboratory (Daley and Barker, 2000). In 1997, ECMWF implemented 4D-Var (Rabier *et al.*, 2000) and so did Météo-France in 2000 (Gauthier and Thépaut, 2001). A considerable amount of research was necessary to achieve these operational implementations. Courtier *et al.* (1994) pointed out that a direct implementation of 4D-Var requires a computational time exceeding the capacity of even the most powerful computers. The incremental formulation of 4D-Var was proposed in which a simplified model is used to perform inner iterations followed by an integration of the full model based on the updated initial conditions. This *outer iteration* provides an updated evaluation of the innovations and of the reference trajectory required to define the simplified tangent linear model. In this context, an operational implementation of 4D-Var is possible. In recent years, experimentation with now operational 4D-Var systems indicates that it is necessary to

R. Swinbank et al. (eds.), Data Assimilation for the Earth System, 167–176.

make the simplified model to agree more closely with the complete high-resolution model both for its dynamics and physical parameterizations. This question is the object of current research regarding the nature of the simplified physical parameterizations that need to be used (Janisková *et al.*, 1999; Mahfouf, 1999).

In the context of the data assimilation cycles, the background error statistics should reflect the information gained from past observations, which is implicitly contained in the background state. Fisher and Andersson (2001) report recent results of experiments with a reduced rank Kalman filter, used to provide flow dependent background error covariances to 4D-Var. Their results did not show substantial improvements in the forecasts. This topic is being investigated using different types of simplified Kalman filters.

In this paper, the focus will be on the variational formulation of the data assimilation problem. In section 2, the incremental formulation will be presented and discussed first in the context of 3D-Var. Section 3 presents the incremental form of 4D-Var. Results with a simple barotropic model are presented to illustrate the capabilities and limitations of the approach. Section 4 reviews some recent results obtained from experimentations by several groups (e.g., ECMWF, Météo-France). Section 5 discusses some avenues being explored in current research. This includes the ensemble Kalman filter (Evensen, 1997; Houtekamer and Mitchell, 2001) or representer algorithms (Bennett and Thorburn, 1982; Xu and Daley, 2002).

2. The incremental formulation of variational assimilation

The variational assimilation problem is expressed here as

$$J(X) = \tfrac{1}{2}(X - X_b)^T B^{-1}(X - X_b) + \tfrac{1}{2}(H(X) - y)^T R^{-1}(H(X) - y) \qquad (2.1)$$

where X is the model state, X_b is the background state, B represents the background-error covariances, y is the observation vector, H, the observation operator while R represents the observation error covariances. To precondition the minimization, the

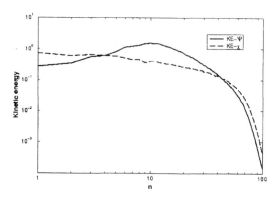

Figure 1. Kinetic energy spectra for the rotational and divergent component for the autocorrelation of background-error statistics at 500 hPa.

control variable $\xi = B^{-1/2}(X - X_b)$ is introduced so that $X \equiv X(\xi) = X_b + B^{1/2}\xi$ and (2.1) can be rewritten as

$$J(\xi) = \tfrac{1}{2}\xi^T\xi + \tfrac{1}{2}\big(H(X(\xi)) - y\big)^T R^{-1}\big(H(X(\xi)) - y\big) . \qquad (2.2)$$

In 3D-Var, the analysis increment $\delta x = B^{1/2}\xi$ has an effective lower resolution that is dictated by the background-error covariances. For example, Fig.1 shows the spectrum of the rotational and divergent kinetic energy of correlations. The use of such covariances will lead to an analysis increment with a resolution that cannot exceed 200 km.

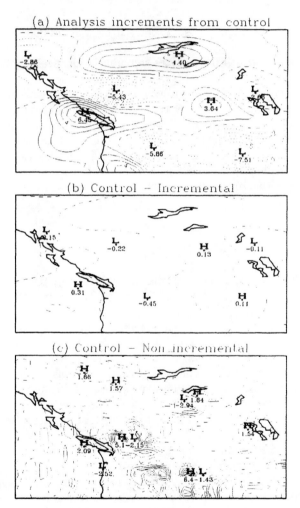

Figure 2. a) Analysis increments of dew point depression at 700 hPa from the control experiment valid at 1200 UTC 24 September 1997. b) Analysis increment of a) minus that of the incremental experiment. c). Analysis increments shown in a) minus that of the non-incremental experiment (from Laroche *et al.,* 1999)

However, it is beneficial to compute the innovations $y' = y - H(X_b)$ using the background state at its full resolution. Following Courtier *et al.* (1994), we approximate $H(X) \cong H(X_b) + (\partial H/\partial X)(X_b)\, \delta \mathbf{x} \equiv H(X_b) + H'(X_b)\, \delta \mathbf{x}$ so that (2.2) is now approximated by

$$J_L(\xi) = \tfrac{1}{2}\xi^T\xi + \tfrac{1}{2}\left(H'(X_b)\delta x(\xi) - y'\right)^T R^{-1}\left(H'(X_b)\delta x(\xi) - y'\right) \qquad (2.3)$$

with $\delta x(\xi) = B^{1/2}\xi$. As pointed out in Courtier *et al.* (1994), J_L has a form very similar to the original problem except that the observation operator has been linearized around X_b and the resulting Jacobian, $H'(X_b)$, is used instead of the nonlinear form of the observation operator. Fig.2 from Laroche *et al.* (1999) shows the analysis increment obtained by using the original cost function at the full resolution of the model and differences between this analysis increment and that obtained by using (2.3) (Fig.2b). The differences indicate that the analyses are virtually identical. Fig.2c however shows the difference between the reference analysis increment and that obtained by solving (2.1) but at a lower resolution. The differences are significant and stress the importance of computing the innovations with respect to the full resolution of the background field.

This approach has been used at the Canadian Meteorological Centre (Laroche *et al.*, 1999) and Météo-France (Desroziers *et al.*, 2003) to produce regional analyses for a variable resolution model while the analysis increment remains global and at a coarser resolution. The small scales features found in the analysis are therefore produced by the model itself in response to changes brought by the analysis to the large-scale components.

3. The incremental form of 4D-Var

The reasons why low-resolution increments are sufficient in 4D-Var are different than those presented above for 3D-Var. In Thépaut and Courtier (1991) and later on in Tanguay *et al.* (1995) and Laroche and Gauthier (1999), it is shown that for the large-scale dynamics, 4D-Var adjusts the more energetic large-scale components first. To determine the analysis increments of 4D-Var within the inner loop, Courtier *et al.* (1994) then proposed to use the tangent linear model (TLM) and its adjoint (Adj) at a coarser resolution with simplified physical parameterizations. However, as for the 3D-Var case, the innovations must be computed with respect to a high-resolution trajectory generated by the high-resolution model with its complete set of physical parameterizations. Outer loops are however needed to update the reference trajectory that defines the TLM and Adj.

Two questions are then raised. First, is it sufficient for the minimization to consider only the large-scale components of the gradient of the cost function and, second, can the evolution of these large scale components be correctly predicted by a simplified model? Fig.3 from Laroche and Gauthier (1999) summarizes results from several experiments carried out with a simple barotropic model on a β-plane. The experiments were cast as identical twins with observations generated from a reference trajectory provided by the assimilating model and random noise representative of observation error was added to those *synthetic* observations. No background term is included. Fig. 3 shows the correla-

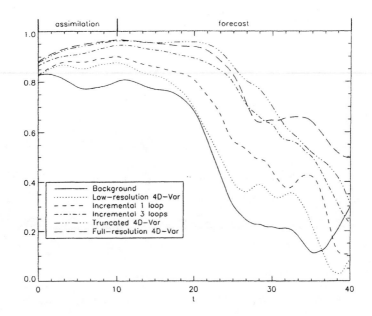

Figure 3. Correlation between the reference vorticity field and those obtained from various variational formulations. The synthetic observations are at low resolution and random noise has been added (from Laroche and Gauthier, 1999).

tion between the reference trajectory (or *nature run*) and the results from several experiments. The assimilation takes place over the first 10 nondimensional time units and forecasts at full resolution are made up to t = 40. The full resolution 4D-Var shows the best that can be obtained from the low-resolution observations. Given the limit of predictability, a reasonable forecast can be obtained up to t ≅ 30. The truncated 4D-Var experiment is one in which the adjoint model is used at the full resolution but the resulting gradient is truncated at a lower resolution (reduced by a factor of 4). In this case, Fig.3 shows that the truncated 4D-Var compares to the results obtained with the full resolution 4D-Var.

The low-resolution 4D-Var experiment is a complete 4D-Var based on the model at low-resolution. Fig.3 indicates that it only marginally improves the fit to the reference trajectory compared to what is obtained from the starting point of the minimization. Finally, the incremental form of 4D-Var shows that with only one outer loop, the results are only slightly improved compared to the low-resolution 4D-Var. However, if three outer loops are considered, then the incremental formulation yields to a reasonable approximation of the truncated 4D-Var. These results show the crucial role of updating the trajectory by performing outer iterations of 4D-Var.

Another question raised by 4D-Var is the impact of model error. It is implicitly assumed that any misfit to the observations is the result of error in the initial conditions. Experience shows that bad forecasts are often caused by errors in the model itself. An example is presented in the context of the barotropic model used in the experiments presented above. To mimic phase errors that occur with numerical weather prediction models that do not displace meteorological systems at the correct phase speed, synthetic

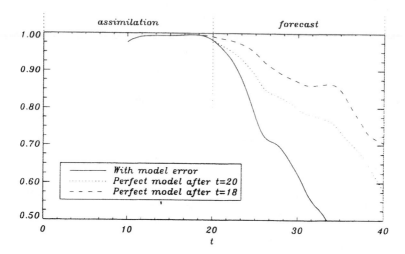

Figure 4. Solid line shows the correlation between the nature run obtained by using β = 0.5 and the assimilation/forecast of a full 4D-Var based on a model using β = 0.4. Using the perfect model to perform the forecast, the dashed (dotted) line show the correlation obtained when using the initial conditions at t = 18 (t = 20).

observations were generated from a nature run obtained with β = 0.5 while β was set to 0.4 in the model used in the assimilation. Fig.4 shows the correlation between the true state and the assimilation/forecast based on the erroneous model: the assimilation window here is 10 < t < 20. It shows that the best solution is obtained not at the end of the assimilation window but at some earlier time. To test the quality of the analysis, the true model (β = 0.5) was used to make the forecast using the 4D-Var analysis at t = 20 (dotted line) and at t = 18 (dashed line) where the maximal correlation to the true state was obtained. It shows that the latter case yields to a substantially better forecast. So, even though these experiments used perfect observations at all grid points and at all times, the presence of model error cannot translate the information from the observations into a better forecast.

In operational systems, model error is often associated with weaknesses in the numerous physical parameterizations used in the model. The incremental formulation of 4D-Var introduces a simplified set of physical parameterizations that should be consistent with those of the complete model. The development of a simplified set of physical parameterizations (deep convection, stratiform precipitation, surface and gravity-wave drag, vertical diffusion and radiation) is presented in Janisková *et al.* (1999) and Mahfouf (1999). As discussed in Mahfouf and Rabier (2000), this translated in a better fit to the observations within the inner loop of the 4D-Var. In Barkmeijer *et al.* (2001) and Puri *et al.* (2001), it is shown that the inclusion of a simplified physics in the TLM/Adj for the computation of the singular vectors used in their ensemble prediction system, has a significant impact on the spread of the resulting ensemble of forecasts.

Finally, it is important to stress that the 4D-Var analysis and forecast are better balanced with respect to the internal balance of the model. Fig.5 shows the average precipitation rates (over 24-h) as a function of forecast time. It clearly shows that,

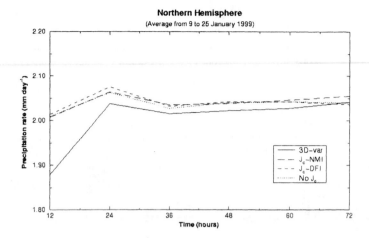

Figure 5. Average precipitation rates (in mm/day) as a function of forecast time. Those have been averaged over the Northern hemisphere for the two-week period extending from 9 to 25 January 1999. (see Gauthier and Thépaut, 2001 for details)

compared to 3D-Var, the 4D-Var analyses do not show a significant imbalance in the first hours of the forecast. This *spin-up* process is often associated with the presence of spurious gravity waves that need to be removed by an initialization process such as nonlinear normal mode initialization or a digital filter. The experiments of Gauthier and Thépaut (2001) show that even without any constraint to suppress these fast oscillations (No J_c experiment), the 4D-Var analysis does not lead to an appreciable precipitation spin-up at the initial time.

4. Experimentation with 4D-Var

The introduction of the time dimension in the assimilation allows information to be extracted from a time series of observations. For instance, the 4D-Var assimilation of measurements related to passive tracers provides information about the winds (Andersson *et al.,* 1994). Moreover, the 4D-Var analysis increments have a baroclinic structure that can be related to the fastest growing perturbations (Thépaut *et al.,* 1996). , Based on the extensive experimentation carried out at ECMWF in preparation for the implementation of 4D-Var, Rabier *et al.* (1998, 2000) report that the main differences between the 3D-Var and 4D-Var analysis increments were observed in those regions identified as the most sensitive to perturbations in the initial conditions. The sensitive regions were determined from the singular vectors. Their results also indicate that the gain obtained with 4D-Var lies in reducing the number of missed forecasts due to rapid cyclogenesis. This indicates that 4D-Var is able to determine the changes to the initial conditions that will trigger or not the development of synoptic systems.

The advantage of 4D-Var over 3D-Var is that it makes it possible to assimilate data at the observation time, which results in an increase of the volume of data that can be assimilated. Moreover, information contained in the temporal variation can also be extracted. However, Järvinen *et al.* (1999) present some difficulties, encountered in the assimilation of surface pressure data from stations reporting every hour. In regions where the real orography differs from that of the model, a bias can be introduced in the surface pressure data. When all hourly reports are assimilated, this results in a significant negative impact. This problem is more acute for isolated stations since there are no surrounding data to weigh against this biased estimate. Removing the effect of this bias from the data can be addressed in different ways and Järvinen *et al.* (1999) introduced a time-correlation in the observation error that manages to alleviate the problem. The net effect is to focus the analysis more on the *variation* of surface pressure than on the mean value of surface pressure. It is the surface pressure tendency that is more closely linked to developing baroclinic systems (Bengtsson, 1980).

5. Conclusion

Any variational problem raises the issue of the convergence of the minimization. One iteration of the minimization being particularly expensive in 4D-Var, its practical implementation has to limit the number of iterations to a rather low number, typically less than a hundred. Fisher and Andersson (2001) proposed a preconditioning of the minimization by approximating the Hessian of the 4D-Var cost function and this significantly improved the convergence. Problems in the conditioning could also explain a case of poor convergence in a 4D-Var experiment reported by Andersson *et al.* (2000). Convergence of the minimization is therefore still an issue and given the nonlinear nature of the model, so is the existence of multiple minima (Tanguay *et al.*, 1995).

Currently, a lot of research is going on to address the importance of model error in 4D-Var and data assimilation in general. In particular, in the context of 4D-Var, the model is assumed to be perfect and even more, in the incremental formulation, the simplified model should be a good approximation of the more accurate high-resolution model with sophisticated physical parameterizations. As discussed in Bouttier (2001), extending the assimilation window to 12-h has accentuated the importance of these differences that could be diagnosed from the differences between observation departures computed with respect to the simplified and high-resolution models. Extending 4D-Var to longer assimilation windows may be more difficult than was thought initially due to the importance of model errors.

It was mentioned briefly that cycling 4D-Var requires that the background-error covariances reflect their flow-dependent nature. Results presented by Fisher and Andersson (2001) indicate that there is a long way from theory to practice. The reduced-rank Kalman filter used in their experiments provided a flow-dependent estimate of the covariances but this only had a marginal impact on the resulting forecasts.

Other avenues are being explored in 4D data assimilation. In particular, the ensemble Kalman filter (EnKF) (Evensen, 1997; Houtekamer and Mitchell, 2000) has been proposed recently to obtain a practical way of implementing a Kalman filter for complex models without having to develop the adjoint of a numerical model. The

Monte-Carlo approach that supports the EnKF then raises some questions about the required size of the ensemble. Up to now, the direct estimate is often noisy and some assumptions must be introduced to address the rank deficiency problem. It also makes it more difficult to introduce non-Gaussian error statistics for errors, and to maintain dynamical balance in the analysis increments.

The 4D-Var algorithm imposes the model as a strong constraint, which has some limitations. Bennett and Thorburn (1992) introduced the weak constraint formulation in which the model is imposed only as a weak constraint. Because it uses the complete model trajectory and not only the initial conditions, this approach is much more demanding than 4D-Var. Recently, Xu and Daley (2002) introduced the *accelerated representer algorithm* which is referred to as *4D-PSAS* in Courtier (1997). It corresponds to the dual formulation of 4D-Var and can be built from the same basic operators (e.g., model integrations, observation operators, and their tangent linear and adjoint, covariance models). In Lagarde *et al.* (2001), a graphical representation has been introduced to represent a wide class of assimilation algorithms altogether with their dual representations. Their analysis show that if some care is taken in developing a data assimilation system, it would be possible to reuse the same basic *building blocks* to obtain new algorithms. This would be an advantage for operational centres that must adapt quickly to new advances in a rapidly evolving field.

Acknowledgement. The author would like to thank Stéphane Laroche and Jean-Noël Thépaut for their comments on the manuscript.

References

Andersson, E., M. Fisher, R. Munro and A. McNally, 2000: Diagnosis of background errors for radiances and other observable quantities in a variational data assimilation scheme, and the explanation of a case of poor convergence. *Quart. J.R. Meteor. Soc.*, **126**, 1455-1472.

————, E., J. Pailleux, J.N. Thépaut, J.R. Eyre, A.P. McNally, G.A. Kelly and P. Courtier, 1994: Use of cloud-cleared radiances in 3D and 4D variational data assimilation. *Quart. J.R. Meteor. Soc.*, **120**, 627-653.

Barkmeijer, J., R. Buizza, T.N. Palmer, K. Puri and J.F. Mahfouf, 2001: Tropical singular vectors computed with linearized diabatic physics. *Quart. J.R. Meteor. Soc.*, **127**, 685-708.

Bengtsson, L., 1980: On the use of a time sequence of surface pressures in four-dimensional data assimilation. *Tellus*, **32**, 189-197.

Bennett, A.F. and M.A. Thorburn, 1992: The generalized inverse of a nonlinear quasigeotrophic ocean circulation model. *J. Phys. Ocean.*, **22**, 213-230.

Bouttier, F., 2001: The development of 12-hourly 4D-Var. *ECMWF Tech. Memo.*, No.348, 21 pages.

Cohn, S., A. da Silva, J. Guo, M. Sienkiewicz and D. Lamich 1998: Assessing the effects of data selection in the DAO Physical-Space Statistical Analysis System. *Mon. Wea. Rev.*, **126**, 2913-2926.

Courtier, P., E. Andersson, W. Heckley, J. Pailleux, D. Vasiljevic, M. Hamrud, A. Hollingsworth, F. Rabier and M. Fisher, 1998: The ECMWF implementation of three dimensional variational assimilation (3D-Var). Part I: formulation. *Quart. J.R. Meteor. Soc.*, **124**, 1783-1808.

————, 1997: Dual formulation of four-dimensional variational data assimilation. *Quart. J.R. Meteor. Soc.*, **123**, 2449-2461.

————, J.N. Thépaut and A. Hollingsworth, 1994: A strategy for operational implementation of 4D-Var, using an incremental approach. *Quart. J.R. Meteor. Soc.*, **120**, 1367-1387.

Daley, R. and E. Barker, 2000: *NAVDAS Source Book: NRL atmospheric variational data assimilation system.* Naval Research Laboratory, Monterey, U.S.A., 151 pages.

Desroziers, G., G. Hello and J.-N. Thépaut, 2003: A 4D-Var reanalysis of the FASTEX experiment. (to appear in *Quart. J.R. Meteor. Soc.*)

Evensen, G., 1997: Advanced data assimilation for strongly nonlinear dynamics. *Mon. Wea. Rev.*, 125, 1342-1354.

Eyre, J.R., 1989: Inversion of cloudy satellite sounding radiances by nonlinear optimal estimation. I: Theory and simulation of TOVS. *Quart. J.R. Meteor. Soc.*, 115, 1001-1026.

Fisher, M. and E. Andersson, 2001: Developments in 4D-Var and Kalman filtering. *ECMWF Tech. Memorandum*, No.347, 36 pages.

Gauthier, P., C. Charette, L. Fillion, P. Koclas and S. Laroche, 1999: Implementation of a 3D variational data assimilation system at the Canadian Meteorological Centre. Part I: The global analysis. *Atmosphere-Ocean*, 37, 103-156.

———— and J.-N. Thépaut, 2001: Impact of the digital filter as a weak constraint in the preoperational 4DVAR assimilation system of Météo-France. *Mon. Wea. Rev.*, 129, 2089-2102.

Houtekamer, P.L. and H.L. Mitchell, 2001: A sequential ensemble Kalman filter for atmospheric data assimilation. *Mon. Wea. Rev.*, 129, 123-137.

Janisková, M., J.-N. Thépaut and J.-F. Geleyn, 1999: Simplified and regular physical parameterizations for incremental four-dimensional variational assimilation. *Mon. Wea. Rev.*, 127, 26-45.

Järvinen, H., E. Andersson and F. Bouttier, 1999: Variational assimilation of time sequences of surface observations with serially correlated errors. *Tellus*, 51A, 468-487.

Lagarde, T., A. Piacentini and O. Thual, 2001: A new representation of data assimilation methods: the PALM flow-charting approach. *Quart. J.R. Meteor. Soc.*, 127, 189-207.

Laroche, S., P. Gauthier, J. St-James and J. Morneau, 1999: Implementation of a 3D variational data assimilation system at the Canadian Meteorological Centre. Part II: the regional analysis. *Atmosphere-Ocean*, 37, 281-307.

———— and P. Gauthier, 1999: A validation of the incremental formulation of 4D variational data assimilation in a nonlinear barotropic flow. *Tellus*, 50A, 557-572.

Lorenc, A.C., S. P. Ballard, R. S. Bell, N. B. Ingleby, P. L. F. Andrews, D. M. Barker, J. R. Bray, A. M. Clayton, T. Dalby, D. Li, T. J. Payne and F. W. Saunders, 2000: The Met. Office Global three-dimensional variational data assimilation scheme. *Quart. J.R. Meteor. Soc.*, 126, 2991-3012.

————, 1986: Analysis methods for numerical weather prediction. *Quart. J.R. Meteor. Soc.*, 112, 1177-1194.

Mahfouf, J.F., 1999: Influence of physical processes on the tangent-linear approximation. *Tellus*, 51A, 147-166.

———— and F. Rabier, 2000: The ECMWF operational implementation of four dimensional variational assimilation. Part II: experimental results with improved physics. *Quart. J.R. Meteor. Soc.*, 126, 1171-1190.

Parrish, D.F., and J.C. Derber, 1992: The National Meteorological Center's spectral statistical interpolation analysis system. *Mon. Wea. Rev.*, 120, 1747-1763.

Puri, K., J. Barkmeijer and T.N. Palmer, 2001: Ensemble prediction of tropical cyclones using targeted diabatic singular vectors. *Quart. J.R. Meteor. Soc.*, 127, 709-734.

Rabier, F., J.-N. Thépaut and P. Courtier, 1998: Extended assimilation and forecast experiments with a four-dimensional variational assimilation system. *Quart. J.R. Meteor. Soc.*, 124, 1861-1887.

————, H. Järvinen, E. Klinker, J.-F. Mahfouf and A. Simmons, 2000: The ECMWF operational implementation of four dimensional variational assimilation. Part I: experimental results with simplified physics. *Quart. J.R. Meteor. Soc.*, 126, 1143-1170.

Talagrand O. and P. Courtier, 1987: Variational assimilation of meteorological observations with the adjoint vorticity equation. I. Theory. *Quart. J. R. Met. Soc.*, 113, 1311-1328.

Tanguay, M., P. Bartello and P. Gauthier, 1995: Four-dimensional data assimilation with a wide range of scales. *Tellus*, 47A, 974-997.

Thépaut, J.-N. and P. Courtier, 1991: Four-dimensional data assimilation using the adjoint of a multilevel primitive equation model. *Quart. J.R. Meteor. Soc.*, 117, 1225-1254.

————, P. Courtier, G. Belaud and G. Lemaître, 1996: Dynamical structure functions in a four-dimensional variational assimilation: a case study. *Quart. J. R. Met. Soc.*, 122, 535-561.

Xu, L. and R. Daley, 2002: Data assimilation with a barotropically unstable shallow water system using representer algorithms. *Tellus*, 54A, 125-137.

QUALITY CONTROL: METHODOLOGY AND APPLICATIONS

PIERRE GAUTHIER, CLÉMENT CHOUINARD
AND BRUCE BRASNETT
Meteorological Service of Canada
Dorval, Québec, Canada

1. Introduction

All assimilation methods rely on data collected from many sources. There are the conventional data sources like surface observations, radiosonde data, aircraft and ship data to which is now added an increasing amount of satellite data (see also the chapters *Observing the atmosphere* by R. Swinbank and *Assimilation of remote sensing observations in Numerical Weather Prediction*, by J.N. Thépaut). Each instrument is prone to error that could be systematic due to an incorrect calibration or random, reflecting the accuracy and representativeness of the measurement. The assimilation methods presented during this course assume that the data used in the assimilation have unbiased errors and are devoid of any serious error due to a malfunction of the instrument. Such errors are referred to as *gross errors*.

There are several sources of gross errors. These can be associated with the calibration of the instrument, transmission problems when data are disseminated to operational centres, pre-processing of the data at reception, etc. Methods had to be developed to detect such errors to reject all data that have a high probability of being in gross error. The background field, x_b, reflects our *a priori* knowledge of the current state of the atmosphere gained from the information of past observations. To check whether the observations, represented here as a vector y, are within a reasonable range of their expected value, they are compared against the background state equivalent of the observations, $H(x_b)$, where H stands for the observation operator. The innovation vector, $y - H(x_b)$, is therefore a measure of the departure of each observation against a common atmospheric state. This information is extremely useful to assess whether any type of gross error does not contaminate the new data. These ideas can be formalised by using conditional probabilities to assess the probability of a datum y being correct given a background state x_b with known (or assumed) error probability distribution. Quality control will be presented in this context using a Bayesian perspective to data assimilation.

In this lecture, innovations will be used to diagnose potential problems with an instrument through the monitoring phase of observations, which is done routinely at operational centres to assess whether something wrong is going on with an instrument.

R. Swinbank et al. (eds.), Data Assimilation for the Earth System, 177–187.

After a brief introduction to conditional probabilities and Bayes' theorem, the background check procedure and variational quality control will be presented.

2. Bayesian approach to data assimilation

In Lorenc (1986), it is shown that the statistical estimation problem of data assimilation can be approached using conditional probabilities and Bayes' theorem. This section briefly introduces this approach and the reader is referred to this paper and to Rodgers (2000) for more details. The *joint probability distribution* of x, the atmospheric state vector, and y, the observation vector, being $p(x,y)$, one can define the marginal probability densities $P(x)$ and $P(y)$ as

$$P(x) = \int p(x, y)dy, \quad P(y) = \int p(x, y)dx$$

where $P(x)$ and $P(y)$ are obtained by integrating over all possible values of y and x respectively. Assuming the background error probability distribution to be Gaussian, then

$$P(x) = \frac{1}{C}\exp\left\{-\frac{1}{2}(x - x_b)^T B^{-1}(x - x_b)\right\}, \tag{1}$$

where B is the background error covariance matrix and C, a normalization constant. In absence of any other information, $P(x)$ represents our *a priori* knowledge of the state of the atmosphere contained within the background state x_b. In absence of any other information, the most probable state would have to be x_b. Similarly, $P(y)$ represents our *a priori* knowledge about the true value of the observation. For example, in the Gaussian case, it would correspond to

$$P(y) = \frac{1}{C_2}\exp\left\{-\frac{1}{2}(y - y_o)^T R^{-1}(y - y_o)\right\} \tag{2}$$

where R is the observation error covariance matrix, C_2, a normalization constant and y_o, the actual value of the measurement.

The *conditional probability* $p(y \mid x = x_0)$ stands for the probability of y taking a particular value given that $x = x_0$. This can be expressed as

$$p(y \mid x = x_0) = \frac{p(x_0, y)}{\int p(x_0, y)dy} \equiv \frac{p(x_0, y)}{P(x_0)}. \tag{3}$$

Similarly,

$$p(x \mid y = y_0) = \frac{p(x, y_0)}{\int p(x, y_0)dx} \equiv \frac{p(x, y_0)}{P(y_0)}. \tag{4}$$

From (3) and (4), one gets that

$$p(x = x_0, y = y_0) = p(x \mid y = y_0)P(y_0) = p(y \mid x = x_0)P(x = x_0) \tag{5}$$

Taking x_0 to be the true value x_t of the atmospheric state while y_0 is associated with the actual values of the observations, then (4) and (5) imply that

$$p(x = x_t \mid y = y_o) = \frac{p(y = y_o \mid x = x_t)P(x = x_t)}{P(y = y_o)}. \tag{6}$$

This is *Bayes' theorem* on conditional probabilities expressing the probability distribution that x is the true value of the atmospheric state given that y_o has been observed. The crux of this argument lies in the fact that since $x = x_t$ it means that $y_t = H(x_t)$, then the probability that $y = y_o$ corresponds to, in the Gaussian case,

$$p(y = y_o \mid x = x_t) = \frac{1}{C_2} \exp\left\{ -\frac{1}{2}(y_o - H(x_t))^T R^{-1}(y_o - H(x_t)) \right\}. \tag{7}$$

Here, we assume that R includes the representativeness error. This conditional probability is then entirely described by the observation error probability distribution with its mean centered at $H(x_t)$. Bayes' theorem therefore implies that

$$J(x) = -\ln\left(p_{a\ posteriori}(x) \right) \equiv -\ln p(x \mid y) = -\ln p(y \mid x) - \ln P(x) + C \tag{8}$$

with $p(x \mid y)$, the *a posteriori* probability distribution of x being the true value. This states that the most probable state, the *maximum likelihood* estimate, is obtained by minimizing (8). In the Gaussian case, (1) and (7) apply and we get that

$$J(x) = \frac{1}{2}(x - x_b)^T B^{-1}(x - x_b) + \frac{1}{2}(H(x) - y)^T R^{-1}(H(x) - y). \tag{9}$$

This corresponds to the variational form of the statistical estimation problem when the error distributions are Gaussian. However, (8) makes it possible to consider non-Gaussian probability distributions. This will be discussed in section 4.

Remark. In the case where H is linear, (9) implies that the *a posteriori* probability distribution is also a Gaussian distribution with x_a being its mean and the covariances correspond to $B^{-1} + H^T R^{-1} H \equiv P_a^{-1}$.

3. Monitoring the observations and the background check quality control

The background field used in the assimilation being our *a priori* estimate of the atmosphere, it offers a common ground against which all observations can be compared. The innovations are such that $y - H(x_b) \cong \varepsilon_o - H'\varepsilon_b$ where $H(x_t) = y_t$ and $H' = \partial H/\partial X$ is Jacobian of the observation operator. It is assumed here that ε_o includes the representativeness error. Averaging the innovations by observation types will reveal whether the observation and background error are unbiased. This phase is the

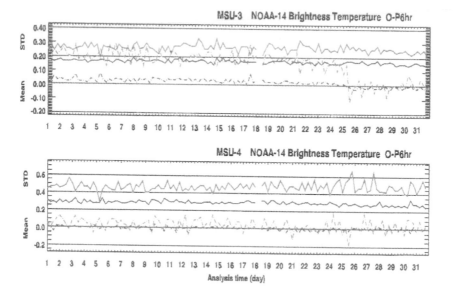

Figure 1. Time series of innovations for ATOVS Level 1d radiance. Innovations (O-F departures) are plotted in red and the departure to the analysis (O-A) are plotted in blue. Both the bias (dashed line) and standard deviations (solid line) are plotted.

monitoring of the observations and allows one to detect serious problems with the data. Fig. 1 shows a timeseries of innovations computed here for radiance data over four microwave channels of ATOVS level 1-d data. The innovations (plotted in red) have been averaged over all data received for the assimilation, and are shown for the month of January 2001. This figure shows that channel MSU-3 had a bias (dashed curve) that was at the same level as the standard deviations (solid line). Such data can be very harmful to the analysis and must be rejected. When a new instrument is added, the monitoring is very useful to characterize the measurement errors and can be used for the calibration.

Assuming now that both the observation and background error are unbiased, the innovations are such that

$$\left\langle (\varepsilon_o - H\varepsilon_b)(\varepsilon_o - H\varepsilon_b)^T \right\rangle = R + HBH^T$$

and represents the error covariances associated with the probability distribution of the innovations. Assuming those to be Gaussian, one has that

$$p(y - H(x_b)) = \frac{1}{C} \exp\left\{ -\frac{1}{2} (y - H(x_b))^T (R + HBH^T)^{-1} (y - H(x_b)) \right\}. \tag{10}$$

For each datum, this implies that, at the observation location, the innovation is distributed as

$$p(y - H\mathbf{x}_b) = \frac{1}{\left(2\pi(\sigma_o^2 + \sigma_b^2)\right)^{1/2}} \exp\left\{-\frac{1}{2}\left(\frac{(y - H\mathbf{x}_b)^2}{\sigma_o^2 + \sigma_b^2}\right)\right\}. \tag{11}$$

Based on this, one concludes that it is very unlikely that the innovation be such that

$$(y - H\mathbf{x}_b)^2 > \lambda(\sigma_o^2 + \sigma_b^2)$$

with λ being defined as the rejection criterion which may depend on observation type and geographical location. For instance, if $\lambda = 2$, the probability of such an event occurring is less than 5%. This *background check* procedure is used to flag data that have a high probability of being in error. It is important to note that it is best to use the background check to only eliminate data that are obviously in error. In this case, λ should be chosen rather high.

4. Variational quality control

Observation error associated with measurement accuracy and representativeness is random by nature and can often be described as normally distributed with mean y_t and variance σ_o^2: this will be denoted as $N(y_t, \sigma_o)$. However, gross errors also occur but have a probability that is independent of the true value y_t. As discussed in Lorenc and Hammon (1988), gross errors can be assumed to be equally probable over a range of admissible values $\hat{y} - D/2 \le y \le \hat{y} + D/2$. Following Dharssi *et al.* (1992) and Ingleby and Lorenc (1993), the probability distribution for observation error is taken to

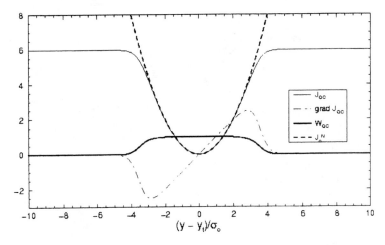

Figure 2. Variational QC-Var cost function (Jo), its gradient and the associated weight WQC represented as a function of the normalized departure from the estimated true value yt = ŷ of the observation value y. The probability of gross error has been set to P = 0.01

be of the form

$$p_o(y) = P\, p_G(y) + (1-P)\mathrm{N}\,(\hat{y},\sigma_0) \tag{12}$$

with P being the overall probability of having a gross error while $p_G(y) = 1/D$, a constant. Finally, $\mathrm{N}\,(\hat{y},\sigma_o) = (\sigma_o\sqrt{2\pi})^{-1}\exp\!\left(-(y-\hat{y})^2/(2\sigma_o^2)\right)$ and it is assumed that $\sigma_o/D \ll 1$. At this stage, $\hat{y} = H(x)$ stands for the estimate of the true value of the observation based on our current knowledge of the atmosphere denoted here by x.

In the Bayesian formulation, the argument is that $p(y|\,x) = p_o(y - Hx)$. Assuming those observation errors are uncorrelated, then

$$J_o(x) = -\ln p_o(y - Hx) = -\ln\!\left(\gamma + (1-P)\mathrm{N}\,(Hx,\sigma_o)\right)$$

with $\gamma = P/D$. As in Andersson and Järvinen (1999), it is convenient to rewrite this in terms of the Gaussian form for $J_o^N = \tfrac{1}{2}(H(X) - y)^2/\sigma_o^2$, which leads to

$$J_o(x) = -\ln\!\left(\frac{(1-P)}{\sigma_o\sqrt{2\pi}}\left[\exp\!\left(-J_o^N(x)\right) + \gamma\right]\right) \tag{13}$$

$$= -\ln\!\left(\exp\!\left(-J_o^N(x)\right) + \gamma\right) + Const.$$

where $\gamma = (P\sigma_o\sqrt{2\pi})/(1-P)D$ while its gradient is

$$\nabla_x J_o(x) = \frac{\exp(-J_o^N)}{\gamma + \exp(-J_o^N)}\nabla_x J_o^N(x) \equiv W_{QC}\nabla_x J_o^N(x)\,. \tag{14}$$

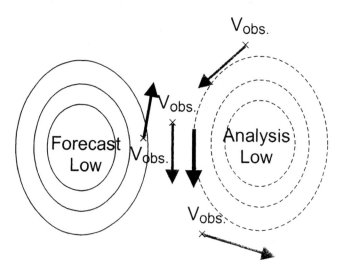

Figure 3. Schematic of a situation in which a comparison of 5 wind observations with the backgound would lead to a rejection of 4 good observations. With QC-Var, the datum that agrees the most with the background state ends up being rejected.

In this case, observation error is considered uncorrelated so that J_o^N can be computed as in the Gaussian case. It then suffices to modify the value of J_o using (13) and the gradient using (14) for each data. This is referred to as the QC-Var cost function that can be implemented very easily with slight modifications to the cost function associated with Gaussian error statistics. Fig. 2 illustrates the associated cost function, gradient and weight W_{QC} as a function of $y' = (y - \hat{y})/\sigma_o$: this figure is similar to Fig.2 of Andersson and Järvinen (1999). For departures exceeding $\sim 4\sigma_o$, the gradient becomes quite flat and the observation has little influence on the minimization.

The background-check procedure introduced earlier presumes that the background state yields a reasonable model equivalent of the observation and any significant departure is an indication that the observation is in error. However, situations do occur where it is the background state itself that is at fault and can disagree significantly with the observation. As illustrated in Fig.3, the background-check procedure would lead to the rejection of good data and the acceptance of bad ones. This figure is a schematic that emphasises a situation where the forecast has wrongly placed a low-pressure system while most of the data in the area would tend to reposition the system eastward. If the comparison is made against the background state, all data would end up being rejected except the one located closest to the forecast low. However, performing a preliminary analysis based on all available data would lead to a repositioning of the system. In this case, the observation departures with respect to this analysis would be the most important for the datum that initially agreed the most with the background state.

In the context of *optimal interpolation* (OI), Lorenc (1981) proposed a quality control procedure using a reduced analysis involving only a small number of observations $y_1, y_2, \ldots, y_K$. A set of K analyses $x^{(k)}$ is obtained based on all data except y_k. If the observation is such that $(y_k - H(x_a^{(k)})) > \lambda (\sigma_o^2 + \sigma_a^2)^{1/2}$ with λ being a pre-set criterion, then the data is rejected. This requires the computation of the analysis error variance that can be done when few observations are used. This procedure is therefore indirectly comparing the consistency among a subset of observations and rejects the data that do not concur with the rest of the data. A formal treatment of this procedure can be found in Ingleby and Lorenc (1993). It will be referred to as the *OI quality control*.

In the QC-Var algorithm, this is achieved by allowing the estimate to benefit partly from the information contained from the ensemble of the observations (Andersson and Järvinen, 1999). A limited number of iterations are therefore completed by turning off QC-Var for a number of iterations at the beginning of the minimization to let the iterate build the main features of the analysis based on all observations. At this point, the QC-Var is activated and the observation departures $y - H(x)$ as introduced in the QC-Var cost function use an *a posteriori* estimate of the model state to assess whether data should be implicitly rejected or not. If QC-Var were activated at the beginning of the minimization, little weight would be given to the data with large departures from the background. This would give more confidence to observations closest to the background state and implicitly rejects those that strongly disagree with it. This is why it is important to first let the minimization proceeds for a number of iterations to let the bulk of the data reposition the analysis. If a hundred iterations are needed for convergence, experimentation shows that turning off QC-Var for the first 30 iterations or so is sufficient. When QC-Var is turned on, the bad data would then be given less weight, as it should be.

5. Experimentation with QC-Var

The results presented here were obtained with the 3D-Var data assimilation system of the Canadian Meteorological Centre (CMC), which includes both a background-check and a variational quality control. This system has been implemented in December 2001 (Gauthier *et al.* 1998, 1999). It is shown here how the probability distribution for observation errors has been estimated. A comparison is presented of the amount of data that have been rejected compared to the rejection rates obtained from the OI quality control.

Previously, the CMC 3D-Var relied on an *OI quality control* (OI-QC) inherited from the previous optimal interpolation analysis used at CMC (Mitchell *et al.*, 1996). The QC-Var does not categorically accept or reject data. Here, data were considered *rejected* by QC-Var if its *a posteriori* weight is less than 0.25. Table 1 gives the rejection rates

Table 1. Comparison of rejection rates and limits for the quality control used in the CMC optimal interpolation analysis and QC-Var which includes also the background-check. The period extends from 19 January 2001 to 31 January 2001.

Obs. Type	Obs. Quantity	Rejection Ratio (%)		Approximate Rejection Limits	
		QC-Var	OI-QC	QC-Var	OI-QC
SYNOP	Pressure (height)	2.7	1.9	3.6 hPa	n/a
	(T- T_d)	0.3	0.0	8.5 K	22 K
	Temperature	2.2	1.2	6.6 K	16.6 K
SHIP	Wind	7.6	0.5	8 m/s	19 m/s
	Pressure (height)	2.3	3.5	8.5 hPa	n/a
	(T- T_d)	0.3	0.0	9.5 K	26 K
	Temperature	1.5	0.9	5.7 K	11.7 K
DRIBU	Pressure (height)	2.8	3.1	6.6 hPa	n/a
	Temperature	3.1	2.4	5.8 K	6.2 K
TEMP	Wind	2.7	0.4	8 - 14 m/s	11 - 20 m/s
	(T- T_d)	1.8	0.0	5 - 16 K	14 - 22 K
	Temperature	3.0	1.3	2.1 - 6.6 K	3.4 - 9.4 K
AMDAR	Wind	1.0	0.4	11 m/s	15 m/s
	Temperature	0.7	0.5	4.0 K	5.0 K
SATOB	Wind	1.3	0.2	13 - 27 m/s	16 - 36 m/s
AIREP	Wind	5.2	1.0	13 m/s	29 m/s
	Temperature	1.7	0.8	5.7 K	9.2 K
ACARS	Wind	2.3	1.0	10 m/s	14 m/s
	Temperature	1.6	2.1	4.0 K	5.0 K

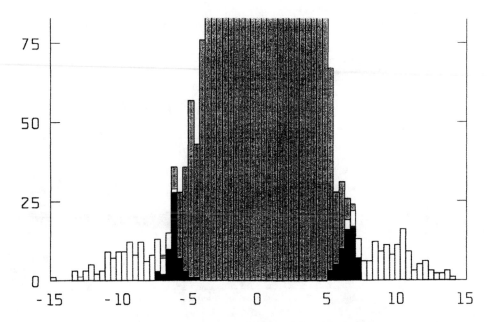

Figure 4.   Histogram of observation departures from the background for AIREP temperatures over the
period of March-April 2002. Number of accepted data is 31,926 (in gray), while 103 data were rejected
by the background check (in white) and 303 were rejected by QC-Var (in black)

and limits used by the background-check and QC-Var (weight W_{QC} evaluated at
convergence) and the OI-QC which corresponds to the procedure of Lorenc (1981)
described above.

Table 1 shows the rejection limits defined as the largest difference between
observation and analysis, computed from samples of accepted observations. Note that it
is the QC-Var algorithm that determines the rejection limit since the definition specifies
the departure from the analysis, not the background. These results indicate that QC-Var
rejects more of almost every type of data. Among the most noteworthy differences are
those of the wind data that are rejected twice as much by QC-Var than by the OI-QC.
This is appropriate since, at various times over the past ten years, meteorologists at
CMC had noted that incorrect wind reports were too often allowed in the analysis.
Another significant difference between the two systems is in the rejection rates of
dewpoint depression (T- T_d). The OI-QC rejects less than 0.1% of these while QC-Var
typically rejects between 0.3% and 1.8%. The rejection limits in Table 1 indicate that, in
the OI-QC, some dewpoint depression observations that differ by more than 20°C from
the analysis were not rejected while the QC-Var does not tolerate such large departures.

Fig. 4 shows the distribution of the number of data as a function of observation
departure from the background (mean value has been removed). It indicates that the
data rejected by QC-Var lie in that region where the Gaussian distribution differs from
the actual distribution of observation error, which includes the probability of gross
error. Fig. 5 represents the probability distribution resulting from having a probability of

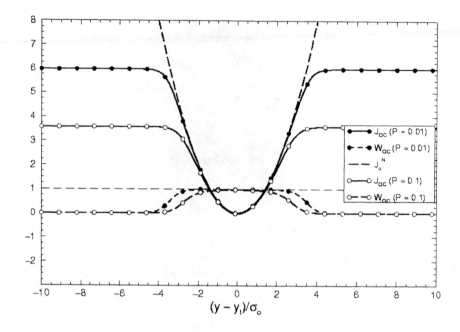

Figure 5. Comparison of the QC-Var cost functions and of W_{QC} when the probability of having a gross error is $P = 0.01$ (black dots) and when $P = 0.1$ (white dots).

gross error set to 10% and 1%. As discussed in Andersson and Järvinen (1999), the probability of gross error can be estimated from the distribution of the innovations ($y - Hx_b$). As expected the region where the weight W_{QC} is close to 1 narrows as the probability of gross error increases.

6. Conclusion

In this chapter, a short introduction to quality control has been presented. It has been shown in particular that a careful monitoring of the observations is necessary to detect systematic biases in the measurements that can often be related to the physical characteristics of the instrument or its environment. Monitoring is a useful and necessary step for new instruments, satellite instruments in particular, during the calibration phase before the data are assimilated (see the chapter *Assimilation of remote sensing observations in Numerical Weather Prediction*). There are numerous cases in which error in the new data could be related to instrument parameters (e.g., pointing angle of an antenna on board a satellite). All operational centres that are involved in data assimilation can then see in near real time any problem with the data they are receiving.

In the first stage, the data pass through a number of crude checks associated with the transmission and encoding/decoding phase. A comparison against the background state is then performed to eliminate a good part of data that are obviously in error. However, comparing against the background state can lead to rejection of good data because the forecast itself is in error. To address this problem, the OI-QC performs a preliminary analysis against which all data are individually compared. Data can then be flagged as rejected and a new analysis without the rejected data is performed. This is the method presented in Lorenc (1981). An alternative is to use a variational quality control that implicitly compares the data individually against a preliminary analysis corresponding to the current iterate of the variational analysis (Ingleby and Lorenc, 1993; Andersson and Järvinen, 1999).

The presentation here has been kept deliberately simple. However, it must be stressed that there are numerous complexities that need to be considered. For instance, as discussed in Andersson and Järvinen (1999), the QC-Var algorithm becomes much more complex when the observation error is correlated in space or even in time, a situation that can occur in 4D-Var (Järvinen et al., 1999).

Acknowledgement. The authors would like to thank Peter Lynch for his comments on the manuscript.

References

Andersson, E. and H. Järvinen, 1999: Variational quality control. *Quart. J.R. Meteor. Soc.*, **125**, 697-722.

Dharssi, I., A.C. Lorenc and N.B. Ingleby, 1992: Treatment of gross errors using maximum probability theory. *Quart. J.R. Meteor. Soc.*, **118**, 1017-1036.

Gauthier, P., C. Charette, L. Fillion, P. Koclas and S. Laroche, 1999: Implementation of a 3D variational data assimilation system at the Canadian Meteorological Centre. Part I: The global analysis. *Atmosphere-Ocean*, **37**, 103-156.

————, M. Buehner and L. Fillion, 1998: Background-error statistics modelling in a 3D variational data assimilation scheme: estimation and impact on the analyses. Proceedings of the *ECMWF Workshop on diagnosis of data assimilation systems*, Reading UK, p. 131-145.

Ingleby, N.B., and A.C. Lorenc, 1993: Bayesian quality control using multivariate normal distributions. *Quart. J.R. Meteor. Soc.*, **119**, 1195-1225.

Järvinen, H., E. Andersson and F. Bouttier, 1999: Variational assimilation of time sequences of surface observations with serially correlated errors. *Tellus, **51A**, 468-487.

Lorenc, A.C., 1986: Analysis methods for numerical weather prediction. *Quart. J.R. Meteor. Soc.*, **112**, 1177-1194.

————, A.C., 1981: A global three-dimensional multivariate statistical interpolation scheme. *Mon. Wea. Rev.*, **109**, 701-721.

Mitchell, H.L., C. Charette, R. Hogue and S.J. Lambert, 1996: Impact of a revised analysis algorithm on an operational data assimilation system. *Mon Wea. Rev.*, **124**, 1243-1255.

Rodgers, C.D., 2000: *Inverse methods for atmospheric sounding: theory and practice.* Series on atmospheric, oceanic and planetary physics, vol.2, World Scientific Ed., New York, 238 pages.

STATISTICAL ASSIMILATION OF SATELLITE DATA: METHOD, ALGORITHMS, EXAMPLES

O.M. POKROVSKY
Main Geophysical Observatory
Karbyshev str.7, St.Petersburg, 194021, Russia

1. Introduction

There were several reasons that caused an intensive development of the 4-D methods for the assimilation of remote sensing data in meteorology. One reason is that the satellite data are continuous in time, in contrast to conventional network observations carried out at prescribed standard terms. But the most part of remote sensing data arrives just between such terms. Another reason is that the remotely sensed meteorological parameters have to be derived from the solution of the ill-posed inverse problems. This last occasion is of special attention, because instead of conventional direct and, therefore, point-wise parameter measurements, in the case of remote sensing we have to deal with some functional of spatial field for this parameter. For example, in the case of atmospheric thermal remote sensing we cannot to retrieve temperature magnitudes at some vertical levels or at some spatial points, but rather averaged values related to some not fully certain weight functions. The third reason is that, actually, the operator of the inverse remote sensing problem does not maintain some constant magnitudes in spatial and temporal coordinates, but, really, it is subjected by disturbances originated from permanent changes occurred in atmospheric optical properties.

The first reason is the most familiar to investigators involved in the 4-D assimilation studies. The two others are less familiar to community and needed to be considered. We will discuss these subjects in three consecutive chapters. The aim of the first chapter is to describe an approach based on the statement and solution of the 4-D inverse problem provided the continuous spatial-temporal analysis of the meteorological fields which is supplied by the radiance data coming from satellite sensors. We illustrate our method facilities by examples of temperature and height field analysis over various areas and based on remote sensing data acquired from different satellite sensors. Despite the absence of the forecasting hydrodynamic model in our scheme, we came to 4-D assimilation method in KBF formulation by means of special transformations of covariance structure functions of the meteorological field treated as a stochastic field.

R. Swinbank et al. (eds.), Data Assimilation for the Earth System, 189–200.
© 2003 *Kluwer Academic Publishers. Printed in the Netherlands.*

2. Statement of the Problem

We assume that the direct data are originated from conventional sounding observations and the indirect data acquired from satellite remote sensing measurements. We suppose that both direct and indirect data are attributed by an arbitrary set of spatial points and temporal instances. It is known that the inverse problem solution can be formulated by means of algebraic system:

$$y = A \cdot x + \varepsilon . \tag{2.1}$$

Here y is a vector of radiance measured by satellite sensor and x is a vector of meteorological parameter vertical profile values. Mostly common known is a problem of thermal remote sensing. In this case x is the vector of temperature vertical profile. In (2.1) A is a radiance transfer operator describing the relationships between the temperature variations and the related radiance fluctuations. Equation (2.1) is valid for an arbitrary spatial point (pixel). Our goal is to obtain the estimates of meteorological parameters at some regular grid of spatial area G. Therefore, let us consider a set of M direct and N indirect measurement performed within the area G. We assume that these measurements are discrete and attributed by some fixed time instant. Therefore, we will consider models of the continuous and discrete observing systems. In general, equation (2.1) components depend on the time coordinate t. So, (2.1) may be rewritten as

$$y(t) = A(t) \cdot x(t) + \varepsilon(t), \tag{2.2}$$

where $A(t)$ is a matrix of $m \times n$ dimension, $y(t)$ and $x(t)$ are vectors of m and n size, respectively. For simplicity in formulas scripting, but without any losses in generality, we assume that the mean values of $y(t), x(t), \varepsilon(t)$ in (2.2) are zeroes. The structure of stochastic field X is completely determined by its first and second moments:

$$\bar{x}(t) = \bar{x}(z(t), t); \quad \Sigma_x\left(t_1, t_2\right) = \Sigma_x\left[z_s\left(t_1\right), t_1, z_s\left(t_2\right), t_2\right], \tag{2.3}$$

where $z_s = z_s(t)$ are spatial coordinates of a pixel observing by a space borne sensor at time t. The corresponding radiance field moments are described by the relationships:

$$\bar{y}(t) = A(t)\bar{x}(t), \quad \Sigma_Y\left(t_1, t_2\right) = A\left(t_1\right)\Sigma_x\left(t_1, t_2\right)A^T\left(t_2\right) + \Sigma_\varepsilon\left(t_1, t_2\right). \tag{2.4}$$

Let us formulate the main problem. We assume that the measurements are available on the time interval $[t_0, T]$. It is necessary to obtain the best linear estimate of meteorological parameter x at arbitrary time instant $t=s$, having in disposal the measured function $y(t)$:

$$\bar{x}(s) = \int_{t_0}^{T} B(s,t)y(t)dt . \tag{2.5}$$

The undetermined matrix function $B(s,t)$ should be obtained from the condition of minimization of a functional related to the residual vector $\delta\tilde{x}(s) = x(s) - \tilde{x}(s)$. Assume that this functional is a variance of the vector. In this case by a simple transformation one can derive an expression for the covariance matrix of the residual vector:

$$
\Sigma_{\delta\tilde{x}} = \int_{t_0}^{T}\int_{t_0}^{T} B(s,t)\left\{A(t)\Sigma_x(t,\tau)A^T(\tau) + \Sigma_\varepsilon(t,\tau)\right\} \cdot B^T(s,\tau) dt\, d\tau +
$$

$$
+ \Sigma_x(s,s) - \int_{t_0}^{T} B(s,t)A(t)\Sigma_x(t,s) dt - \int_{t_0}^{T}\Sigma_x(s,t)A^T(t)B^T(s,t) dt . \quad (2.6)
$$

By using the standard Euler technique, one can come to the set of the Fredholm integral equations of the first kind:

$$
\int_{t_0}^{T} B(s,t)\left\{A(t)\Sigma_x(t,\tau)A^T(\tau) + \Sigma_\varepsilon(t,\tau)\right\} dt = \Sigma_x(s,\tau)A^T(\tau). \quad (2.7)
$$

This equation should be solved with respect to the unknown weight matrix function $B(s,t)$ on the time interval $[t_0, T]$. The relationship (2.7) is known as the Wiener-Hopf filter (WHF) equation (Wiener, 1949). This equation admits an analytical solution only in stationary case, when Fourier analysis technique is applicable. In general, there are no efficient methods for solving this problem. One simplified assumption may be accepted in this case. It is due to the discrete kind of any information arriving from sensors. In the case of remote sensing it is a pixel-wise of input data. This assumption allows us to obtain a solution in the general case of non-stationary field X. It is necessary to note that, from the other hand, there is another way for the WHF problem solution, but under severe assumption of the Markov property of stochastic processes $x(t)$ (Bharucha-Reid, 1960). This property assumes that the estimated process at any given time instant depends only on its distribution at one, latest of the preceding time instants. This assumption leads to the Kalman-Bucy filter (KBF) (Kalman and Bucy, 1961) and provides an estimation problem solution for consecutive time instants. In fact, it is very restrictive assumption with respect to the atmospheric processes closely linked both in time and in space. For example, it is unwise to assume that for the sequence of temperature or pressure fields, which are strongly correlated on the time interval of 1 hour, the 2, 3, 4-th hour correlations might be neglected.

3. Method

Under the assumptions made about the covariance function of measured stochastic process $y(t)$, the relationship (2.4) is valid (Kuznetsov et al., 1965). An expression for cross-covariance function of $x(t)$ and $y(t)$ may be written in the form:

$$
\Sigma_{xy}(s,\tau) = \Sigma_x(s,\tau)\cdot A^T(\tau). \quad (3.1)
$$

We assume that the observations are performed at the consecutive time instances t_1, t_2, ..., t_N $(t_0 \leq t_1 \leq t_2, ..., \leq t_N \leq T)$ and employ the system (2.7). Let us introduce the following designations: $B_p(s)=B(s, t_p)$; $\Sigma_y(p,r)=\Sigma_y(t_p, t_r)$, $\Sigma_{xy}(s,r)=\Sigma_{xy}(s,t_r)$ $(p, r = 1, ..., N)$. This allows us to rewrite (2.7) as a of set of the following N algebraic equations:

$$\sum_{p=1}^{N} B_p(s) \cdot \Sigma_y(p,r) = \Sigma_{xy}(s,r), \quad (r = 1, ..., N) \tag{3.2}$$

with respect to the unknown matrix $B_p(s)$ $(p=1, ..., N)$. Each of them is of $n \times m$ size. The covariance matrices $\Sigma_y(p,r)$ and $\Sigma_{xy}(s,r)$ are of $m \times m$ and $n \times m$ sizes, respectively. Therefore, system (3.2) has $N \times m \times n$ equations. Hence, the more measurement points N, the more size of the related system (3.2). The main problem in numerical solution of (3.2) is the size of the matrix to be inverted. Our aim is to show how to transform system (3.2) in matrix equation sequences of lower sizes without any restricting assumption (Pokrovsky, 1974). Our approach is based on the construction of the canonic basis $\{Y_i(t)\}(i = 1, 2, ..., N)$ for the stochastic process $y(t)$. Let us consider the first equation of (3.2). We start transformations by multiplying this equation $(r=1)$ from the right side by the matrix $[\Sigma_y(1,1)]^{-1}$. Let $Y_1(p)=\Sigma_y(p,1) \cdot [\Sigma_y(1,1)]^{-1}$, $C(s,1)=\Sigma_{xy}(s,1) \cdot [\Sigma_y(1,1)]^{-1}$, then one can rewrite this equation in form:

$$\sum_{p=1}^{N} B_p(s) \cdot Y_1(p) = C(s,1).$$

It follows from above that $Y_1(1) = I$ (I is the unit matrix). As a result of this step we obtained the first term $Y_1(p)$ of the canonic basis $\{Y_i(t)\}$. Now it is necessary to perform in both sides of (3.2) a term by a term subtraction of the following expression:

$$\left(\sum_{p=1}^{N} B_p(s) \cdot Y_1(p)\right) \cdot \Sigma_y(1,r).$$

We come to the transformed equations of system (3.2) for the numbers $r = 2, 3, ..., N$:

$$\sum_{p=1}^{N} B_p(s) \cdot \Sigma_y(p,r/1) = \Sigma_{xy}(s,r/1), \tag{3.3}$$

where

$$\Sigma_y(p,r/1) = \Sigma_y(p,r) - Y_1(p)\cdot\Sigma_y(1,r),$$

$$\Sigma_{xy}(s,r/1) = \Sigma_{xy}(s,r) - \Sigma_{xy}(s,1)\cdot\Sigma_y^{-1}(1,1)\cdot\Sigma_y(1,r).$$

By multiplying the both sides of (3.3) for r=2 by $\left[\Sigma_y(2,2/1)\right]^{-1}$, we come to the relationship: $\sum\limits_{p=1}^{N} B_p(s)Y_2(p) = C(s,2)$, where $C(s,2) = \Sigma_{xy}(s,2/1)\left[\Sigma_y(2,2/1)\right]^{-1}$. It is necessary to note that for the resolved matrix function $Y_2(p) = \Sigma_y(p,2/1)\left[\Sigma_y(2,2/1)\right]^{-1}$ the equalities $Y_2(1) = 0$, $Y_2(2) = I$ (0 is the zero matrix) are valid. As a result of this step we obtained the second term $Y_2(p)$ of the canonic basis $\{Y_i(t)\}$. At the third step we are to subtract the expression $\left(\sum\limits_{p=1}^{N} B_p(s)Y_2(p)\right)\Sigma_y(2,r/1)$ from both sides of (3.3) for r=3,...,N and come to the transformed system written in the form:

$$\sum_{p=1}^{N} B_p(s)\Sigma_y(p,r/2) = \Sigma_{xy}(s,r/2), \tag{3.4}$$

where designations $\Sigma_y(p,r/2) = \Sigma_y(p,r/1) - \Sigma_y(p,2/1)\left[\Sigma_y(2,2/1)\right]^{-1}\Sigma_y(2,r/1)$ and $\Sigma_{xy}(s,r/2) = \Sigma_{xy}(s,r/1) - \Sigma_{xy}(s,2/1)\left[\Sigma_y(2,2/1)\right]^{-1}\Sigma_y(2,r/1)$ are used. Let us multiply (3.4) for r=3 at the right hand by $\left[\Sigma_y(3,3/2)\right]^{-1}$ and come to a new transformed system:

$$\sum_{p=1}^{N} B_p(s)Y_3(p) = C(s,3), \tag{3.5}$$

where $C(s,3) = \Sigma_{xy}(s,3/2)\left[\Sigma_y(3,3/2)\right]^{-1}$ and $Y_3(p) = \Sigma_y(p,3/2)\left[\Sigma_y(3,3/2)\right]^{-1}$. It is easy to find out that the following equalities are valid $Y_3(1) = Y_3(2) = 0$, $Y_3(3) = I$. The latter proved that we built the third member of the canonic bases $\{Y_i(t)\}$ for the stochastic process y(t). One can obtain a sequence of matrix functions $Y_r(p)$ by the described algorithm. From the algebraic point of view we transformed a block system of equations (3.2) to lower triangular form. Therefore, at the last step (r=N) one can obtain the simple equation: $B_N(s) = C(s,N)$, since $Y_N(k) = 0$ for k<N and $Y_N(N) = I$. This is a starting point for backward steps of the algorithm. It needs to get all matrix operators $B_r(s)$ (r =1,..,N). In fact, using latter equation and taking into account the equations

$Y_{N-1}(N-1) = I$ and $Y_{N-1}(p) = 0$ for $p < N-1$, one can obtain the second equation for $B_r(s)$: $B_{N-1}(s) = C(s, N-1) - C(s, N)Y_{N-1}(N)$. It is easy to continue this backward process and obtain all the matrix operators $B_r(s)$. Therefore, the matrix block $B(s) = (B_1(s), \ldots, B_N(s))$ is built and the problem solution is completed.

The main advantage of the algorithm just described is that, instead of the inversion of $N \times m \times m \times N$ matrix, it provides the step-wise procedure using the inversion of $m \times m$ matrix at each of N steps. It is important in cases when N achieves a considerable value. Another remark concerns the stability of the inversion process. Implementation of step-wise procedure provides some improvement in stability of inversion process, due to reducing in dimension. Another essential factor is the presence of the additive term $\Sigma_\varepsilon(t)$ in the matrix $\Sigma_y(1,1) = A(1)\Sigma_x(1,1) \cdot A^T(1) + \Sigma_\varepsilon(1)$ to be inverted at the first step of the forward pass. It is necessary to note that this regularized term is presented at each step of the described recurrent procedure. For example, at the second step the matrix $\Sigma_y(2, 2/1)$ to be inverted contains the term: $\Sigma_y(2, 2/1) = \left[A(2) \cdot \Sigma_x(2,2) \cdot A^T(2) - \Sigma_y(2,1) \cdot \Sigma_y^{-1}(1,1) \cdot \Sigma_y(1,2) + \Sigma_\varepsilon(2) \right]$.

4. Algorithms

There are two cases of interest to be considered: $N=1$ and $N=2$. These cases are linked to some other known methods. When $N=1$, we come to the method of statistical regularization which provides a solution of the inverse problem for remote sensing at the single point to perform a vertical sounding of the atmosphere (Pokrovsky, 1984). The system (3.2) gets a simple form of the single matrix equation: $B_1(s) \cdot \Sigma_y(1,1) = \Sigma_{xy}(s,1)$.

That leads to the solution:

$$B_1(s) = \Sigma_x(s,1)A^T \left(A\Sigma_x(1,1)A^T + \Sigma_\varepsilon(1) \right)^{-1}. \tag{4.1}$$

When $s = t$, (13) provides a solution of the inverse problem for this point. Otherwise, (4.1) gives an extrapolation formula at the arbitrary point $s \neq t$, which does not coincide with an observing point. Another important case is $N=2$, when measurements are carried out at two consecutive time instants and should be linked and adjusted. Our aim is to show how to come to the KBF formulas in this particular case. The reason is that the KBF is just intended for implication in such cases. In fact, the discrete KBF is formed by a system of matrix equations:

$$x_{k+1} = D_k x_k + \eta_k \ , \ y_k = A_k x_k + \varepsilon_k \quad (k = 1,2,\ldots), \tag{4.2}$$

where the former is a state equation and the latter is an equation of the observing system. In (4.2) D_k is a transition (or propagation) matrix and η_k is a model noise, which is assumed to be statistically independent for different values of k. Also, it is assumed that

$\overline{\eta}_k = 0$ and η_k is independent of x_k. Let us denote the covariance matrix by $\Sigma_x(k) = \Sigma_x(t_k)$, $(k = 1,2,...)$. By taking into account the known results on Gauss conditional expectation, one can get a simple stochastic equation for a forecasting value of x_{k+1}, based on x_k:

$$x_{k+1} = \Sigma_x(k+1,k)\Sigma_x^{-1}(k)x_k + \eta_k, \tag{4.3}$$

where η_k is a random vector attributed by zero mean values and covariance matrix:

$$\Sigma_\eta(k) = \Sigma_x(k+1/k) = \Sigma_x(k+1) - \Sigma_x(k+1,k)\Sigma_x^{-1}(k)\Sigma_x^T(k+1,k). \tag{4.4}$$

System (4.3) is satisfied by all conditions of the KBF state equation (4.2). Really, one can get that $E\left(x_k\eta_k^T\right) = 0$, $(k = 1,2,...)$. Also, if we introduce usual designation $x_{k+2/k+1} = D_{k+1}x_{k+1}$, one can obtain $x_{k+2} = \eta_{k+1} + x_{k+2/k+1}$ from (4.2). Then, $E\left(\eta_{k+1}x_{k+2/k+1}^T\right) = 0$. Hence, $E\left(\eta_{k+1},\eta_k^T\right) = 0$ is valid. Therefore, the KBF recurrent estimates may be applied to the system of two equations: (4.3) and the second equation taken from system (4.2). In this case the solution of the problem may be described by the recurrent two-step procedure: prediction $x_{k+1/k}$ and analysis $x_{k+1/k+1}$ carried out by update new observations y_{k+1}:

$$x_{k+1/k+1} = (I - R_{k+1}A_{k+1})x_{k+1/k} + R_{k+1}y_{k+1},$$
$$x_{k+1/k} = D_k x_{k/k-1} + D_k R_k\left(y_k - A_k x_{k/k-1}\right),$$
$$R_k = \Sigma_x(k/k-1)A_k^T\left[A_k\Sigma_x(k/k-1)A_k^T + \Sigma_\varepsilon(k)\right]^{-1} \tag{4.5}$$

with the initial condition $x_{0/-1} = \overline{x}(t_0)$. Also, it is necessary to add a recurrent equation for the conditional covariance matrix $\Sigma_x(k+1/k) = D_k\Sigma_k(k/k)D_k^T + \Sigma_\eta(k)$, where $\Sigma_x(k/k) = \Sigma_x(k/k-1) - \Sigma_x(k/k-1)A_k^T\left(A_k\Sigma_x(k/k-1)A_k^T - \Sigma_\varepsilon(k)\right)^{-1}A_k\Sigma_x(k/k-1)$ with the initial condition $\Sigma_x(0/-1) = \Sigma_x(t_0)$.

Now we consider how to derive the optimal filter for $N=2$ in the framework of our approach and compare the result with the KBF formulas obtained above. Having in mind (16), let us introduce a conditional covariance matrix for the observing data:

$$\Sigma_y(k+1/k) = A_{k+1}\Sigma_x(k+1/k)A_{k+1}^T + \Sigma_\varepsilon(k+1).$$

Assume that $N=2$, $t = t_k$, t_{k+1}, $s = t_{k+1}$. By using (3.1)-(3.5), after simple transformations one can get a set of expressions for the filter operators:

$$B_k(k + 1) = \Sigma_x(k + 1/k)A_{k+1}^T\left(A_{k+1}\Sigma_x(k + 1/k)A_{k+1}^T + \Sigma_\varepsilon(k + 1)\right)^{-1}, \tag{4.6}$$

$$B_{k+1}(k + 1) = \left[\Sigma_{xy}(k + 1, k) - \Sigma_{xy}(k + 1, k + 1)\Sigma_y^{-1}(k + 1/k)\Sigma_y(k + 1, k)\right]\Sigma_y^{-1}(k, k). \tag{4.7}$$

Formula (4.6) gives an exact analog of the representation (4.5) for R_{k+1}. If we neglect the second term in the right-hand side of (4.7), it is easy to come to a formula for the predictor $D_k R_k$. It is necessary to note that the second term in the right-hand side of (4.7) is responsible for a correction in the cross-covariance matrix $\Sigma_{xy}(k + 1, k)$, due to the new measurement data $y(t_{k+1})$. So, even in the case $N=2$ our approach is more flexible to innovated measurement information than the KBF and allows us to obtain more correct estimates. A question arises: how does our result differ form the KBF recurrent estimates? The answer is that this difference is determined by neglecting all the non-diagonal and the first upper-diagonal blocks in the block matrix $\Sigma_y(p, r)(p, r = 1, \ldots, N)$. Here we mean matrices $\Sigma_y(r, r)$ and $\Sigma_y(r, r + 1)$. Therefore, a difference in results should be considerable.

5. Illustrative Examples

Atmospheric thermal remote sensing was the main area for application of the proposed method (for details see (Belyavsky *et al.*, 1983; Pokrovsky and Ivanykin, 1978; Pokrovsky, 1984). We carried out several series of experiments with the satellite data obtained from the spectrometers SIRS, ITPR and IRIS boarded at different platforms: Nimbus-5, Nimbus-6 and Meteor over areas supplied by collocated radiosonde data to evaluate this method and to compare it with other approaches. The major subjects of study were the temperature and height fields at standard atmospheric levels as well as their vertical profiles. Let us start from the evaluation of the temperature vertical profile retrieval by using the above filter of different orders (values of N). Vertical profiles of the discrepancy of the retrieved temperature values from the collocated radiosonde data for Central Europe are presented in Fig.1. The left side of Fig.1 figures out two layers of high discrepancy values: tropopause and boundary layer. It is a known feature of the retrieval by the 1-D inversion method. These data were distributed by means of the international WMO network and called the SATEM code. The results of the same retrieval, but obtained by the proposed assimilation method, are presented in the left side of Fig.1.

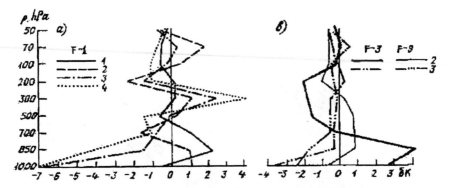

Figure 1. Deviations of retrieved temperature profiles (Nimbus-5 orbit over Central Europe, 10.05.1974) from collocated radiosonde data (1 – 35880, 2 - 27595, 3- 27196, 4 - 27037): a) SATEM retrievals, b) estimates provided by assimilation methods for filter operators of the first F-1, third F-3, and ninth F-9 orders.

This shows that the retrieval errors become less. The tropopause maximum noticed for the SATEM data has disappeared. Near the surface the maximum is reduced two times. It is easy to notice that the higher the order of the filter, the smaller the retrieval errors. It turned out that the most efficient is the filter of the order $N=9$. At least, Fig.1 demonstrates a monotonous decreasing of the deviation scale as far as the filter number N is increased. It is strong horizontal and vertical atmospheric temperature correlation that is the major cause of the proposed filter method efficiency (Pokrovsky and Ivanykin, 1978). In fact, as it was told above, the optical features of the atmosphere are subjected to substantial disturbances originated from natural fluctuations of its absorption compounds: water vapour, ozone, aerosol etc. (also see Pokrovsky, 1984). It leads to noticeable variation in values of the operator A entered in all the above filter equations. But, if one retrieve the vertical profiles for each spatial point in independent mode, fluctuations in A would give a local effect in the temperature estimates.

Therefore, as it is shown in Fig.2 (curve 3), the temperature horizontal profile (i.e., the temperature field cross-section obtained along the satellite orbit sub-track) is fluctuated in accordance with the corresponding atmospheric transmittance function variation. In contrast to this finding, a horizontal temperature profile derived by application of the assimilation filter (curve 2) looks more similar to the reference objective analysis (curve 1) acquired for the same area. Small and medium scale temperature disturbances originated from the SATEM data might become a cause of the instability after their implication in the numerical forecasting. In fact, our experiments prove that the filtered fields derived by the assimilation method do not contain any high frequency oscillations. This important conclusion is confirmed by the results of a similar experiment carried out for the data obtained from the Soviet satellite "Meteor" over Great Britain. In this case the outgoing radiance was measured by the Fourier spectrometer, i.e. by the instrument of other type than was at "Nimbus-5" board. This result proves the same relationship between the SATEM and assimilated temperature fields. We have considered cross-sections of horizontal temperature fields at standard levels. Now let us come to investigation of the field analysis maps (Fig.3). We did not consider the SATEM data in spatial analysis because of low quality results which might

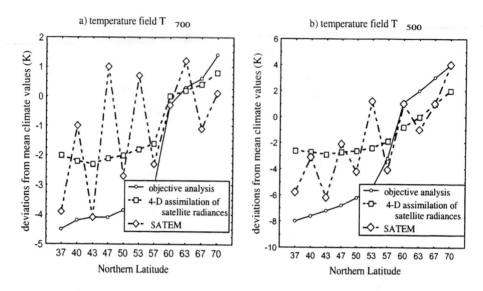

Figure 2. Latitude cross-section of anomaly fields of temperature for "Nimbus-5" satellite orbit at 10.05.74 over Central Europe: a) 700 mb surface, b) 500 mb surface.
Notations: 1- radiosonde data objective analysis, 2- direct assimilation of radiance data in 3D analysis, 3 – SATEM

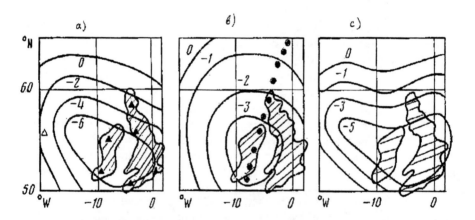

Figure 3. Analysis of H500 field anomaly for "Meteor" satellite orbit at 22.08.77 over Great Britain:
a) radiosonde data (triangles indicate site locations), b) data from satellite orbit
(circles indicate measurement points), c) joint data set

be expected after the above consideration. In this case we used the corrected transmittance functions for all channels of satellite instrument (for details, see other papers in this book). That allows us to obtain an anomaly height field (Fig.3b) which is a very similar to the results of the objective analysis based on the radiosonde data (Fig.3a). In general, the low atmospheric pressure anomaly, obtained by using only the radiance data, agrees very well with the synoptic analysis. However, the actual anomaly is to the west, and only by using the both data sets is this feature obtained (Fig.3c). After the comparison of Fig.3a,b,c, one can come to the conclusion that the assimilation scheme produces the information impact of different kinds of data which is consistent with its accuracy. Fig.3c demonstrates that the radiosonde data dominates in the joint analysis case. The above case corresponds to the radiance measurements in clear sky conditions.

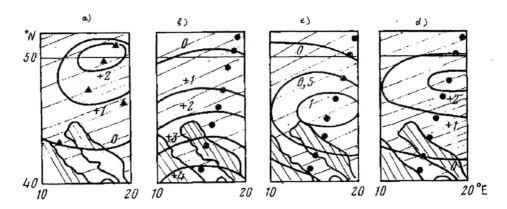

Figure 4. Analysis of T$_{500}$ field anomaly for "Meteor" satellite orbit at 22.08.77 over Central Europe: a) radiosonde data (triangles indicate site locations), b) data from satellite orbit (circles indicate measurement points): 8 points, c) 5 points, d) 6 points.

The cloudiness is a disturbing factor of major importance in remote sensing. In Fig.4 we consider cloudiness situation. A reference field of T$_{500}$ (Fig.4a) computed by radiosonde data shows a positive temperature anomaly, while the field computed by all available radiance data (Fig.4b) indicates a false southward temperature anomaly originated from the cloudiness impact. The withdrawal of radiance data attributed 3 southern orbit points obscured by cloudiness allows us to come to a more realistic field (Fig.4c, 4d).

References

Belyavsky, A.I., Pokrovsky, O.M., Spankuch, D., and Guldner, J. (1983) Numerical analysis of temperature and geopotential fields from satellite radiometric Measurements, *Izvestiya, Acad.of Sciences, USSR, Atmos. and Ocean. Physics* **19**, 870-877.

Bharucha-Reid, A.T. (1960) *Elements of the Theory of Markov Processes and Their Applications*, Mc-Graw Hill, N.Y.

Kalman, R.E. and Bucy, R.S. (1961) New results in linear filtering and prediction theory, *Trans.ASME*, ser.D **83**, 95-108.

Kuznetsov, P.I., Stratonovich, L.A. and Tikhonov, V.I. (1965) *Non-Linear Transformations of Stochastic Processes*, Pergamon Press, N.Y.

Pokrovsky, O.M. (1974) An assimilation of conventional and satellite data in 3-D analysis of meteorological fields, *Soviet Meteorology and Hydrology*, **6**, 33-39.

Pokrovsky, O.M. (1974) An optimal 4-D assimilation of conventional and satellite data for meteorological field analysis, *Soviet Meteorology and Hydrology*, **8**, 29-36. (Allerton Press Inc.,NY).

Pokrovsky, O.M. and Ivanykin, E.E. (1978) Spatial analysis of temperature and height fields on the basis of data from remote sounding of the atmosphere, *Z. fur Meteorologie* **1**, 3-23.

Pokrovsky, O.M. (1984) *An Optimization of Meteorological Remote Sensing of Atmosphere from Satellites* Hydrometeoizdat, Leningrad.

Wiener, N. (1949) *The Extrapolation, Interpolation and Smoothing of Stationary Time Series*, J.Wiley, N.Y.

THEORETICAL IMPACT ASSESSMENT OF SATELLITE DATA ON WEATHER FORECASTS

O. M. POKROVSKY
Main Geophysical Observatory
Karbyshev str.7, St.Petersburg, 194021, Russia

1. Introduction

The global meteorological observing system is extremely expensive and in the present economical situation some conventional observations such as radiosondes begin to be severely reduced. At the same time improved satellite systems become available (Kondratyev *et al.*, 1996). The operational observing network, which uses both conventional and satellite measurements, influences the weather forecast accuracy through the initial atmospheric state uncertainty (Beliavsky and Pokrovsky, 1983; Ghil *et al.*, 1979; Pokrovsky, 1984; Pokrovsky, 2000). Therefore, there is an urgent necessity to investigate the importance of different observing subsystems on numerical weather forecasting performance (Epstein, 1969; Kondratyev *et al.*, 1996; Pokrovsky and Denisov, 1985).

In contrast to some other studies, based on observing system numerical experiments with operational forecast models, this work is based on the estimation of the covariance matrices of analysis and forecast errors. We suppose that the 4-D assimilation is a tool to combine observed and predicted values of the state variables of a system in an optimal way to produce the best possible estimate of the true state (Pokrovsky, 1984). This requires the knowledge of the corresponding covariance matrices of observational and short-term forecast errors. The former matrix is relatively easy to specify, since it depends primarily on the error characteristics of measuring instruments. However, there is a substantial difference between conventional and satellite measurement errors related to the presence of spatial correlation and cross-correlation components in the remote sensing data (Pokrovsky, 1984). Estimating the covariance matrix of the short-term prediction error is more difficult (Pokrovsky and Beliavsky, 1983).

It is well known that for a linear system the Kalman-Bucy filter (KBF) (Kalman and Bucy, 1961) provides the best linear estimates of the state of a system. A stochastic prediction approach in meteorology based on ideas of KBF was developed in the past and documented in (Denisov and Pokrovsky, 1980). A generalised approach for a non-linear meteorological model to be incorporated in KBF was proposed in (Pokrovsky and Beliavsky, 1983).

In this paper we present a methodology and description of some results. We introduce the conventional and satellite observing system impact indexes. The

R. Swinbank et al. (eds.), Data Assimilation for the Earth System, 201–211.

relationship between these indexes at stages of analysis and forecast is a central question of discussion.

2. Direct Sensitivity Method

The KBF formulation of 4-D assimilation has an evident advantage for the observing subsystem impact studies. Instead of routine investigation of a few meteorological fields, the KBF approach allows to investigate the arbitrary size ensemble of such fields described by the mean field and the covariance matrix. This approach would lead to the sensitivity estimates for different observing systems. It could become a background for optimal design of complex observing systems. There are two novel issues discussed below. The first one is that the quadratic term was incorporated in KBF. The second, the procedure for model stochastic term estimation was evaluated for FGGE data.

2.1. METHODOLOGY

We consider a standard approach based on two equations describing the state variable evolution of a system and the updating measurement information, respectively:

$$\begin{cases} dx(t)\big/_{dt} = \mathbf{A}(t) \cdot \mathbf{x}(t) + \mathbf{b} + \mathbf{C}\big[\mathbf{x} \cdot \mathbf{x}^T\big] + \mathbf{G} \cdot \mathbf{w}(t) \\ \mathbf{y}(t) = \mathbf{B}(t) \cdot \mathbf{x}(t) + \varepsilon(t). \end{cases} \tag{2.1}$$

Unlike the KBF formulation, we introduce a quadratic term in (1):

$$\mathbf{C}(\mathbf{P}) = \sum_{j,k} c_{ijk} \, p_{jk} \, . \tag{2.2}$$

This term (2.2) was chosen in accordance to the Lorenz minimal meteorological equations (Lorenz, 1960). We also assume the presence of a stochastic term in (2.1), which accounts for the lack of parametrization. For simplicity, we assume that it is similar to colour noise and described by: $\mathrm{cov}(\mathbf{w}(t), \mathbf{w}(\tau)) = \Sigma_{\mathbf{w}} \cdot \delta(t - \tau)$. The same assumption is made for observational error: $\mathrm{cov}(\varepsilon(t), \varepsilon(\tau)) = \Sigma_{\varepsilon} \cdot \delta(t - \tau)$ and the initial state uncertainty: $\mathrm{cov}(\mathbf{x}(t_0), \mathbf{x}(t_0)) = \Sigma_{\mathbf{x}}(t_0)$. The corresponding evolution equations for the trajectory of ensemble mean and error covariance matrix (see Pokrovsky and Beliavsky, 1983) are:

$$
\begin{cases}
\dfrac{d\hat{x}}{dt} = b + (A - \Sigma_x \cdot B^T \cdot \Sigma_\varepsilon^{-1} \cdot B)\hat{x} + C\left[\hat{x} \cdot \hat{x}^T\right] + C\left[\Sigma_x\right] + \Sigma_x \cdot A^T \cdot \Sigma_\varepsilon^{-1} \cdot y \\
\dfrac{d\Sigma_x}{dt} = (A + R)\Sigma_x + \Sigma_x (A + R)^T - \Sigma_x \cdot B^T \cdot \Sigma_\varepsilon^{-1} \cdot B \cdot \Sigma_x + G \cdot \Sigma_w \cdot G^T \\
\hat{x}(t_0) = \bar{x}_0; \quad \Sigma_x(t_0) = \Sigma_0.
\end{cases}
\tag{2.3}
$$

We use the standard notation T for the transpose matrix, $\bar{x}$ is a mean field. System (2.3) differs from the standard KBF equations (Kalman and Bucy, 1961) by the presence of C and R terms that are explained by the quadratic part in (2.1). We use the following notation:

$$
R(t) = \{r_{ij}(t)\} = \sum_k \bar{x}_k(t)\left(c_{ijk} + c_{ikj}\right).
$$

The covariance matrix evolution depends on the sign of the difference between the third and the forth terms, which are responsible for the observational system contribution and the model stochastic error influence. The third term is known as the Fischer information matrix. The fourth term describes the model error growth due to stochastic effects. The more measurement data impact contribution, the lower trend in covariance error matrix. It is easy to understand that the observing system contribution mainly depends on the error covariance matrix: its values and the structure determined by cross-correlation elements.

2.2. MODEL PERTURBATION STOCHASTIC TERM ESTIMATION

We have to estimate the model error mean μ_w and its covariance matrix Σ_w for the time interval from t_0 to T. For simplicity, we consider the case $G = I$. Let us suppose that we have both the observed and predicted state variable values for $t \in [t_0, T]$. Then it is possible to compute the differences $\Delta x^{(i)}(T) = x_{obs}^{(i)}(T) - x_{pred}^{(i)}(T)$ and the corresponding sample mean $\mu_{\Delta x}$ and the covariance matrix $\Sigma_{\Delta x}$. It had been shown that the unknown μ_w, Σ_w are determined from the system of equations

$$
\begin{cases}
\mu_{\Delta x} = A^{-1} \cdot \left(e^{AT} - I\right) \cdot \mu_w \\
\Sigma_{\Delta x} = e^{AT}\left\{\displaystyle\int_0^T e^{-At} \cdot \Sigma_w \cdot e^{-At}\, dt\right\} e^{A^T \cdot T}
\end{cases}
\tag{2.4}
$$

derived from the system of state equations (2.1).

3. Observing Data Impact Research

The KBF may be divided into an analysis step and a forecast step. In the analysis step, the state of the system is estimated as a combination of observed and forecast values, and the covariance matrix for the forecast errors is modified to give the covariance matrix of analysis error. The discrete recursive update procedure is presented below:

$$\mathbf{x}(i/i) = \mathbf{x}(i/i-1) + \mathbf{\Sigma_x}(i/i-1) \cdot \mathbf{B}^T \cdot \left[\mathbf{B} \cdot \mathbf{\Sigma_x}(i/i-1) \cdot \mathbf{B}^T + \mathbf{\Sigma_\varepsilon}\right]^{-1} \cdot \left[\mathbf{y}(i) - \mathbf{B} \cdot \mathbf{x}(i/i-1)\right], (3.1)$$

$$\mathbf{\Sigma_x}(i/i) = \mathbf{\Sigma_x}(i/i-1) - \mathbf{\Sigma_x}(i/i-1) \cdot \mathbf{B}^T \cdot \left[\mathbf{B} \cdot \mathbf{\Sigma_x}(i/i-1) \cdot \mathbf{B}^T + \mathbf{\Sigma_\varepsilon}\right]^{-1} \cdot \mathbf{B} \cdot \mathbf{\Sigma_x}(i/i-1). \quad (3.2)$$

To quantify the observing system impact we introduce some indexes of the covariance matrix $\mathbf{\Sigma_x}(t)$. These are several indexes describing the weather forecast features: $\xi(t) = \lambda_{\min}(\mathbf{\Sigma_x}(t))/\lambda_{\max}(\mathbf{\Sigma_x}(t))$ is the ill-conditioned index (ICI), $\eta(t) = tr(\mathbf{\Sigma_x}(t))$ is the root mean square (RMS) uncertainty, $L_S(t) = (\eta_A(t) - \eta_{A+S}(t))/\eta_A(t)$ describes the subsystem (S) data impact (SDI) with respect to the conventional (sounder (A)) network contribution.

At the forecast step, both the analysed mean state and the covariance matrix of analysis error are propagated forward in time (see (2.3)) to give the forecast state and covariance matrix for the next analysis cycle. To quantify the relaxation rate of updated information at the forecast step we introduce the ratio index $Q(t) = \eta(t)/\eta(t_0)$ which describes the relative error growth (REG).

In this study we use a simplified barotropic spectral model based on empirical orthogonal functions (EOF) of the height field H_{500} (Fechner, 1975; Rinne, 1971) in medium and high latitudes of Northern Hemisphere. This model includes non-linear terms and allows us to investigate their influence. On the other hand, a set of 30 EOF's provides a satisfactory level of approximation and the system of such a dimension can easily be integrated as well as the corresponding Ricatti matrix equation for the covariance matrix. It should be noted that this approach is an alternative to the well known Ensemble Prediction System (EPS) based on multiple runs from initial fields produced by a perturbation procedure built on a set of singular vectors of the covariance matrix $\mathbf{\Sigma_x}$. The EPS trajectory spread allows one to get only approximate information on distortion of uncertainty covariance matrix $\mathbf{\Sigma_x}$. An important advantage of our approach is its ability to provide a better tracking of the error covariance matrix ICI values during the forecasting cycles. But surely EPS is less demanding in computer time.

The results presented above could be considered as a background for the study results described below. We will consider two aspects of the impact problem: the evaluation of the impact of the initial field on short-term forecasting, and the 4-D assimilation scheme that assumes a continuous satellite and a discrete conventional observation updating.

3.1. INITIAL FIELD IMPACT

The aim of this study was to reveal the most important features of observing systems, which affect more the forecast REG (Pokrovsky, 2000). The spatial error correlation related to the satellite data and its main features were investigated in (Bartello and Mitchell, 1992; Beliavsky and Pokrovsky, 1983; Pokrovsky 1984). Our past contribution (Pokrovsky and Beliavsky, 1983) was aimed at finding the origins of a substantial increase in the error correlation (remote sensing retrieval error correlation and uneven location of ground-based network). This problem was investigated both at the analysis and forecasting stages. The Northern Hemisphere sounder network including 680 radiosonde stations is considered as a reference. Our task is to estimate the relative impact of additional observation subsystems. We model two regional sonde networks of 6 stations over data sparse areas (A1- near the Arctic coast of Siberia, A2- over the North Atlantic) in order to compare with a satellite measurement data contribution (Tab.1). We considered only the impact of thermal remote sensing data (TOVS type). The hypothetical Atlantic radiosonde network A1 (Pokrovsky and Denisov, 1985) provides three times less informative contribution than the satellite subsystem, but it existed really only in 1979 (FGGE period) or before. Moreover, the satellite impact value slightly increases with time while SDI (see above) values $L(t)$ for A1, A2 networks decrease. The investigation of the ICI values for the covariance matrix at analysis stage (Tab l.) clarifies the SDI dependence on the ICI values. The lower ICI values correspond to lower SDI patterns, and the higher ICI values correspond to an increase of the SDI patterns. The Arctic coastal network is six times less informative than the ship network. Generally, the satellite subsystem displays a clear forecasting impact advantage due to a more even network location. It is occurred despite an influence of the error spatial cross-correlation. On the other hand, the relationship between the informative contribution of coastal and ship networks indicates that each sounder station of the set A1 is six times less informative than that of A2.

Table 1. Supplement (to A) observational subsystem data impact $L(t)$, (%)

Supplement subsystem	ICI	Forecasting Day			
		0	1	2	3
S(satellite)	0.27	18.6	9.2	19.9	20.6
A1(coastal)	0.09	1.1	1.0	0.9	0.8
A2(ship)	0.13	6.3	6.1	5.4	5.1

This result urges us to continue our study. We trace the REG $Q(t)$ temporal patterns and its correspondence to the ICI values. Now our goal is to compare these characteristics for the reference network A and for the satellite subsystem S. The temporal evolutions of $Q(t)$ for A and S as well as the related ICI time trajectories are presented in Fig. 1 and Fig. 2. We assume that initial height fields are approximated by EOF expansions. The Root Mean Square (RMS) of the EOF approximation is equal to 15 m (Fechner, 1975). It is an estimation of the residual term obtained from a 30 EOF expansion. The first 10 EOF set is called the long-wave (LW) component, while the last 10 EOF set is called the short-wave (SW) component. The objective analysis RMS obtained for the system A is about 30 m at high and mid latitudes. The same value for the subsystem S lies in interval 60-80 m. Comparative analysis of Fig.1 and Fig.2 shows that the REG values do not depend on the initial field RMS but on the observational system ICI determined by the network configuration. The higher ICI, the lower REG. Therefore, an advantage of the satellite subsystem for weather forecasting is evident in

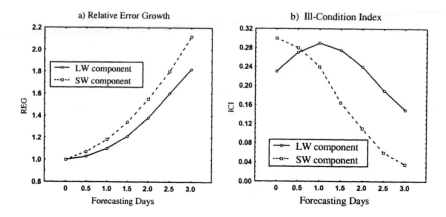

Figure 1. Relationship between forecast error growth and ICI for initial fields, derived from aerological network data

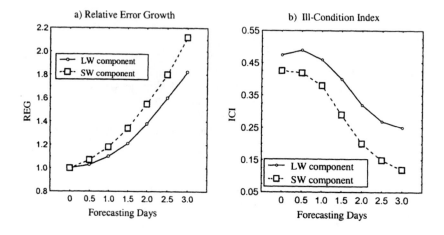

Figure 2. Relationship between forecast error growth and ICI for initial fields derived from radiosonde network data

sense of lower relative error growth. Furthermore, the REG values are lower for the LW component than for the SW. The consideration of the ICI values indicates the same relationships between the LW and SW components. The ICI values related with the LW component are always higher than that of the SW component. That implies faster acceleration of small-scale disturbances. Therefore, the LW components are forecasted with a lower error than the SW. As differences between Q_{LW} and Q_{SW} are similar for A and S observational systems we can come to the conclusion that the REG is mainly affected by the non-linear terms of equations (2.3).

3.2. CONTINUOUS ASSIMILATION IMPACT

An assimilation cycle includes two stages: the analysis and the forecast. We supposed that satellite data are assimilated once per hour and are coming from one or two polar orbiting satellites. Spatial and cross correlation patterns have been obtained by the technique described above (Pokrovsky, 1984). A generalised error correlation index (ECI) was introduced to investigate the error correlation effect on REG. For interpretation simplicity this ECI varies from 0 to 1. ECI zero value implies an absence of satellite retrieval error correlation. Our training study (Epstein, 1969) concluded that, in fact, ECI values responded the remote sensing data locates between 0.7 and 0.9. It was assumed that the conventional sounder information is assimilated twice a day at 00 Z and 12 Z. The covariance matrix for model disturbance term **Gw** was estimated from FGGE statistics for 1979 by technique (2.4). The continuous assimilation and the impact of renewing the input data led to changes in the REG time pattern behaviour. Here REG $Q(t)$ becomes a decreasing function in time. The REG sharply decreases after one day, especially around 0.5 day (see Tab.2). Then the REG reaches the asymptotic constant values which are the highest for A+S system, intermediate for A, and the lowest for satellite S+S (two satellites). This result means that the satellite information input is mostly effective in an assimilation scheme after several cycles. The impact values SDI for the remote sensing system S are decreasing also with time (Fig.3). The higher ECI values, the lower satellite data impact index SDI. The initial impact value (at $t=0$) mainly depends on the retrieval error. Our SDI estimates (13-18%) obtained from simulations are similar to the results of other authors (Ghil et al., 1979) based on real measurement data which could be explained by the completeness of main noise factors taken into account in our study. We derived the RMS error estimate 80 m and ECI value 0,7 at the analysis stage for S system (no ground-based data).

Table 2. REG values $Q(t)$ (at the stage of analysis)

Observing system	ECI	RMS (m)	Forecasting Day					
			0	0,5	1,0	1,5	2	3
A	0	12.9	1.0	0.68	0.62	0.61	0.60	0.59
A+S	0.7	11.5	1.0	0.70	0.65	0.64	0.63	0.63
S+S	0.9	21.0	1.0	0.64	0.56	0.53	0.52	0.52

208

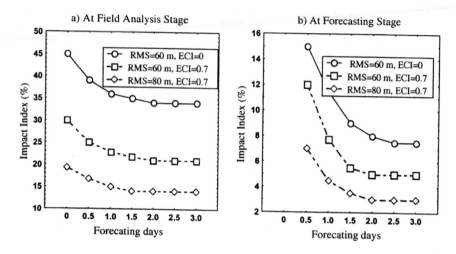

a) At Field Analysis Stage

b) At Forecasting Stage

Figure 3. Remote Sensing Impact Index

At the forecast stage the covariance error matrix increases more rapidly. This leads to a more moderate decrease of the REG function Q (t) (see Tab.3). The accumulated effect is kept above 0,85 for A and above 0,90 for A+S. But it goes down to 0,57 for S+S. This confirms our previous conclusion about a more effective satellite data impact after several assimilation cycles. Also the accumulation effect of the satellite data impact lasts longer than that of the ground-based data (Tab.2 and Tab.3) at both stages (analysis and forecast). The impact value SDI starts from a low initial value between 7 and 12% (Fig.3b). It could be explained by the additional effect of the error perturbation term $\mathbf{G \cdot w}$ propagated by the model.

Table 3. REG values $Q(t)$ at the stage of forecast

Observing system	ECI	RMS (m)	Forecasting Day					
			0,5	1,0	1,5	2,0	2,5	3,0
A	0	21.0	1.0	0.88	0.86	0.85	0.85	0.85
A+S	0.7	19.9	1.0	0.91	0.90	0.90	0.90	0.90
S+S	0.9	32.0	1.0	0.70	0.62	0.59	0.58	0.57

4. Adjoint Sensitivity Technique

The mathematical tools obtained by the introduction of the adjoint techniques to observing system impact studies, together with the mathematical clarity of Kalman filter formulation of data assimilation problem, provide the opportunity for significantly improving on understanding of peculiarities of different observing system data information content. This technique can easily include observations of different origins (radiosondes, surface and satellite data). In general, the calculation of adjoint linearized model equations involves the evaluation of more terms than in the original non-linear model and requires the increase in computer time by a factor of two at least. An assimilation cycle was equal to 6 hours. It was implemented in a three-day model forecast. Therefore, this approach should be rather used in a parallel way at low resolution (with simple physical parametrizations).

Taking into account the linear type of linearized state variable model and observation system (2.1), one can represent any linear functional of the solution x (t) at the moment of forecast $t = T$ by linear operations applied to the data of discrete and continuous observations [12]. This fact may be expressed by the relationship:

$$f(\mathbf{x}(t)) = \mathbf{\alpha}^T \cdot \mathbf{x}(T) = -\int_{t_0}^{T} \mathbf{p}^T(t) \cdot \mathbf{y}(t) dt + \mathbf{p}_0^T \cdot \mathbf{x}_0 . \tag{4.1}$$

We suppose that $\mathbf{\alpha}$ is a known vector, $\mathbf{p}$ (t) and p0 are a vector-function and a vector to be determined. The vector $\mathbf{p}_0^T$ describes the sensitivity of the forecasted variables to the accuracy of the data of the system of discrete observations $\mathbf{X}_0$ (conventional). Let $\mathbf{p}_0^T$ be the value of information (VOI) of discrete observations. The function $\mathbf{p}^T(t)$ in (4.1) describes the sensitivity of the predicted parameter $\mathbf{\alpha}^T \cdot \mathbf{x}(t)$ to the errors in the prescribed continuous (satellite) observation data y(t). Let us call $\mathbf{p}^T(t)$ the VOI for continuous information.

With a system of continuous observations defined by (2.1), one can overcome the limitations of the problem in the case of discrete systems of observations. In this case $\mathbf{B}(t) \neq 0$, $\mathbf{p}(t) \neq 0$ are valid. Let us consider an auxiliary adjoint system of inhomogeneous equations:

$$d\mathbf{x}^* \Big/ dt = -\left(\mathbf{A}^T + \mathbf{B}^T \cdot \Sigma_\varepsilon^{-1} \cdot \mathbf{B} \cdot \mathbf{S}\right) \cdot \mathbf{x}^*, \quad \mathbf{x}^*(T) = \mathbf{\alpha} . \tag{4.2}$$

Equation (4.2) additionally includes in its right side the observation system impact (second) term. To minimise the STD of estimate (4.1) under the constraint (2.1) it is necessary to use the technique of the canonical Euler-Lagrange equations. Upon some transformations (see (Pokrovsky, 2000)), we come to the solution of two equations. First is the Ricatti matrix equation which is integrated forward in time (Pokrovsky, 1984):

$$d\mathbf{S} \Big/ dt = \mathbf{S} \cdot \mathbf{A}^T + \mathbf{A} \cdot \mathbf{S} - \mathbf{S} \cdot \mathbf{B}^T \cdot \Sigma_\varepsilon^{-1} \cdot \mathbf{B} \cdot \mathbf{S} + \mathbf{G} \cdot \Sigma_\mathbf{W} \cdot \mathbf{G}^T, \quad \mathbf{S}(t_0) = \Sigma_{\mathbf{x}_0} . \tag{4.3}$$

When equation (3.1) is solved, the vector-function p (t) is determined from the equations:

$$\mathbf{p}_0 = \mathbf{x}^*(t_0), \quad \mathbf{p}(t) = \Sigma_\varepsilon^{-1} \cdot \mathbf{B} \cdot \mathbf{S}(t) \cdot \mathbf{x}^*(t) . \tag{4.4}$$

Thus, the problem of an optimal design of the continuous system of observations requires the solution of not only the direct and adjoint equations obtained from the classical method, but also of auxiliary equation (4.3) and (4.4).

5. Sensitivity Analysis

There are two versions of the linear functional $\mathbf{a}^T \mathbf{x}(T)$ for a model that provides a short-term forecast of the height field in the Northern Hemisphere (Beliavsky and Pokrovsky, 1983). The first version uses the input vector $\mathbf{a}$ related to the specified height field averaged over some spatial range. The second version $\mathbf{a} = \mathbf{a}_s$ assumes a determination of individual spectral components of the EOF expansion of the hemispherical geopotential field. When considering the problem of VOI computation aiming at forecasting the averaged height fields $\mathbf{a}^T \mathbf{x}(T)$, two types of models have been studied: (i) the deterministic dynamic model (1) in which $\mathbf{G} = 0$; (ii) the dynamic-stochastic model (2.1) in which the perturbation term $\mathbf{G}\cdot\mathbf{w}$ was obtained from the FGGE data (Beliavsky and Pokrovsky, 1983). For simplicity of the VOI analysis, $p(s, t)$ is presented as a product $p(s,t) = pr(t) \cdot ps(s,t)$ of the temporal change $pr(t)$ and spatial distribution $ps(s, t)$. Note, that the simplicity of such a presentation is determined by normalising the function $ps(s, t)$ in s for each value of t, according to the rule: max $p(s, t)$ = 1. The decrease of $pr(t)$ (from $t = T$ to $t = t_0$) illustrates the loss of the satellite information with time. In the case of a pure dynamical prognostic model the value of VOI for satellite data halves with time growing from zero to 96 hours. For a more realistic dynamic-stochastic model, the values of VOI $pr(t)$ decrease by a factor of 28 in the same time interval. During the first 24 hours the value of satellite information for the dynamic-stochastic model decreases by a factor of 1.5 to 2, after two days – by a factor of 3, after three days - by a factor of 7. Thus, a real inadequacy of the prognostic model determines the necessity to renew the satellite information every six hours. For this purpose a remote sensing system of at least two satellites is needed.

To determine the preferable zones of satellite data collection, we consider the spatial distribution of VOI, $ps(s, t)$. We shall discuss, first, the cases of the deterministic prognostic mode. For $t = 0$, the maximum VOI values are concentrated in the forecast region. This squared region located at the central part of European Russia (Moscow). An increase of the forecast range to two days leads to the appearance of great values of VOI in the areas corresponding to the satellite orbits over the North area of Atlantic Ocean. With $t = 72$ hours there is a need of satellite measurements over the Eastern Siberia. Considerable natural fluctuations of the height field (Siberian Anticyclone) and a sparse sounder network explain the choice of the region of Eastern Siberia. The VOI increases more in the Atlantic Ocean and less in the Pacific. The extension of t to two days determines a considerable growth of VOI over the Atlantic Ocean. In the range interval from 2 to 3 days the VOI starts increasing more uniformly in all the regions of the Northern Hemisphere which indicates the limitation of this method for designing a remote sensing observing network. Simultaneously with the computation of VOI, calculations were made of the sounder VOI, p_0. At 72 hours range one can point out two regions of extremely high values of sounder VOI: (i) North-eastern region of North

America, in the interval 60°-80°N; (ii) Eastern Siberia and Far East.

Our main goal was to consider a non-linear KBF and the related adjoint equation technique to develop a rigorous approach, in order to study the VOI determined for conventional and satellite observation systems. First, we showed a fundamental value of ICI (Ill Conditioned Index) in impact studies. We took into account most of the input information error sources and introduced a relevant short-term forecast model noise term obtained from a FGGE training data set. The specification of correlation patterns of remote sensing data provides some new findings of a main reason of its low information content (with respect to expected values). We introduced its correlation index which determines lower value of covariance matrix ICI for a joint conventional / remote-sensing observational system. This helps to obtain a more realistic impact estimate. The study clarifies also the reasons of the relatively low impact of the data from a second satellite and the importance of adding ships and coastal stations located in data sparse areas. The latter would supply a substantial improvement of covariance matrix ICI values. Another way of improving the remote sensing data quality is to perform a new optical correction procedure.

References

Bartello, P. and Mitchell, H.L. (1992) A continuous three-dimensional model of short-range forecast error covariance, *Tellus* **44A**, 217-235.

Beliavsky, A.I. and Pokrovsky, O.M. (1983) Contribution effect of the remote sensing data on the numerical analysis of geopotential fields, *Soviet Meteorology and Hydrology* 1 , 14-21.

Denisov, S.G. and Pokrovsky, O.M. (1980) Analysis of the information content of the system of the remote sensing of the height fields, *Proc. of Acad. of Sci. of USSR, ser.Phys.Atmos. and Ocean* **16**, 582-590.

Epstein, E.S. (1969) Stochastic dynamic prediction, *Tellus* **21**, 39-759.

Fechner, H. (1975) Darstellung der Geopotentials der Winterlichen Nordhalbkugel Durch Naturaliche Ortogonalfunctionen, Berichte Inst. fur Meereskunde, Kiel, N5.

Ghil, M., Halem, M. and Atlas, R. (1979) Time-continuous assimilation of remote sensing data and its effect on weather forecasting, *Mon.Weath. Rev.* **107**, 140-171.

Kalman, R.E. and Bucy, R.S. (1961) New results in linear filtering and prediction theory, *Trans.ASME, ser.D* **83**, 95-108.

Kondratyev, K.Y., Buznikov, A.A. and Pokrovsky, O.M.(1996) *Global Change and Remote Sensing*, J.Wiley and Sons, Praxis Publ, Chichester.

Lorenz, E.N. (1960) Maximum simplification of dynamic equations, *Tellus* **12**, 243-254.

Pokrovsky, O.M. and Beliavsky, A.I. (1983) On a 4-D assimilation of satellite information based on the non-linear model, *Soviet Meteorology and Hydrology* **8**, 44-54. (Allerton Press Inc.,NY).

Pokrovsky, O.M. and Denisov, S.G. (1985) The information content of the oceani network for an objective analysis of NH geopotential fields, *Soviet Meteorology and Hydrology* **10**, 37-43.

Pokrovsky, O.M. (1984) *An Optimization of Meteorological Remote Sensing of Atmosphere from Satellites*, Hydrometeoizdat, Leningrad. (in Russian).

Pokrovsky, O.M. (2000) Direct and adjoint sensitivity approach to impact assessment of ground-based and satellite data on Weather Forecasting. *Proceedings of the Second CGC/WMO Workshop on the Impact of Various Observing Systems on Numerical Weather Prediction.* World Weather Watch Technical Rep.N 19 (WMO/TD N1034), WMO, Geneva, 99-118.

Rinne, J. (1971) Investigation of the forecasting error of a simple barotropic model with the aid of empirical orthogonal functions, *Geophysica (Helsinki)* **11**, 11-32.

THE CORRELATION FEATURES OF THE INVERSE PROBLEM SOLUTION IN ATMOSPHERIC REMOTE SENSING

O. M. POKROVSKY

Main Geophysical Observatory
Karbyshev str.7, St.Petersburg, 194021, Russia

1. Introduction

There are two different approaches to retrieve the air temperature and humidity fields from remote sensing data. One approach is a conventional one and it based on 1-D inverse problem solution. This approach assumes the two-stage procedure: vertical profile retrieval from radiance data and optimal interpolation of retrieval values. Another approach is based on the irradiance data direct assimilation within the objective analysis of these fields. The first approach is more widely used because of the international distribution of SATEM data through the meteorological data network. But, unfortunately, the retrieved temperature and atmospheric constituents profiles contain highly correlated noise components that prevents the increase of the satellite data information content along with sensor sensitivity improvement.

Atmospheric parameter retrieval is based on the solution of the Fredholm integral equation of the first kind (Pokrovsky, 1984). The kernel of this equation is determined by so called atmospheric transmittance functions which are uncertain due to their dependence on such atmospheric absorbing constituents as water vapour, ozone, aerosol and some others. These constituent vertical profiles are varied both in space and time. Therefore, these profiles and transmittance functions are not known exactly. Hence, there are uncertainties in transmittance functions which generate additional errors affecting the inverse problem solution. The aim of this chapter is to trace these errors and their influence on the assimilation process in different methods.

2. The 1-D Retrieval Solution Correlation Features

The 1-D retrieval techniques provide the temperature and humidity vertical profiles in atmosphere. These profiles (well known as SATEM) contain the errors with specific correlation structure (Beliavsky and Pokrovsky, 1982; Schlatter and Branstator, 1978). Both spatial and temporal correlation is mainly due to large-scale perturbations of the atmosphere optical model. These perturbations are linked with fluctuation of aerosol, ozone and water vapor content and have different spatial scales. Large-scale perturbation is responsible, primarily, to fluctuations of aerosol and ozone constituents. Small-scale component of retrieval error is mainly determined by variability of water vapor and

R. Swinbank et al. (eds.), Data Assimilation for the Earth System, 213–223.

cloudiness impact (Schlatter and Branstator, 1978). An example displayed this kind of perturbation is presented in Fig.1. The comparison of temperature field cross-sections derived by back-ground objective analysis, standard 1-D retrieval procedure and 3D-satellite radiance assimilation method (Pokrovsky and Beliavsky, 1983, Beliavsky and Pokrovsky, 1982) demonstrates substantial improvement gained by removing the intermediate 1-D retrieval stage. But, there are some other error sources originated in ill-condition nature of remote sensing estimate (Pokrovsky and Ivanikin, 1978, Pokrovsky and Beliavsky, 1983). The equation of remote sensing system is described by means of approximate matrix equation of the first kind:

$$y = B \cdot x + \varepsilon.$$ (2.1)

It is well known that the optimal solution of (2.2) is described by the Gauss-Markov formula:

$$\hat{x} = \bar{x} + \Sigma_x \cdot B^T \cdot \left(B \cdot \Sigma_x \cdot B^T + \Sigma_\varepsilon \right)^{-1} \cdot \left(y - B \cdot \bar{x} \right).$$ (2.2)

We will illustrate our analysis of error correlation features of inverse problem solution by examples related to atmospheric temperature sounding in 15μm CO_2 absorption band. We performed a study for matrix B in (2.1) corresponding to the spectral band set of the spectrometer ITPR (Pokrovsky, 1984).

2.1. CROSS-COVARIANCE PATTERNS FOR TRUE PROFILES

The matrix described the cross-correlation between the true profile x and its departure δx from retrieved profile may be derived from (2.1) (see Beliavsky and Pokrovsky, 1982) and written as:

$$- \Sigma_{\delta x} = \Sigma_x - \Sigma_x \cdot B^T \cdot \left(B \cdot \Sigma_x \cdot B^T + \Sigma_\varepsilon \right)^{-1} \cdot B \cdot \Sigma_x$$ (2.3)

Formula (2.3) shows that the cross-correlation between true profile and estimation uncertainty is of negative sign. We suggest a simple explanation for this phenomenon. The vector of the true temperature values x belongs to an ellipsoid of probability distribution in multidimensional space described by the covariance matrix Σ_x. The regularised solution $\hat{x}$ (in (2.2)) is a more smooth function than the true profile x. Therefore, $\hat{x}$ is located more closely to the ellipsoid centre than x. Hence, the deviations $\hat{\delta} = \hat{x} - x$ and $x - \bar{x}$ are of opposite signs. Our computation shows that these cross-correlations reach considerable values: 0,7-0,8 and should be taken into account. Vertical distribution of cross-correlation for temperature sounding (Fig.1) is non-uniform. Maximal correlation values are attributed to tropopause level, where the remote sensing method is least reliable. Evident correlation increasing could be found at the land surface level in the case of cloudy atmosphere. This result leads to the conclusion that the less uncertainty of input data contribution, the less cross-correlation between the true profile **x** and its departure δ**x**. We should note that this uncertainty includes atmospheric optical model fluctuations which are due to cloudiness and principal absorber constituent content changes. Therefore, improvement of information on current optical atmospheric model is a basis for the remote sensing data accuracy increasing and its impact to analysis and forecast.

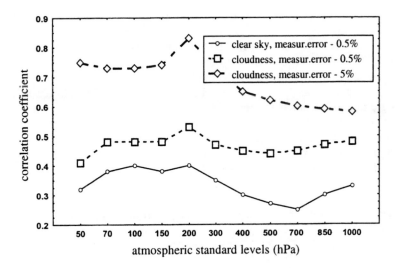

Figure 1. Cross-correlation between satellite temperature retrieve errors and corresponding true temperature values at standard levels

2.2. SPATIAL ERROR CROSS-CORRELATION

Spatial correlation of retrieval errors due to an arbitrary pair of points i and j has also a fundamental value and may be calculated from the expression (Beliavsky and Pokrovsky, 1982):

$$\Sigma_{\delta_{ij}} = \Sigma_{\delta_i} \cdot \Sigma_{x_i}^{-1} \cdot \Sigma_{x_{ij}} \cdot \Sigma_{x_j}^{-1} \cdot \Sigma_{\delta_j} . \tag{2.4}$$

The notations in (2.4) are evident. A matrix having one index is a covariance, and a two-index set matrix is a cross-covariance. This correlation component is of importance due to well-known strong spatial correlation of temperature and humidity fields. This function is monotonous decreasing of distance and its behaviour is similar to temperature correlation function. We performed some computations of horizontal correlation function dependence and found out that our theoretical estimates are in good agreement with empirical values obtained by Schlatter and Branstator (1978). Later, these results were confirmed in papers of many other investigators. The most substantial cross-correlation values are found out for remote sensing in cloudy conditions. These values exceed 0,5 threshold. At most studies the cross-correlation component (2.3) is neglected and spatial component (2.4) is taken into account only in few papers. Our studies had shown that error correlation components cause a rapid decreasing of covariance matrix ill-condition index (ICI) value (Pokrovsky, 1974a; Pokrovsky, 1974b). It turns out that ICI is closely linked to the forecast error growth.

2.3. ERROR CROSS-COVARIANCE FOR RETRIEVAL VERTICAL PROFILE

It should be noted that the above-described property is also valid for grid field estimates obtained by optimal interpolation and other objective analysis techniques. It could be easily explained by the fact of using the linear model (2.1). Furthermore, model (2.1) is also used in spectral analysis schemes (Pokrovsky, 1984). Therefore, these finding features are rather fundamental ones. Formula (2.4) as well might be used for deriving the cross-correlation between temperature estimations attributed by different vertical levels. It shows that such cross correlations should be similar to covariance matrix for temperature vertical profile. Computations prove that theoretical values agree with empirical data of Schlatter and Branstator (1978).

3. Correction Procedure for Atmospheric Transmittance Functions

3.1. TRANSMITTANCE FUNCTION RETRIEVAL

We start from physical background of the inverse problem solution provided the vertical profile retrieval from the remote sensing data. We suppose here the vertical profiles of temperature and main atmospheric constituents (water vapour, ozone, aerosol and small gaseous constituents). The integral form of the radiation transfer equation valid in infrared and microwave spectral bands:

$$I_{\Delta v} = J_v[T(p_s)]\tau_{\Delta v}(p_s) - \int_0^{p_s} J_v[T(p)] d\tau_{\Delta v}(p) \qquad (3.1)$$

is the main relationship related the outgoing radiance $I_{\Delta v}$ registered by a satellite remote sensor and vertical profile of temperature $T(p)$, which enters in the expression for the transmittance functions $\tau_{\Delta v}(p)$. Here p is an atmospheric pressure, p_s is a pressure at the Earth surface. Let us consider equation (3.1) in the variation form:

$$\delta I_{\Delta v} = \frac{\partial J_v}{\partial T}[T(p_s)]\tau_{\Delta v}(p_s)\delta T(p_s) - \int_0^{p_s} \frac{\partial J_v}{\partial T}[T(p)]\delta T(p) d\tau_{\Delta v}(p) + \int_0^{p_s} \delta\tau_{\Delta v}(p) dJ_v[T(p)] \quad (3.2)$$

in order to come to formulation of the inverse problem mentioned above. Here $\bar{T}(p)$ is the mean (climatic) temperature profile, $\bar{\tau}_{\Delta v}$ is the theoretical transmittance function for the spectral band Δv. Equation (3.2) allows us to make the insight in the factor set determined the radiance variations which are registered by a satellite borne sensor in both time and space coordinates.

Most of considered inverse problems are formulated for the IR range (5-50 μm). Here are principal absorbing atmospheric constituents: carbon dioxide, water vapour, ozone and aerosol. Carbon dioxide is considered as the uniformly mixed atmospheric compound and its variation is neglected. Therefore, the total transmittance function might be determined as a product of constituent transmittances:

$$\tau_{\Delta v} = \tau_{\Delta v}^{H_2O} \cdot \tau_{\Delta v}^{O_3} \cdot \tau_{\Delta v}^{CO_2} \cdot \tau_{\Delta v}^{A} .$$

Above we assume that the water vapour transmittance should be presented as a product of continuous and spectral line absorption $\tau_{\Delta v}^{H_2O} = \tau_{\Delta v}^{L} \cdot \tau_{\Delta v}^{c}$. Hence, finally we get the following expression for the total transmittance variation:

$$\delta\tau_{\Delta v} = \delta\tau_{\Delta v}^{H_2O} \cdot \tau_{\Delta v}^{O_3} \cdot \tau_{\Delta v}^{CO_2} \cdot \tau_{\Delta v}^{A} + \tau_{\Delta v}^{H_2O} \cdot \delta\tau_{\Delta v}^{O_3} \cdot \tau_{\Delta v}^{CO_2} \cdot \tau_{\Delta v}^{A} + \delta\tau_{\Delta v}^{A} \cdot \tau_{\Delta v}^{H_2O} \cdot \tau_{\Delta v}^{CO_2} \cdot \tau_{\Delta v}^{O_3} . \quad (3.3)$$

Therefore, considered inverse problem (3.1) is a four component inverse problem formulated with respect to vertical profiles of temperature variation δT and transmittance variations of $\delta\tau_{\Delta v}^{H_2O}$ -water vapour, $\delta\tau_{\Delta v}^{O_3}$ -ozone, $\delta\tau_{\Delta v}^{A}$ -aerosol. Let us introduce a general expression for the transmittance function

$$\delta\tau_{\Delta v}(\bar{u}) = \frac{\partial\tau_{\Delta v}}{\partial\bar{u}(p)} \frac{1}{g} \int_0^p \left(\frac{p'}{p_0}\right)^N \left(\frac{T}{T_0}\right)^M \delta q(p') dp' , \quad (3.4)$$

and for the related atmospheric optical path

$$\bar{u}(p) = \frac{1}{g} \int_0^p \left(\frac{p}{p_0}\right)^n \left(\frac{T}{T_0}\right)^m q(p) dp , \quad (3.5)$$

where m and n are empirical constants, T_0 and p_0 temperature and pressure assigned to homogeneous atmospheric optical path, $q(p)$ the vertical profile of atmospheric constituent in search. Equations (3.1)-(3.5) may be used to retrieve the unknown profiles $q(p)$. But our main target is to retrieve the temperature profile δT. Therefore, we need to reduce the transmittance uncertainty magnitudes $\delta\tau_{\Delta v}$ for each spectral channel Δv in operational mode. To develop an algorithm for $\delta\tau_{\Delta v}$ evaluation let us rewrite (3.2) in vector form:

$$\delta I_{\Delta v} = \mathbf{a}_{\Delta \tau}^{\mathrm{T}} \cdot \delta \mathbf{T} + \mathbf{b}_v^{\mathrm{T}} \cdot \delta \tau_{\Delta v} . \tag{3.6}$$

The vectors entered in (3.6) have the forms: $\delta \mathbf{T} = \left(\delta T(p_1), \ldots, \delta T(p_s) \right)^T$, $\delta \tau_{\Delta v} = \left(\delta \tau_{\Delta v}(p_1), \ldots, \delta \tau_{\Delta v}(p_s) \right)^T$. Equation (3.6) is a background for the two-component inverse problem statement. A standard formulation employs the assumption that $\delta \tau_{\Delta v} = 0$. But, actually, it is not the case. So, we have to state a problem of the variation $\delta \tau_{\Delta v}$ retrieval from (3.6). In this case we should assume that the temperature profile is known from additional more reliable source, e.g. radiosonde data. It means that $\delta T = 0$. So, we came to the inverse problem of the second kind:

$$\widetilde{y}_{\Delta v} = \mathbf{b}_v^{\mathrm{T}} \cdot \delta \tau_{\Delta v} + \widetilde{\varepsilon}_{\Delta v} \tag{3.7}$$

which is stated with respect to $\delta \tau_{\Delta v}$. The noisy term $\widetilde{\varepsilon}_{\Delta v}$ in (3.7) represents a sum of satellite radiometric and radio sounding errors. We assume that these noise components are statistically independent and get the transmittance variation estimates in the form:

$$\delta \widehat{\tau}_{\Delta v} = \Sigma_\tau \cdot \mathbf{b}_v \cdot \widetilde{y}_{\Delta v} \Big/ \left(\mathbf{b}_v^{\mathrm{T}} \cdot \Sigma_\tau \cdot \mathbf{b}_v + \sigma_{\widetilde{\varepsilon}}^2 (\Delta v) \right). \tag{3.8}$$

Here Σ_τ is a covariance matrix of the transmittance variation $\delta \tau_{\Delta v}$, $\sigma_{\varepsilon}^2 (\Delta v)$ the variance of the sum noise component $\widetilde{\varepsilon}_{\Delta v}$. So, one can perform the transmittance function correction having a pair of collocated measurements: radiance δI_v and radio sounding $\delta T(p)$. This method is recommend to be applied under the cloudless condition. In the case of several collocated pairs are to be available the recurrent version of formula (3.8) might be implemented. The recurrent technique is more relevant to operational mode when an information flow but not a single measurement is available.

We suggested (Pokrovsky and Beliavsky, 1983) a modified inversion technique (3.8) to make a solution class of the inverse problem (3.2) to be more compact. In fact, there are two physical restrictions on the transmittance functions: its monotonous behaviour as vertical coordinate function and its equality to one at the top of the atmosphere. Therefore, the constraint $\sum\limits_{i=1}^{N} d\tau^i{}_{\Delta v} = 1$ is valid for any Δv. Here i is a layer index: $\left(\delta \tau^i{}_{\Delta v} = \delta \tau_{\Delta v}(p_i) \right)$. Another feature provides a validity of the following constraints:

$$\sum_{i=1}^{N} \delta \left(d\tau^i{}_{\Delta v} \right) = \sum_{i=1}^{N} d \left(\delta \tau^i{}_{\Delta v} \right) = 0 .$$

The above restrictions on the admissible class of inverse problem solution permit to use more sophisticated method. We applied a modified algorithm (3.8) to minimise a residual in (3.2) under restrictions mentioned. This modification allows us to obtain a physically interpretable solution for the transmittance functions in search.

3.2. RESIDUAL ANALYSIS

The correction problem for the transmittance function outlined above may be considered as an inverse problem of the second kind. In fact, we assume that the temperature profile, which in the conventional inverse problem is considered to be retrieved, is here assumed to be known from radiosonde data. On the other hand, the transmittance function, which is assumed to be known in inverse problem of the first kind, is a solution of the inverse problem of the second kind. In both problems the spectral dependence of the radiance residuals $\Delta(\delta I_{\Delta v}) = \delta I_{\Delta v} - \delta \tilde{I}_{\Delta v}$ is an input data for inversion.

It is a difference between measured $\delta \tilde{I}_{\Delta v}$ and calculated $\delta I_{\Delta v}$ radiance values. Let us consider several implication examples. First experiment series was performed for Nimbus-5 (N-5) data set for the year 1973. We used a set of collocated measurements for 34 points. Second series was completed for Nimbus-6 (N-6) data set for 1975. Here we employed a sample of simultaneous satellite and radiosonde measurements for 18 points. The third series was carried out for Meteor (M) satellite data for 1977. In this case we have 23 collocated pairs at our disposal. To explore the residual spectral dependence we split the spectral area 667-750 cm^{-1} (15 μm carbon dioxide absorption band) into three sub-areas: (I) 677-698 cm^{-1}, (II) 699-730 cm^{-1} и (III) 730-750 cm^{-1} in accordance to the atmosphere absorption rate. It is necessary to note that measurements obtained at sensor channels, which attributed I band, give the information on temperature at higher troposphere and lower stratosphere. In contrast, measurements, related III spectral sub-area, provide the information on temperature at lower troposphere layers. This is a reason which may explain why the radiance residual function $\Delta(\delta I_{\Delta v})$ in both spectral areas has a regular behaviour as a function of spatial coordinates. In contrast its behaviour is irregular at II area, where the impacts of different atmospheric layers are superimposed and lead to frequent changes in signs of the radiance residual $\Delta(\delta I_{\Delta v})$. It is a fortune that most of spectral channels, related the thermal remote sensing system, enters at 1-th and 3-th bands. Despite of differences in residual behaviour, some joint features had been found out. The radiance residual $\Delta(\delta I_{\Delta v})$ at I spectral band maintains a constant (positive or negative) sign along both western and eastern continent coasts, where large-scale ozone variations are observed. Positive sign in residual at III spectral band was noticed at ocean area located closely to continents, where the water vapour concentration is increased considerably. In contrast, negative residual values for III band are observed at Arctic and polar area (Eastern Siberia, Spitzbergen, Novaja Zemlya islands) and also at Central Asia, where the water vapour concentration was turned out to be the below climatic values.

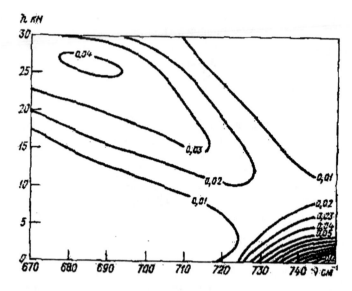

Figure 2. Atmospheric transmittance function variability at 15 μm carbon dioxide absorption band due to changes in concentrations of water vapour, ozone and aerosol

To perform an application of the above algorithm we carried out a preliminary study on theoretical variability of the transmittance function at 15 μm carbon dioxide absorption band (see (3.3), (3.4)). We investigated the variation $\delta\tau_{\Delta v}$ in spectral and vertical coordinates, which is related to changes in vertical profiles of water vapour, ozone and aerosol. In this study we used the "LOWTRAN" algorithm. Here we employed several climatic models related to different seasons and latitude belts. Generalised structure of the variation $\delta\tau_{\Delta v}$ in spectral and vertical coordinates is presented in Fig.2. It shows that there are two ranges of high sensitivity: (i) stratospheric at I spectral band and (ii) lower tropospheric at III spectral band. Our study proved that aerosol and ozone are greatly responsible for the variation $\delta\tau_{\Delta v}$ at stratospheric layers, while water vapour substantially impacts on $\delta\tau_{\Delta v}$ at lower troposphere. At higher troposphere the ozone impact might be compared with the water vapour contribution. It follows from Fig.2 that the spectral band II plays a role of a transition zone, where stratospheric and tropospheric factors are superimposed.

At the next step we performed computation of corrections $\delta\tau_{\Delta v}$ for the transmittance functions. We consider $\delta\tau_{\Delta v}$ as a function of vertical and spectral coordinates. As we used several collocated data sets (N-5, N-6, M), some sample of $\delta\tau_{\Delta v}$ might be obtained. In fact, we obtained three samples related to different geographic areas: 1) European (middle latitude), 2) Central and Arctic Asia, 3) Pacific and Indian oceans (tropical belt). Mean values of $\delta\tau_{\Delta v}$ in each sample were computed. We obtained three pictures similar to Fig.2. All these figures have the same structure: two anomaly zones: stratospheric located in I band and lower tropospheric at III band. It turned out that the figure $\delta\tau_{\Delta v}$ for European area was most similar to the theoretical computation results (Fig. 2). But, there is a difference in ascending of III band maximum location at level of 3 km. This discrepancy might be explained by possible impact of weak cloudiness, which was not taken into account in the theoretical computation of $\delta\tau_{\Delta v}$. The second $\delta\tau_{\Delta v}$ sample (related to tropical belt) demonstrates a stronger lower tropospheric anomaly at I band, while the third $\delta\tau_{\Delta v}$ (Asian and Arctic) sample displays a stronger stratospheric anomaly at III band.

The residual $\Delta(\delta I_{\Delta v})$ analysis performed for each satellite orbit or band of orbit showed that there is a systematic component in $\Delta(\delta I_{\Delta v})$ for each spectral channel (see Fig.3). Proposed correction technique allows us to reduce it (see curve 2 in Fig.3). But also there is a random component in the residual, which may arise by two reasons: space borne sensor measurement noise and radiosonde observing errors. The first reason may cause a disturbances of about 0.5-1.0 w/sr*m^2*cm^{-1}, while the second may cause a deviation of 1.0-2.0 w/sr*m^2*cm^{-1}. These reasons can explain the radiance residual behaviour and its magnitudes presented in Fig.3. To evaluate our technique effectiveness we proposed an implementation procedure based on serial correlation criteria developed by Anderson (Anderson, 1958). The coefficient of serial correlation may be determined by the equality:

$$R = \frac{1}{\sigma^2} \sum_{k=1}^{m} \Delta(\delta I_{\Delta v}(s_k))\Delta(\delta I_{\Delta v}(s_{k+1})),$$

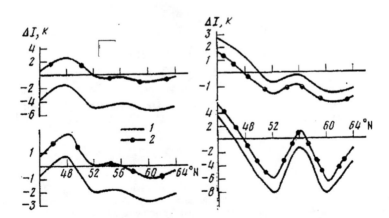

Figure 3. Residuals for different radiometer channels (N-5, I orbit): a) 668,3 cm^{-1}, b) 689,5 cm^{-1}, c) 713,8 cm^{-1}, d) 747,0 cm^{-1}.

Table 1. Serial correlation statistics σ and R

Transmittance functions	Spectral channel, cm⁻¹					
	747		713,8		668,3	
	σ	R	σ	R	σ	R
	I orbit					
Theoretical model	8,0	0,66	2,2	0,57	5,2	0,53
First correction	3,1	0,61	0,89	0,51	0,43	0,29
Second correction	2,3	0,43	0,71	0,32	0,31	0,21
Third correction	2,0	0,31	0,6	0,2	0,25	0,16
Transmittance functions	Spectral channel, cm⁻¹					
	747		668,3			
	σ	R	σ	R	σ	R
	II orbit					
Theoretical model	7.4	0.62	1.6	0.53	3.9	0.49
First correction	3.2	0.58	1.5	0.49	0.52	0.27
Second correction	2.6	0.38	1.2	0.38	0.49	0.22
Third correction	2.2	0.29	1.0	0.31	0.42	0.19

where s_k is the index measurement point at the considered orbit or orbit band, and σ^2 the related variance:

$$\sigma^2 = \frac{1}{m-1} \sum_{k=1}^{m} \Delta(\delta I_{\Delta v}(s_k))^2 \ .$$

Anderson provided a table of R_α critical values for the probability function $P(R > R_\alpha) = a$ for a set of a and m values. Therefore, σ^2 and R completely describe the effectiveness of the correction procedure. Our procedure is stepwise. It assumes that after first step we should split the considered orbit into bands at which innovated residual function maintains a constant sign. The next step should be carried out separately for each band.

References

Anderson, T. (1958) Introduction to Multivariate Statistical Analysis, Wiley and Sons, N.Y.

Beliavsky, A.I. and Pokrovsky, O.M. (1982) On the covariance features of inverse problem solution relate to atmospheric remote sensing, Soviet Meteorology and Hydrology 2, 23-28.

Pokrovsky, O.M. (1974a) An assimilation of conventional and satellite data in 3-D analysis of meteorological fields, Soviet Meteorology and Hydrology 6, 33-39.

Pokrovsky, O.M. (1974b) An optimal 4-D assimilation of conventional and satellite data for meteorological field analysis, Soviet Meteorology and Hydrology 8, 29-36.

Pokrovsky, O.M. and Ivanykin, E.E. (1978) Spatial analysis of temperature and height fields on the basis of data from remote sounding of the atmosphere, Zeit. fur Meteorologie 1, 3-23.

Pokrovsky, O.M. and Beliavsky, A.I. (1983) On the technique of operational correction of an optical atmospheric model when solving the problem of thermal remote sensing, *Izvestiya, Acad. of Sciences, USSR*, Atmos.and Ocean. Physics **19**, 455-461. (Edition of AGU)

Pokrovsky, O.M. (1984). *An Optimization of Meteorological Remote Sensing of Atmosphere from Satellites*, Hydrometeoizdat, Leningrad. (in Russian).

Schlatter ,T.W. and Branstator, G.W. (1978) Errors in Nimbus-6 temperature profiles and their spatial correlation, Preprint NCAR Ms 78/0501-1.

ASSIMILATION OF REMOTE SENSING OBSERVATIONS IN NUMERICAL WEATHER PREDICTION

JEAN-NOËL THÉPAUT
European Centre for Medium Range Weather Forecasts
Shinfield Park, Reading, Berkshire, RG2 9AX, U. K.

1. Introduction

Over the last few years, satellite data have progressively become a major (if not the predominant) source of information assimilated in Numerical Weather Prediction (NWP) models. This has been made possible thanks to a substantial enhancement of the remote sensing instruments measuring various atmospheric quantities but also largely to the improvements in data assimilation techniques to better exploit the information contained in such data. The advantage of satellite data is that they provide a uniform spatial and temporal coverage of the atmosphere. This advantage is however balanced by a general poor vertical resolution of the instruments currently used, and the difficulty to handle clouds, precipitations and surface contributions to the information content of the data. The future improvements of NWP models and a better handling of new observing techniques (radio-occultation, passive limb soundings, active sensors) in data assimilation schemes may overcome some of these limitations.

This lecture briefly stresses how satellite data have become an essential part of current data assimilation schemes (section 2). The lecture also reviews what remote sensing measurements are actually measuring (section 3) and the cons and pros of directly assimilating raw measurements versus retrieved products (section 4). Section 5 aims at describing the foreseen evolution of the observing system from space and the opportunities and challenges this will entail for operational NWP data assimilation systems.

2. The importance of satellite data in Numerical Weather Prediction

It is now well recognised by all the NWP centres that satellite data play a major role in improving the accuracy of the forecasts. An illustration of this can be given in Fig.1 after Simmons and Hollingsworth (2001). This figure presents running annual-mean anomaly correlation of 500 hPa height for ECMWF's operational three-, five- and seven-day forecasts for the extratropical northern and southern hemispheres for the period from January 1980 to November 2002. The first remark to make about this figure is the general upward trend of the curves (indicating a progressive improvement of the forecast quality over the covered period). Another striking feature is the higher rate of

R. Swinbank et al. (eds.), Data Assimilation for the Earth System, 225–240.
© 2003 *Kluwer Academic Publishers. Printed in the Netherlands.*

improvement in the forecasts in the southern hemisphere. In twenty-two years, the skill of medium range weather forecasts in the southern hemisphere has reached a level now comparable to the one in the northern hemisphere. Bearing in mind the relatively poor data coverage provided by the conventional observing system (in particular over the oceans), this result is a strong indication of an improved usage of satellite data in the ECMWF assimilation system.

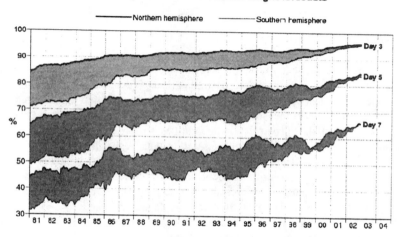

Figure 1. Anomaly correlations of 500hPa height for 3-, 5- and 7-day forecasts for the extratropical northern and southern hemispheres, plotted in the form of annual running means of scores for the period from January 1980 to October 2002.

This general trend has been observed by most of the NWP centres (NCEP and the Met Office in particular). Recently, the impact of satellite data on the improvement of numerical forecasts has been confirmed by Bouttier and Kelly (2001), in the context of Observing System Experiments. Fig. 2 summarizes some key results of their data impact studies accumulated over two three week periods. This figure represents 500 hPa height anomaly correlation for the extratropical northern hemisphere averaged over 43 day-3/day-5/day-7 forecasts for various assimilation experiments. The first column corresponds to the control experiment (all data in), the second column to the experiment when no aircraft data are used, the third column to the experiment when no sondes data are used, and the fourth when no upper-air satellite data are used. What is particularly striking is that (probably for the first time in the NWP literature), clearly the impact of the satellite data is now similar to, if not larger (as can be seen at day 7 on Fig.2), than the impact of conventional sondes **in the northern hemisphere.** This result may be partially explained by the progressive reduction of the radiosonde observing network for the last twenty years. Nevertheless, it demonstrates that the investment in methodologies for exploiting remote sensed observations in NWP is paying off. Moreover, most of the atmospheric sounding instruments are currently not used over land due to the difficulty to handle the surface contribution to the radiative transfer equation (see next section). One can then measure the range of improvement expected from satellite observations with a better treatment of surface conditions.

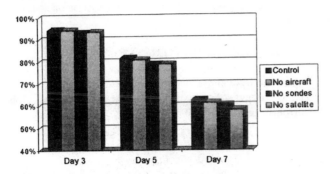

Figure 2. 500 hPa height anomaly correlations for the extratropical northern hemisphere averaged over 43 day-3/day-5/day-7 forecasts from various assimilation experiments (see text for details)

3. What do remotely sensed instruments actually measure?

Before considering how to assimilate satellite data in NWP data assimilation schemes, it is useful to identify what is specific to the satellite observations and in particular what they are actually measuring. It is indeed important to realize that the measured quantities by satellite instruments do not relate directly to geophysical quantities. Satellite instruments do not measure temperature, do not measure humidity, and do not measure wind. Satellite instruments (active and passive) measure essentially the radiance L that reaches the top of the atmosphere at frequency v. The radiance is related to geophysical parameters through the radiative transfer equation. The reader is invited to read Liou (1980) and Stephens (1994) for a very good introduction to atmospheric radiation and radiative transfer equation. In short, the radiative transfer equation can be summarized as follows:

$$L(v) = \int_0^\infty B(v, T(z)) \left[\frac{d\tau(v)}{dz} \right] dz$$

+ (surface emission)
+(surface reflection)
+(surface scattering)
   +(cloud/rain contribution)   (1)

where $B(v, T(Z))$ is the Planck radiance for a scene temperature T at altitude Z, and $\tau(v)$ the altitude Z to space transmittance.

Eyre (2000) provides an excellent overview of the different instrument technologies commonly used to observe the atmosphere from space, and a brief summary is given here. By selecting radiation at different frequencies (or **channels**), a satellite instrument can provide information on a range of geophysical variables. A distinction has to be made between passive and active instruments. Passive instruments sense radiation emitted by the surface and/or atmosphere (or the solar radiation reflected by it). Active instruments emit radiation and measure how much of it is reflected or back-scattered by the surface and/or atmosphere.

3.1 DIFFERENT WAYS OF SENSING THE EARTH/ATMOSPHERE

As stated above, depending on the frequency, the measured radiance will be sensitive to different geophysical variables (upper air temperature, surface parameters, clouds,...). In general the channels currently used for NWP applications maybe considered as one of 3 different types.

3.1.1 Atmospheric sounding channels from passive instruments
These channels are located in parts of the infrared and microwave spectrum for which the main contribution to the measured radiance is described by the first term of the right hand side of Eq 1:

$$L(v) = \int_0^\infty B(v, T(z)) \left[\frac{d\tau(v)}{dz} \right] dz \qquad (2)$$

These channels avoid frequencies for which surface radiation or cloud contribution are important. These channels are primarily used to obtain information about atmospheric temperature and humidity (Atmospheric sounding channels from the HIRS (High resolution Infrared Sounder) and AMSU (Advanced Microwave Sounding Unit) on board NOAA satellites fall into this category).

3.1.2 Surface sensing channels from passive instruments
These channels, called "imaging" channels, are located in atmospheric "window" regions of the infra-red and microwave spectrum at frequencies where there is very little interaction with the atmosphere and the main contribution to the measured radiance in this case is:

$$L(v) = Surface\ emission\ [T_{surf}, \varepsilon(wind,...)] \qquad (3)$$

where T_{surf} is the surface temperature and ε the surface emissivity. These channels are primarily used to obtain information on surface temperature and quantities that influence the surface emissivity such as wind (through the roughness over sea) and vegetation (land). They can also be used to obtain information on cloud top (in the Infrared) and rain (in the microwave). In addition, sequences of Infrared images from geostationary satellites can be used to track the cloud movements and indirectly derive wind information.

3.1.3 *Surface sensing channels from active instruments*

These instruments (e.g. scatterometer) emit microwave radiation towards the surface in the atmospheric window parts of the spectrum such that radiance scattered back from the surface is:

$$L(v) = \textbf{\textit{Surface scattering } [\varepsilon(wind)]} \tag{4}$$

These instruments provide information on ocean winds. Some similar-class active instruments such as altimeters and SARS (Synthetic Aperture Radars) provide information on wave height and spectra.

3.2 WEIGHTING FUNCTIONS

We will now insist on a notion that is important to capture to understand how radiance observations are used in data assimilation systems.

Let us consider the simple case of a channel for which the primary absorber is a well mixed gas (oxygen or carbon dioxide). It can be shown that the measured radiance is in this case a weighted average of the atmospheric temperature profile, that is Eq. 1 reduces to:

$$L(v) = \int_0^\infty B(v, T(z))K(z)dz \tag{5}$$

where $K(z) = d\tau / dz$.

The function $K(Z)$ that defines this vertical average is known as a **weighting function**. Fig.3 a) and b) represent two idealized weighting functions. If the weighting function of a channel is a delta-function, the measured radiance is uniquely sensitive to the temperature at a single level in the atmosphere. Conversely, if the weighting function is constant with Z throughout the atmosphere, the measured radiance is sensitive to the mean temperature of the atmospheric profile.

In reality, the shape of the real atmospheric weighting functions are somewhere in between these two idealized cases. The consequences of the shape of the weighting function on the actual radiation emitted at the top of the atmosphere are illustrated in Fig. 4. The surface and the shape of the arrows A B C represent the amount of radiation emitted from the bottom of the arrow and reaching the top of the atmosphere; the corresponding weighting function is displayed on the right of the figure.

Arrow A shows that a lot of radiation is emitted from the dense lower atmosphere (large arrow base) but very little reaches the top of the atmosphere because of absorption. Arrow C indicates that high in the atmosphere, very little radiation is emitted but most will reach the top of the atmosphere. At some level (peak of the weighting function, arrow B), there is an optimal balance between the amount of radiation emitted and the amount reaching the top of the atmosphere. The altitude of the

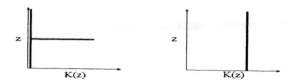

Figure 3. Two idealized weighting functions: The delta function corresponds to a sensitivity of the radiance to a single atmospheric level. The vertically constant weighting function corresponds to a sensitivity of the radiance to the mean temperature between the surface and the top of the atmosphere

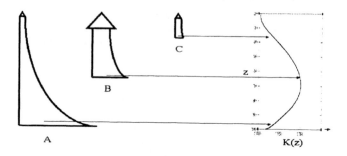

Figure 4. Schematic amount of radiation emitted from different layers of the atmophere as a function of a given weighting function (see text for details)

peak of the weighting function depends on the strength of the absorption for a given channel. Channels in parts of the spectrum where the absorption is strong peak high in the atmosphere (this is the case for channels near the center of CO_2 or O_2 lines). Conversely, channels in parts of the spectrum where the absorption is weak (this is the case for channels in the wings of CO_2 or CO_2 lines) peak low in the atmosphere. By selecting a number of channels with varying absorption strengths, one can sample the atmospheric temperature at different altitudes.

The problem of extracting the atmospheric temperature profile from a set of measured radiances is called the **retrieval** or **inverse** problem. Unfortunately, with a finite number of channels and with weighting functions that are generally quite broad, the inverse problem is generally ill-posed (i. e. an infinite number of different temperature profiles could give the same measured radiance). If one wants to utilize satellite radiances to determine the initial conditions of a NWP model, the role of data assimilation is to solve this ill-posed problem. Any technique requires the use of a prior information, the quality of which will drive the accuracy of the final retrieved product. The review of some of the techniques commonly used in NWP centres to exploit satellite information in data assimilation is provided in the next section.

4. From retrievals to direct assimilation of satellite radiances

4.1 TECHNIQUES FOR SATELLITE DATA ASSIMILATION

Among others, three basic options have been commonly used to use satellite data in NWP data assimilation (see e.g. Simmons 2000):

4.1.1 *Assimilation of retrieved products from space agencies or research institutes*
These retrieved products have the advantage of being of the same nature as meteorological variables. The principle of the retrieval can vary from center to center but it is generally based on a statistical relationship that predicts atmospheric temperature/humidity from measure radiances. This statistical relationship can be built out of a sample of atmospheric profiles (from radiosondes) collocated with a sample of radiance measurements and can be based on regression, library search or neural network techniques. One limitation of this approach for NWP is that it is limited by the statistical characteristics of the training dataset and may miss some extreme but important atmospheric features that are statistically rare in the training sample. Another risk is that the retrieved product will suffer from a poorer prior estimate of the atmosphere than that from an NWP model. Last but not least, the error characteristics of the retrieved products are not easy to assign when the inputs and the chain of processing the product (cloud clearing, limb adjustment, surface corrections,...) is poorly known by the NWP center. Nevertheless, and although they have been widely superceded by more direct methods, the use of retrievals was the original way TOVS radiances were assimilated in NWP and it is still the approach most commonly adopted for the assimilation of wind information from successive geostationary images. The reader will find information on famous retrieval techniques in Chédin et al. (1985) and Reale et al. (1989).

4.1.2 *Locally produced or "1D-Var" retrievals*
The retrievals are produced by the NWP center prior to the main assimilation suite, using background information from a short-range forecast (typically 6 hour). The retrieval is the outcome of an optimal estimation (minimizing for example a cost function or solving the standard optimum interpolation equation) adjusting atmospheric profiles to background atmospheric profiles and measured radiances. In that case, prior information is generally very accurate and contains information about important atmospheric phenomena (such as fronts, tropopause folding,...). In principle, the error characteristics (covariances) of the prior information and resulting retrieval are better known. This ingredient is vital for the subsequent assimilation process. However, the error characteristics of the retrieval may remain complicated due to its correlation with the forecast background that is in fact used twice in the subsequent assimilation. The implications of using cross-covariances between observation and background errors are

discussed in Joiner and Dee (2000). An example of an interactive retrieval procedure can be found in Phalippou (1995).

4.1.3 Direct assimilation of radiances

Variational techniques such as 3D-Var or 4D-Var (Rabier et al. 1998) allow the direct assimilation of radiance observations and therefore avoid the need for an explicit retrieval step. The retrieval step is essentially incorporated within the main analysis by finding the model variables that minimize the following cost function:

$$J(x) = (x - x_b)^T B^{-1}(x - x_b) + (y - H(x))^T R^{-1}(y - H(x)) \qquad (6)$$

where x is the atmospheric state, x_b the prior (or background) estimation of the atmospheric state (with associated error covariance B), y is the observation vector made of measured radiances and all the other data available for the analysis time (with associated error covariance R) and H the observation operator that maps the atmospheric profile to radiance space. A full description of the variational technique is provided by Talagrand and Gauthier elsewhere in this volume. As can be seen from the equation above, the forecast background still provides the prior information to supplement the radiances. However, it is not used twice (as it is in a 1D-Var preprocessor context) and this avoids the problem of assimilating retrievals with complicated error structures. Furthermore, the inversion is further constrained by the simultaneous assimilation of other observations. Note in particular that in 4D-Var, the adjustments forced by radiances at different times of the assimilation window will be consistent with the forecast model physics and dynamics (for example and as shown by Andersson et al. 1994, radiances can cause wind adjustments during the assimilation). Last, the characterization of observational errors in radiance space is much easier. In practice, the approach adopted by NWP centres is pragmatic and observations are used in the space where errors are easier to characterize. As it happens, the "model-to-satellite" approach tends to become the rule, but exceptions exist. For example, atmospheric motion winds derived from cloud tracking are assimilated in all NWP systems because the direct assimilation of cloud information is not mature enough for various reasons discussed in section 5. Stoffelen (2000) also explains the advantages of assimilating ambiguous wind products versus raw signals from scatterometers, claiming that observational errors are better characterized in the wind space (gaussian distribution) than in the sigma-naught space (skewed distribution).

4.2 IMPORTANT ISSUES FOR THE ASSIMILATION OF SATELLITE RADIANCES

Because of all the advantages mentioned above, the direct use of radiances in variational techniques has become the preferred choice to incorporate measured satellite sounder observations in most of the operational NWP centres. However, important issues still need to be addressed to handle properly the treatment of remote sensed observations in data assimilation.

4.2.1 *Biases*

Systematic errors must be removed before the assimilation is performed to avoid propagation of various biases into the analysis (see also the chapter on *Observation monitoring and Quality Control* by P. Gauthier). Biases in radiance data assimilation can originate from various sources: instrument error (calibration), radiative transfer error, cloud/rain screening error, background model error,...In practice, NWP centres address the problem of **bias correction** by **monitoring** departures between forecast background (or analysis) and radiance observations, thus identifying possible systematic differences. The use of several independent instruments is then recommended to disentangle the possible source of these differences. An illustration of this is given in Fig.5 showing the standard monitoring statistics daily produced at ECMWF (http://www.ecmwf.int). This plot shows on a daily basis during April 2001, the global mean departure between the ECMWF model background and channel 5 (temperature channel peaking at 600 hPa) from the HIRS (High Infrared Radiation Sounder) instrument on board NOAA-14 (left panel) and NOAA-16 (right panel). The grey line represents the raw mean departure while the dark thick line represents the bias corrected departure. In this particular comparison, one clearly sees that NOAA-14 satellite has a +2K radiance bias against the model, while NOAA-16 has no radiance bias against the model (the same in both comparisons). This illustrates how powerful and important monitoring procedures and cross calibration techniques are to quantify biases in satellite instruments, although problems are not always as simple as the one illustrated in Fig. 5.

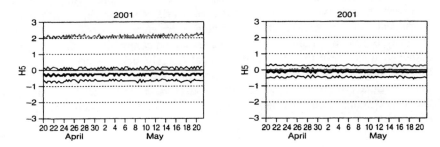

Figure 5. Example of monitoring of channel 5 from the HIRS instrument on board NOAA-14 (left panel) and NOAA-16 (right panel) for April 2001. The grey line represents the non-corrected mean departure between observations and model first-guess. The thick dark line represents the bias corrected mean departure. Thin black lines show one standard deviation around the mean departure.

Indeed, although bias correction is mainly dedicated to remove radiative transfer and observational biases, it is very difficult to make absolutely sure that none of the model biases are removed at the same time. Monitoring is also essential to identify sudden deterioration in instrument performance, as a trigger for blacklisting actions.

4.2.2 *Quality Control*

In the chapter *Observation monitoring and quality control*, Gauthier provides an extensive review of issues related to quality control in data assimilation. Quality control is necessary in general to reject data of "bad" quality but also data that cannot be properly simulated by the NWP model. This is particularly true for satellite radiances that can be difficult to use because of the contribution of clouds, rain, surface emission and other parameters which are not represented correctly in the forecast model. This issue remains certainly one of the most challenging areas of research.

4.2.3 Thinning

As pointed out by Derber (2000), satellite data are specific in that the spacing of the observations is small (a few kilometers). However, the background error covariances used in current NWP models have currently fairly large horizontal scales and therefore forbid the details observed by satellites to be properly analysed. In addition, Liu and Rabier (2002) have shown that when observation errors are correlated, increasing the observation density beyond a certain threshold adds very little information even within an optimal data assimilation scheme. The problem gets worse in a suboptimal context that neglects observational error correlations. Therefore and also due to the additional computational burden of processing high resolution radiance data, satellite observations are thinned to a resolution that is scientifically "digestible" by the error statistics of the assimilation system, and technically compliant with computational constraints.

4.2.4 Observational error characterization

As stated above, the characterization of observational errors is in principle much easier for raw radiances than for retrievals because of far less processing steps involved. However, the observational error covariance matrix R should represent accurately the instrument, radiative transfer and representativeness errors (errors due to the inability of the analysis system to represent scales at a resolution compatible with the observations). Although it is likely that R is not diagonal (merely because errors in the radiative transfer model lead to inter channel correlations), most NWP centres ignore it in practice. Specifying properly R in radiance data assimilation remains a difficult issue that is becoming more important with the new generation of multi-channel advanced sounders.

5. Future challenges in satellite data assimilation

If satellite observations are already playing an important role in improving numerical forecasts today, the combination of increased model resolution and improved physics, enhanced data assimilation techniques, together with advances in satellite instruments over the next decade offer new potential observations available for NWP data assimilation centres. The horizontal and vertical resolution of numerical forecast models has been steadily increasing for the last twenty years in most NWP centres (following this trend and as an example, the horizontal resolution of the ECMWF model is expected to typically reach 10-15km in 2015, with a vertical resolution of around 80m in the boundary layer, 200-300m in the troposphere and 500-600m in the stratosphere).

5.1 EVOLUTION OF THE OBSERVING SYSTEM FROM SPACE

In parallel to model and data assimilation improvements, the evolution of the current and future satellite observing systems clearly indicates:

- Increased spectral resolution in the infrared instruments onboard polar orbiters first (such as AIRS -Advanced Infrared Sounder- on board AQUA or IASI -Infrared Atmospheric Sounding Interferometer- onboard METOP) and later on geostationary platforms (such as GIFTS – Geostationary Imaging Fourier Transform spectrometer-). Those instruments will measure radiation in many thousands of different channels, and therefore provide atmospheric temperature and composition information at a much higher accuracy and vertical resolution than what can be achieved with the current generation of current instruments such as HIRS.

- Improved channel combination and scanning in the microwave. The SSMIS (Special Sensor Microwave Imager/Sounder) instrument will combine the sounding and imaging capabilities of the current sensors to provide an enhanced description of temperature/humidity/rain triplet profiles.

- More instruments providing information on the land surface but also on atmospheric composition (including trace gases). These new observations will not only be useful for NWP application but also for climate, chemistry and environmental research.

- Wind observations from Doppler lidars. An ESA demonstration mission, ADM-Aeolus (Atmospheric Dynamics Mission) is expected around 2008. For the first time, NWP centres will have the opportunity to assimilate globally 3D-wind information.

- New sources of information on clouds and precipitation. Satellite imagery is already providing a lot of information on cloud fields, and rain information can be retrieved from microwave imagers. New missions such as Cloudsat-Calipso (to be flown together with AQUA) and Earthcare will enhance the validation of the cloud parameterizations of NWP models by providing for the first time information on cloud profiles thanks to the use of "active" instruments penetrating atmospheric hydrometeors. Global information on rain will be available for NWP with missions such as the Global Precipitation Mission (GPM) and its European component (E-GPM). A particular effort is already taking place in a number of NWP centres to assimilate cloud and rain products from satellite. A prerequisite to assimilate these observations is the ability of the numerical model to "look like" the observations. Fig. 6 represents the comparison of a METEOSAT-7 infrared image (11μm) on December 10 2000 at 03UTC and the same image as simulated from the operational ECMWF 3-hour forecast.

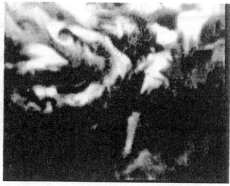

Figure 6. comparison of a METEOSAT-7 infrared image (11µm) on December 10 2000 at 03UTC (left panel) and the same image as simulated from the operational ECMWF 3-hour forecast (right panel). After Chevallier et al., 2001.

- The degree of agreement between the model and the observations is sufficiently good to be convinced that assimilation of such quantities is at hand. Several approaches are being developed to assimilate cloud and rain products in NWP. One of the most promising approaches is probably the direct use of cloudy or rainy radiances in a four-dimensional data assimilation system. This route is currently under investigation at ECMWF. This technique requires an improved simplified physics in the incremental formulation of 4D-Var (see the chapter "Operational implementation of 4D variational assimilation" by P. Gauthier), a fast but accurate radiative transfer able to simulate cloud /rain absorption and scattering and a proper treatment of representativeness errors (in particular subgridscale variability errors). The route is certainly long and intermediate steps (such as interactive 1D-Var as a preprocessor to 4D-Var) may be necessary, but early results suggest that it may be achievable. This approach would have the advantage of providing an optimal coherent treatment of satellite observations under clear, cloudy and rainy areas.

- Radio-occultation techniques using GPS (Global Positioning System). These techniques exploit an opportunity that the GPS constellation (originally designed for other applications) already exists. GPS receivers (such as the GRAS instrument on board METOP) measure the Doppler shift of a GPS signal refracted along the atmospheric limb path. This refraction is proportional to (among other parameters) the density of the atmosphere, and therefore indirectly to temperature and humidity profiles. Provided a sufficient number of receivers are installed on LEO (Low Earth Orbiting) satellites, this technique could offer high vertical resolution (balanced by a somewhat coarse horizontal resolution of ~200-300 km), self-calibrated and "all weather" observations of atmospheric temperature and humidity.

5.2 ASSIMILATION OF ADVANCED SOUNDERS

Among the list of challenges that NWP centres will have to tackle in order to better absorb the massive amount of information available from space within the next decade, the exploitation of advanced sounders is currently probably the topic of most active research. Indeed, the current generation of operational instruments (both infrared and microwave) suffers from a lack of vertical resolution. This is because the radiance measured at a particular frequency (or channel) originates from a deep atmospheric layer and with only a small number of channels it is impossible to resolve fine vertical structure. Advanced infrared sounders measure radiation in many thousands of channels. While individually they only provide a broad layer measurement, their combination may provide significantly higher vertical resolution. Preparations are under way to use these data operationally in NWP and to achieve this, key issues that are being considered:

5.2.1 Monitoring and bias correction of high spectral resolution data

The monitoring system is extremely useful when a new instrument is launched. NWP centres currently play a key role in the early checkout of the AIRS instrument on the NASA AQUA. The departure statistics provide vital information on the quality of the new data and how well the spectral characteristics of the instrument have been determined. The existing tools have been extended to simultaneously visualize several hundreds of channels. The monitoring statistics also provide the input to the bias corrections that will have to be applied and to the characterizations of the observational errors. It is worth stressing that a comprehensive and consistent bias handling system will be a key element for the current and future exploitation of future satellite instruments. In particular, the extraction from advanced sounders of new quantities (such as trace gases) with a low signal-to-noise ratio will only be possible if different bias sources in temperature and water vapour are thoroughly understood and controlled. This requires a capability to disentangle systematic errors inherent to instrument, radiative transfer and NWP model deficiencies.

5.2.2 Channel selection

It is neither feasible nor efficient to assimilate all of the advanced channels (approximately 2400 in the case of AIRS and 8400 in the case of IASI) and a policy of channel selection has to be designed. The challenge is to find a set of channels that is small enough to be assimilated efficiently in a global NWP system (with operational time constraints), but which is still large enough to capture the important atmospheric variability. Most centres are designing channel selection strategies based on optimal theory or on a more pragmatic approach relying on a selection predefined at source (Collard 2000, Rabier et al. 2002, Fourrié and Thépaut 2002).

5.2.3 Data compression

Even if all the advanced sounder channels could be assimilated, current transmission links and archiving resources are under pressure to deal with the huge data volumes involved. To handle these large volumes of data, data compression techniques are also being considered. The purpose of these is to reduce the data volume of the measured spectra without any distinguishable loss of accuracy. One candidate for this is to use a truncated set from the Empirical Orthogonal Functions (or EOFs) of the spectra with a suitably defined reproduction metric (Huang and Antonelli, 2001). However, it must be stressed that any loss of accuracy resulting from data compression is undesirable (even if it is within the anticipated measurement / system noise) and that every precaution not to compromise the information content of the spectra has to be taken.

5.2.4 Cloud detection

A key issue for the successful exploitation of advanced infrared sounders is the detection of the presence of clouds since they have a significant impact on the radiance measurements. Given that important atmospheric structures occur in cloudy areas, it is of utmost importance to take into account the effect of clouds when present and cloud detection is therefore needed to correctly interpret the data. A failure to detect the presence of clouds can result in a (possibly large) signal being wrongly interpreted by the assimilation system and force erroneous adjustments to temperature and humidity. An over-stringent cloud detection system can lead to very little data reaching the assimilation and thus reduce the impact of the new high resolution instrument. A number of strategies are being investigated by different NWP centres. McNally and Watts (2003) propose a technique based on the identification of characteristic spectral signatures or patterns in the observed radiances. The pattern recognition algorithm is used to identify which radiances at a particular location are free of cloud and which are not (it is therefore not designed to identify locations of completely clear spectra). In the longer term, cloud affected radiances could be exploited as they can, in principle, provide unique high-resolution information about the temperature structure near the cloud top (see the discussion above about the assimilation of cloudy and rainy information from satellites). In that respect, advanced sounders may indeed be able to determine the cloud top and cloud amount accurately.

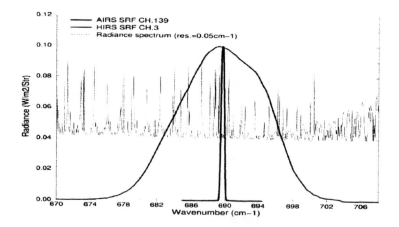

Figure 7. Spectral Response Functions (SRF) of HIRS channel 3 and AIRS channel 139 (longwave CO_2 band) superimposed with a radiance spectrum at a resolution of $0.05 cm^{-1}$.

5.2.5 *Environment opportunities*

Although this issue goes beyond strict NWP applications, it is worth mentioning that advanced sounders will also provide information about minor constituents of the atmosphere (ozone, N_2O, CO, CH_4). It is also hoped that CO_2 information can be extracted from high resolution sounders. Fig.7 displays the Spectral Response Function (SRF) of HIRS (thick broad line) and AIRS (peaked line) in the longwave infrared domain. It is clear that by sampling the IR spectrum at very high resolution, one can measure radiation that is only dependent on temperature and the atmospheric CO_2 concentration (small group of pure lines), while instruments with coarse spectral resolution sample radiation that is a mixture of absorbing species. If we have accurate temperature information (from NWP analyses driven by AMSU for example), it is hoped that the CO_2 signal can be separated out.

6. Conclusions

This chapter has reviewed the increasing importance of satellite data in Numerical Weather Prediction. A special emphasis on what makes the satellite data specific has been presented. The advantages of assimilating raw information versus products have been reviewed. Finally, a number of future challenges to better exploit an increasing amount of information from satellites has been described. Combined with improvements that make up numerical models (e. g. dynamics, physical parameterizations of sub-grid scale processes, heat and moisture exchanges with the surface,...), advanced data assimilation methods will make it possible to extract the maximum information content from existing and future satellite instruments. This combination will be the key to further improvements in NWP systems within the next decade.

As a final comment, the reader may have noticed that one word has been almost absent from the discussion: radiosonde. Indeed, the tone of the lecture may have indicated that satellite data are already and will continue to supercede conventional observing systems, in particular radiosondes. However, as mentioned in the chapter *Observing the Atmosphere* by R. Swinbank, it is worth pointing out that radiosondes will remain essential as a very good independent dataset to calibrate the satellite data and also because satellite data are and will remain difficult to use over land for various reasons such as a good description of soil properties, skin temperature and surface emissivity. In that sense, the conventional and space observing systems are clearly complementary.

Acknowledgements. The author is indebted to the research department at ECMWF, in particular Tony McNally who provided a lot of materials presented in this paper and the associated lecture. Adrian Simmons and Frédéric Chevallier also contributed significantly to this paper.

References

Andersson, E., J. Pailleux, J.N. Thépaut, J.R. Eyre, A.P. McNally, G.A. Kelly and P. Courtier, 1994: Use of cloud-cleared radiances in 3D and 4D variational data assimilation. *Quart. J.R. Meteor. Soc.*, **120**, 627-653.

Bouttier, F., and G. Kelly, 2001: Observing system experiments in the ECMWF 4D-Var data assimilation system. Quart. J.R. Meteor. Soc., 127, 1469-1488.

Chédin, A., N. Scott, C. Wahiche and P. Moulinier, 1985: The Improved Initialization Inversion method: a high resolution physical method for temperature retrievals from the TIROS-N series. J. Climate Appl. Meteor., 24, 124-143.

Chevallier, F. and G. Kelly, 2002: Model clouds as seen from space: comparison with geostationary imagery in the 11 micron region. Mon. Wea. Rev., 130, 712-722.

Collard, A., 2000: Assimilation of IASI and AIRS data: Information Content and Quality Control. Proceedings of the ECMWF seminar on the exploitation of the new generation of satellite instruments for Numerical Weather Prediction. Pp 201-224. available from the ECMWF library.

Derber, J., 2000. Assimilation of TOVS, GOES and ATOVS radiances. Proceedings of the ECMWF seminar on the exploitation of the new generation of satellite instruments for Numerical Weather Prediction. Pp 47-56. available from the ECMWF library.

Eyre, J. R., 2000: Planet Earth seen from space: Basic concepts. Proceedings of the ECMWF seminar on the exploitation of the new generation of satellite instruments for Numerical Weather Prediction. Pp 5-20. available from the ECMWF library.

Fourrié, N. and J.-N. Thépaut, 2002: Validation of the NESDIS Near Real time AIRS channel selection. ECMWF Tech. Mem. 390. available from the ECMWF library.

Huang, H. L., and P. Antonelli, 2001: Application of Principle Component Analysis to High Resolution Infrared Measurement Compression and Retrieval, J. Appl. Meteor., 40, 365-388.

Joiner, J., and D. Dee, 2000: An error analysis of radiance and suboptimal retrieval assimilation, Quart. J.R. Meteor. Soc., 126, 1495-1514.

Liou, K. N., 1980: An introduction to atmospheric radiation. Internal Geophysics Series. Vol 26, Academic Press.

Liu, Z. Q. and F. Rabier, 2002: The interaction between model resolution, observation resolution and observation density in data assimilation: A one-dimensional study. Q. J. R. Meteorol. Soc., 128, 1367-1386.

McNally, A. P. and P. D. Watts, 2003: A cloud detection algorithm for high spectral resolution infrared sounders. submitted to Q. J. R. Meteorol. Soc.

Phalippou, L., 1995: Variational retrieval of humidity profile, wind speed and cloud liquid water path from SSM/I: potential for numerical weather prediction. ECMWF Tech. Memorandum 216.available from the ECMWF library

Rabier, F., J.-N. Thépaut and P. Courtier, 1998: Extended assimilation and forecast experiments with a four-dimensional variational assimilation system. Quart. J.R. Meteor. Soc., 124, 1861-1887.

Rabier, F. , N. Fourrié, D. Chafai and P. Prunet, 2002: Channel selection methods for infrared atmospheric sounding interferometer radiances. Q. J. R. Meteorol. Soc., 128, 1011-1027.

Reale, A., C. Novak, M. Chalfant, H. Drahos and D. Gray, 1989: NOAA/NESDIS sounding products. Proceedings of ECMWF/EUMETSAT workshop on the use of satellite data in operational numerical weather prediction. Pp 153-172. available from the ECMWF library

Rodgers, C. D., 1976: Retrieval of Atmospheric Temperature and Composition from Remote Measurements of Thermal Radiation. Rev. Geophys. and Space Phys., 14, 609-624.

Simmons, A. and A. Hollingsworth, 2002: Some aspects of the improvement in skill of numerical weather prediction. Q. J. R. Meteorol. Soc., 128, 647-687.

Stephens, G. L., 1994: Remote sensing of the lower atmosphere. An introduction. Oxford University Press.

Stoffelen, A., 2000: A generic approach for assimilating scatterometer observations. Proceedings of the ECMWF seminar on the exploitation of the new generation of satellite instruments for Numerical Weather Prediction. Pp 73-100. available from the ECMWF library.

RESEARCH SATELLITES

W. A. LAHOZ

DARC, Department of Meteorology, University of Reading
Reading RG6 6BB, UK

1. Introduction

Our knowledge of the Earth System ultimately comes from observations. Although observations have uncertainties and biases, they are the "truth" against which theories and models must be confronted and evaluated. Predictions of the variability of the Earth System require an understanding of the variability in the observations representing the "truth". This requires observations which are: (a) high-quality (*i.e.*, have small errors and biases), (b) consistent (*i.e.*, there is uniformity in the observing system characteristics), and (c) long-term (*i.e.*, the results have statistical significance). Further requirements include global coverage at temporal and space resolution which allow the representation of significant phenomena (*e.g.* climate change, weather extremes).

2. Observations

Examples of observing platforms include ground-based instruments, sondes, balloons, aircraft and remote sounding satellites. Each platform has advantages and disadvantages (see chapter *Observing the Atmosphere*). For example, satellite observations have low spatial and temporal resolution but good global coverage, whereas ground-based instruments, sondes, aircraft, balloons and sondes have high spatial and temporal resolution but poor global coverage. For observing instruments aboard satellite platforms, nadir sounders have good horizontal resolution but poor vertical resolution, whereas limb sounders have poor horizontal resolution but good vertical resolution (Figure 1). Currently, the observations used by the Met Office for operational numerical weather prediction (NWP) come from: aircraft, nadir-viewing satellites, sondes, and the surface. This chapter concerns satellite data.

The heterogeneity of observations, the changes in observing systems, the lack of a significant global observing system before the International Geophysical Year (IGY) of 1957, and the lack of significant satellite observations before the 1970's, makes it difficult to meet the observational requirements for studying the variability in the Earth system. Concerns about these difficulties have led to the implementation of multi-year re-analyses (currently of the atmosphere) of past observations using a fixed, up-to-date data

R. Swinbank et al. (eds.), Data Assimilation for the Earth System, 241–250.
© 2003 *Kluwer Academic Publishers. Printed in the Netherlands.*

assimilation system which attempts to combine, in an optimal manner, information from observations and a sophisticated model (Bengtsson and Shukla 1988, Trenberth and Olson 1988). These re-analyses are used extensively to study issues such as climate change and stratospheric ozone depletion (see chapter *Reanalysis*). Examples of existing re-analyses include a 15-year re-analyses for the period 1979-1993 by ECMWF (Gibson *et al.* 1997), and a 40-year re-analyses from 1957 by NCEP in collaboration with NCAR, (Kalnay *et al.* 1996). Currently, ECMWF are producing a 40-year re-analysis for the period 1957-2000.

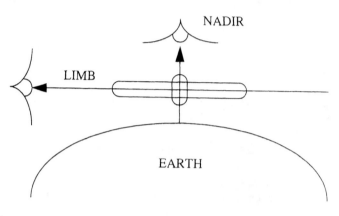

Figure 1. Schematic of nadir and limb observation geometries. The ovals represent the volume associated with typical horizontal and vertical resolutions

It is important to realise that earth observing instruments do not measure directly temperature, ozone or similar geophysical parameters. What they measure is photon counts (level 0 data). Algorithms then transform the level 0 data into radiances (level 1 data). Subsequently, using retrieval techniques (Rodgers 2000), retrievals of profiles or total column amounts are derived (level 2 data). It is the level 2 data that most Earth Observation scientists use as the starting point for their studies. Analyses derived from the assimilation of level 1 and/or 2 data are termed level 3 and/or 4 data.

This level 2 data from a satellite instrument is not a point measurement, but instead represents an observation which is representative of a finite volume in the atmosphere, the dimensions of this volume being determined by the horizontal and vertical resolution of the measurement. How this measurement represents an "average" over this volume is indicated by the averaging kernel of the measurement (Rodgers 2000).

The level 2 data (as well as the level 1 and level 0 data) will have associated with it a number of errors, including random and systematic errors, and the error of representativity. Random errors (sometimes termed precision) have the property that they can be reduced by averaging the data. This is not the case of the systematic error or bias (sometimes termed accuracy). The error of representativity is associated with the extent to which the "average" measurement represents a measurement within any point of the finite volume for which this "average" measurement is appropriate.

3. Satellite data

There are a wide range of satellite data types. Examples include: (a) geostationary satellites located above a fixed spot on the equator (coverage tends to be between 60°N and 60°S, with the focus on the tropics), (b) polar orbiting satellites (coverage tends to be almost global), (c) sun-synchronous satellites (fixed equator crossing time; these are a subset of polar orbiting satellites), (d) non-sun-synchronous satellites (varying equator crossing time). Satellites have instruments which observe the Earth System at a range of frequencies: infrared (IR), visible, ultraviolet (UV) and the microwave. Often, a satellite will have different instruments measuring the same geophysical parameter (*e.g.* ozone, temperature, water vapour) using different frequencies, thereby providing the opportunity to evaluate the observations by instrument intercomparison. This is the case of NASA's UARS and ESA's Envisat satellites.

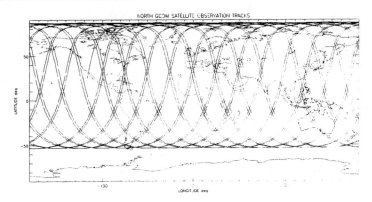

Figure 2: Coverage of the proposed SWIFT instrument (North looking configuration).

Sun-synchronous satellites (*e.g.* ESA's Envisat) have the advantage that the instruments always face away from the sun (and no periodic satellite yaw manoeuvre to avoid sunlight damaging the instruments is necessary), but the disadvantage that they cannot observe the diurnal cycle at a particular location. For example, NO and NO_2, which play a role in determining the distribution of ozone, have strong diurnal cycles. Non-sun-synchronous satellites (*e.g.* NASA's UARS) have the advantage of being able to observe the diurnal cycle at a particular location, but the disadvantage that a periodic satellite yaw manoeuvre is needed to avoid sunlight damaging the instruments. In the case of the SWIFT instrument (Figure 2), this will have the effect that the satellite will either have a North Looking configuration (53°S-87°N) or a South Looking configuration (87°S-53°N).

Satellites can also be divided into operational and research types. Operational satellites provide data in near-real-time (typically within 3 hours of data acquisition), currently the vast majority are nadir viewing (note that ECMWF assimilate retrieved ozone layers from SBUV/2), and are used in NWP. Research satellites provide data off line (typically more than 2 days after data acquisition), use both nadir and limb geometries, and are used in climate studies. Recently, however, research satellite data is beginning to become of interest to the NWP community, mainly for the production of stratospheric ozone analyses.

3.1 OPERATIONAL SATELLITES

Examples of past satellite operational data include the USA TIROS-N, ESSA-N, NOAA-N and GOES-N series which mainly made measurements of temperature and relative humidity. USSR operational satellites have included the COSMOS/METEOR-N series which mainly made measurements of temperature, relative humidity, ozone and clouds. Further details about these satellites can be found in Houghton *et al.* (1984). Examples of current satellite operational data include TOVS/ATOVS, DMSP, ERS-2 and GEOS, which mainly make measurements of temperature, relative humidity and precipitation. The Met Office and the ECMWF operational centres make use of data from these satellites (see chapter *Assimilation of Remote Sensing Observations in Numerical Weather Prediction*).

Examples of future satellite operational data are METOP (including instruments such as IASI, AMSU-A, HIRS/4 and GOME-2) and MSG (including instruments such as SEVIRI and GERB). These satellites will mainly make measurements of temperature, relative humidity, ozone and clouds (*e.g.* for deriving wind information) in the troposphere and lower stratosphere, in some cases over the depth of the troposphere and the stratosphere (IASI, AMSU-A, GOME-2), in other cases at one level (*e.g.* cloud-top from SEVIRI). A notable change in the geophysical parameters which are planned to be measured by operational satellites is the presence of ozone measurements. Interest in ozone is due to: (a) concerns about ozone depletion, (b) recognition that one must develop a coupled climate/chemistry capability to understand, simulate and predict climate change, and (c) the potential to extract wind information in the tropics from tracer observations, including ozone. Currently, ECMWF have an operational ozone product using 4d-var and SBUV/2 ozone (six unevenly spaced layers between 0.1 hPa and the surface) and GOME total column ozone observations (http://www.ecmwf.int).

3.2 RESEARCH SATELLITES

Most of the space agencies in the world: NASA (USA), ESA (Europe), NASDA (Japan) and CSA (Canada) have launched research satellites over the last 10 years, and are involved in plans to launch research satellites over the next 5-10 years. The following list of research satellite missions is meant to be illustrative and is not exhaustive.

3.2.1 NASA satellites
NASA is involved with UARS and EP TOMS (there have been several TOMS instruments, of which EP is one example), and is involved with the EOS series of satellites (Terra, Aqua and Aura). NASA's UARS was launched in September 1991, and

some data are still being received. It carries a suite of instruments making measurements of the stratosphere and mesosphere that are used to infer meteorological and chemical fields. In particular, a number of UARS limb sounder instruments (CLAES, HALOE, HRDI, ISAMS and MLS) have made measurements of temperature, ozone, water vapour, ClO and winds. Measurements of upper tropospheric water vapour have also been made by the MLS instrument. EP TOMS measures total column ozone (*i.e.*, the amount of ozone in a column above a particular location on Earth). The UARS data have been extensively evaluated (UARS special issue in *J. Geophys. Res.*, 1996, Vol. **101**, 9539-10473), and have contributed to our understanding of many aspects of the atmospheric circulation and chemistry (UARS special issue in *J. Atmos. Sci.*, 1994, Vol. **51**, 2781-3105).

EOS Terra was launched in December 1999. It carries on board 5 instruments: ASTER, CERES, MISR, MODIS and MOPITT. EOS Terra provides information on: (1) land surface, water, and ice (ASTER), (2) radiation (CERES), (3) radiation and biosphere parameters (MISR), (4) biological and physical process on land and the ocean (MODIS), and (5) CO and CH_4 in the troposphere, where they are pollution markers (MOPITT). An example of the impact of EOS Terra is provided by MOPITT, which is providing the first global CO (air pollution) measurements from space (http:// www.atmosp.physics.utoronto.ca/MOPITT/home/html).

EOS Aqua (http://eos-aqua.gsfc.nasa.gov) was launched in May 2002. It carries on board 6 instruments: AMSR/E, MODIS, AMSU, AIRS, HSB, CERES. EOS Aqua will provide information on: (1) clouds, radiation and precipitation (AMSR/E), (2) clouds, radiation, aerosol and biosphere parameters (MODIS), (3) temperature and humidity (AMSU, AIRS, HSB), and (4) radiation (CERES). The NWP community is interested in using AIRS temperature and humidity data.

EOS Aura (http://eos-aura.gsfc.nasa.gov) is due for launch in early 2004. It will carry on board 4 instruments: EOS MLS, HIRDLS, TES and OMI. EOS Aura will provide information on: (1) chemistry of the lower stratosphere and upper troposphere, chemistry of the middle and upper atmosphere, upper tropospheric water, and the impact of volcanoes on global change (EOS MLS), (2) temperature and constituents in the upper troposphere, stratosphere and mesosphere (HIRDLS), (3) global maps of tropospheric ozone and its photochemical precursors (TES), and (4) maps of total column ozone (which will continue the TOMS record) and UV-B radiation (OMI). EOS MLS (which builds upon the experience of UARS MLS) and HIRDLS are limb sounders, and TES can be used in both nadir and limb sounder mode. Two of the many innovative aspects of EOS Aura include near-real-time production of ozone data from OMI, and the possibility that limb radiances from HIRDLS may be available for operational use (currently, operational centres only use radiances from nadir instruments).

3.2.2 ESA satellites

ESA is involved with the ERS-2 satellite (which carries the GOME instrument), the ODIN mission, and Envisat. GOME is a nadir sounder which has been making measurements of total column ozone and NO_2 since 1995. Height-resolved ozone information from GOME is now being produced. Early results based on comparison against independent data and assimilation tests suggest the height-resolved ozone product is robust.

ODIN (which also involves the CSA) was launched on February 2001 (http://www.snsb.se/Odin.Odin.html). It carries on board 2 instruments: OSIRIS and SMR. ODIN is providing information on ozone and NO_2 (total columns and profiles).

Envisat (http://envisat.esa.int) was launched on 1 March 2002 (GMT). It carries on board 10 instruments: AATSR, ASAR, DORIS. GOMOS, LRR, MERIS, MIPAS, MWR, RA-2, and SCIAMACHY. Envisat provides information on: (1) temperature, ozone, water vapour and other atmospheric constituents using limb, nadir and occultation geometries (MIPAS, SCIAMACHY, GOMOS), (2) aerosol (AATSR, MERIS), (3) sea-surface-temperatures (AATSR), (4) sea colour (MERIS), (5) land and ocean images (ASAR), (6) land, ice, and ocean monitoring (RA-2), (7) water vapour column and land surface parameters (MWR), and (8) cryosphere and land surface parameters (DORIS). LRR will be used to calibrate RA-2. The broad spectrum of information from Envisat reflects the paradigm that the Earth System should be treated as a whole, and that information from its various components should be integrated.

One of the many innovative aspects of Envisat is the combination of limb and nadir geometries (e.g. in SCIAMACHY itself, or by combining MIPAS limb and SCIAMACHY nadir data), with the potential to extract information on tropospheric constituents such as ozone, which are indicators of pollution and are very difficult to measure. One of the novel features of the evaluation phase of Envisat is the use of data assimilation techniques (either based on GCMs or CTMs) by participants in the ACVT-MA working group.

Other future missions from ESA include CRYOSAT, GOCE and ADM, all of which are part of ESA's Living Planet Programme (http://www.esa.int/export/esaLP). CRYOSAT will measure cryosphere parameters, GOCE will measure the Earth's gravity field, and ADM will measure winds in the troposphere and lower stratosphere. All these missions include novel measurements and are due to fly in the middle of the 2000's.

3.2.3 NASDA satellites

NASDA (http://www.eorc.nasda.go.jp) is involved with the ADEOS mission (launched in 1996), which carried several instruments on board, including ADEOS TOMS (which measured total column ozone) and ILAS (a limb instrument which measured temperature, ozone, water vapour and other atmospheric constituents). Unfortunately, the ADEOS mission only lasted for 10 months.

TRMM (http://www.eorc.nasda.go.jp/TRMM/index_e.htm) is a joint project of Japan (NASDA) and the USA (NASA). It was launched on November 1997. It can observe the rainfall rate in the tropics and its horizontal and vertical distribution, which were not possible by the other measuring methods. Precipitation in the tropics comprises of more than two thirds of the global rainfall. TRMM data can help understand global change and implement environmental policies.

NASDA is also involved with the ADEOS-II and GCOM-A1 missions. ADEOS-II was launched in December 2002. It will carry on board 5 instruments: AMSR, GLI, SeaWinds, POLDER and ILAS-II. It will provide information on: (1) water column, precipitation, and ocean and ice parameters (AMSR), (2) land, ice and biosphere parameters (GLI), (3) winds over the ocean (SeaWinds), (4) radiation parameters

(POLDER), and (5) temperature, ozone and other atmospheric constituents (ILAS-II).

GOSAT (the new name for GCOM-A1) is due to be launched in the late 2000's. The latest information is that it will carry on board 2 instruments: SWIFT (http:// swift.yorku.ca/index.htm) and SOFIS. It will provide information on winds, ozone and other atmospheric chemical species, and carbon dioxide. The University of Reading, in collaboration with Cambridge University, The Met Office (UK) and SERCO, have carried out an Observing System Simulation Experiment (OSSE) to evaluate the potential for science studies of SWIFT ozone and winds measurements (Lahoz *et al.* 2003). Despite their shortcomings (*e.g.* difficulties of interpretation), OSSEs are essentially the only tool available for space agencies to assess the impact of future missions.

3.4 BENEFITS OF RESEARCH SATELLITES

Research satellites have a number of benefits. Because they can have both limb and nadir viewing instruments, they allow the combination of limb/nadir geometries to provide better atmospheric analyses, and provide information on tropospheric ozone (Struthers *et al.* 2002). Because they have instruments which focus on measurements of ozone and of photochemical species which affect the ozone distribution, they provide information for studying ozone depletion, and information which enables the development of coupled climate/chemistry models, and a chemical forecasting capability (important for UV and pollution forecasting). The increasing interest by the operational centres in ozone and chemical forecasting is making research satellites (including the assimilation of limb radiances) more attractive to them.

A number of approaches are implemented to include ozone (and photochemistry) in forecasting models. CTMs have sophisticated photochemistry schemes but, because they are forced by off-line winds, they do not allow feedbacks between the photochemistry, dynamics and radiation. An example of an operational centre which takes this approach is KNMI, which uses GOME observations to make forecasts of total column ozone (http:/ /www.knmi.nl/gome_fd/index.html). Research institutions which also take this approach include BIRA-IASB and NCAR. In contrast, GCMs have sophisticated dynamics and radiation schemes (including feedbacks), but have simple photochemical schemes (*e.g.*, the Cariolle scheme). Examples of operational centres which take this approach include the Met Office, and ECMWF (which incorporates the Cariolle scheme into its operational ozone assimilation system). A research institution which also takes this approach is DARC. One way forward is to couple the dynamics and radiation of a GCM with the sophisticated photochemistry of a CTM. Such an approach is being investigated by a number of research and operational groups in Europe, Canada and Japan. The SPARC Data Assimilation Working Group provides a forum for these efforts (http:// dao.gsfc.nasa.gov/DAO_people/ivanka/report.html).

4. Future prospects

There are a number of activities, all involving the use of data from research satellites, which are likely to become important in the future. These include: (1) the operational use of research satellite data by significant numbers of operational centres. Examples of data which could be used include ozone (already being assimilated operationally by ECMWF) and stratospheric water vapour. (2) the assimilation of limb radiances by research and operational groups. A lot of work is being done on developing fast and accurate forward models and the interface between the forward model and the assimilation. Progress is more advanced in the case of IR radiances than in the case of UV and visible radiances. (3) chemical forecasting, including tropospheric pollution forecasting. (4) an Earth System approach to environmental and associated socio-economic issues. This Earth System approach would incorporate the biosphere and the carbon cycle, and the coupling of all components of the Earth System. These activities have hitherto not been as advanced as atmospheric modelling and ocean modelling.

5. List of acronyms

AATSR: Advanced Along Track Scanning Radiometer
ACVT-MA: Atmospheric Chemistry Validation Team - Modelling and Assimilation
ADEOS: ADvanced Earth Observing Satellite
ADM: Atmospheric Dynamics Mission
AIRS: Atmospheric InfraRed Sounder
AMSR: Advanced Microwave Scanning Radiometer
AMSU: Advanced Microwave Sounding Unit
ASAR: Advanced Synthetic Aperture Radar
ASTER: Advanced Spaceborne Thermal Emission and reflection Radiometer
ATOVS: Advanced TOVS
BIRA-IASB: Belgisch Instituut voor Ruimte Aeronomie (Belgian Institute of Space Aeronomy)
CERES: Clouds and the Earth's Radiant Energy System
CLAES: Cryogenic Limb Array Etalon Spectrometer
CSA: Canadian Space Agency
CTM: Chemistry-Transport Model
DARC: Data Assimilation Research Centre, UK
DMSP: Defense Meteorological Satellite Program
DORIS: Doppler Orbitography and Radiopositioning Integrated by Satellite
ECMWF: European Centre for Medium-range Weather Forecasts
EOS: Earth Observing System
EOS MLS: EOS Microwave Limb Sounder
EP: Earth Probe
ERS: European Research Satellite
ESA: European Space Agency
ESSA: Environmental Survey SAtellite
GCM: General Circulation Model
GCOM: Global Change Observation Mission
GERB: Geostationary Earth Radiation Budget experiment

GLI: GLobal Imager
GOCE: Gravity field and steady-state OCEan circulation
GOES: Geostationary Operational Environmental Satellite
GOME: Global Ozone Monitoring Experiment
GOMOS: Global Ozone Monitoring by Occultation of Stars
GOSAT: Greenhouse gas Observing SATellite
HALOE: Halogen Occultation Sounder
HIRDLS: HIgh Resolution Dynamics Limb Sounder
HIRS/4: High resolution Infrared Radiation Sounder/4
HRDI: High Resolution Doppler Imager
HSB: Humidity Sounder for Brazil
IASI: Infrared Atmospheric Sounding Interferometer
ILAS: Improved Limb Atmospheric Spectrometer
ISAMS: Improved Stratospheric And Mesospheric Sounder
KNMI: Koninklijk Nederlands Meteorologisch Instituut (The Royal Dutch Meteorological Institute)
LRR: Laser RetroReflector
MERIS: MEdium Resolution Imaging Spectrometer
MIPAS: Michelson Interferometer for Passive Atmospheric Sounding
MISR: Multi-angle Imaging SpectroRadiometer
MLS: Microwave Limb Sounder
MODIS: MODerate resolution Imaging Spectroradiometer
MOPITT: Measurements Of Pollution In The Troposphere
MSG: Meteosat Second Generation
MWR: MicroWave Radiometer
NASA: National Aeronautics and Space Administration
NASDA: NAtional Space Development Agency of Japan
NCAR: National Center for Atmospheric Research
NCEP: National Centers for Environmental Prediction
NOAA: National Oceanic and Atmospheric Administration
OMI: Ozone Monitoring Instrument
OSIRIS: Optical Spectrograph and InfraRed Imager System
POLDER: POLarization and Directionality of the Earth's Reflectance
RA-2: Radar Altimeter 2
SBUV/2: Solar Backscatter Ultra-Violet/2
SCIAMACHY: SCanning Imaging Absorption spectroMeter for Atmospheric CHartographY
SEVIRI: Spinning Enhanced Visible and InfraRed Imager
SOFIS: Solar Occultation FourIer transform Spectrometer
SMR: Sub-Millimeter Radiometer
SPARC: Stratospheric Processes And their Role in Climate
SWIFT: Stratospheric Wind Interferometer For Transport Studies
TES: Tropospheric Emission Spectrometer
TIROS: Television and InfraRed Observations Satellite
TOMS: Total Ozone Mapping Spectrometer

TOVS: TIROS Operational Vertical Sounder
TRMM: Tropical Rainfall Measuring Mission
UARS: Upper Atmosphere Research Satellite

References

Bengtsson, L., and J. Shukla, 1988: Integration of space and in situ observations to study global climate change. *Bull. Amer. Meteorol. Soc.,* **69**, 1130-1143.

Gibson, J.K., P. Kållberg, S. Uppala, A. Hernández, A. Nomura, and E. Serrano, 1997: ECMWF Re-analysis Project report. 1. ERA description. European Centre for Medium-range Weather Forecasts, Reading, UK.

Houghton, J.T., F.W. Taylor, and C.D. Rodgers, 1984: *Remote Sounding of Atmospheres.* Cambridge University Press.

Kalnay, E., K. Kanamitsu, R. Kistler, W. Collins, D. Deaven, L. Gandin, M. Iredell, S. Saha, G. White, J. Woollen, Y. Zhu, M. Chelliah, W. Ebisuzaki, W. Higgins, J. Janowiak, K.C. Mo, C Ropelewski, J. Wang, A. Leetma, R. Reynolds, R. Jenne, and D. Joseph, 1996: The NCEP/NCAR 40-Year Reanalysis Project. *Bull. Amer. Meteorol. Soc.,* **77**, 437-471.

Lahoz, W.A., R. Brugge, S. Migliorini, D. Lary, A. Lee, R. Swinbank, D. Jackson, 2003: SWIFT -- Application of Stratospheric Wind Measurements to Process Studies and Meteorology. Final Report, ESTEC Contract No. 15344/01/NL/MM, March 2003.

Rodgers, C.D., 2000: *Inverse Methods for Atmospheric Sounding: Theory and Practice.* World Scientific.

Struthers, H., R. Brugge, W. A. Lahoz, A. O'Neill, and R. Swinbank, 2002: Assimilation of ozone profiles and total column measurements into a global General Circulation Model. *J. Geophys. Res.,* **107**, 10.1029/2001JD000957.

Trenberth, K.E., and J.G. Olson, 1988: An evaluation and intercomparison of global analyses from NMC and ECMWF. *Bull. Amer. Meteorol. Soc.,* **69**, 1045-1057.

THE STRUCTURE AND EVOLUTION OF THE ATMOSPHERE

ALAN O'NEILL
Data Assimilation Research Centre
University of Reading, Reading, U.K.

The application of data assimilation to observations of a physical system, e.g. the atmosphere, demands a good knowledge of the structure and evolution of the system, as well as the factors that govern them. With such knowledge, it is generally possible to make at least an intelligent estimate of quantities that lie at the core of the data assimilation algorithm: the error covariance matrices for the observations and the model background fields. These matrices are (or should be) state dependent.

In a short chapter, it is impossible to give a useful and balanced account of such a diverse subject as atmospheric structure and evolution. It is preferable to refer the reader to some of the excellent references that cover key areas of the subject. These references, supplemented by data resources available on the web, provide much of the background knowledge needed to turn to the data assimilation problem intelligently.

For a first-class introduction to the fundamentals of atmospheric science – covering composition and structure, governing equations, fluid dynamics and simple atmospheric chemistry – the book by Salby (1996) is hard to beat. Another good text, covering similar ground but somewhat more briefly is Andrews (2000). Basic dynamical meteorology is treated in the classic text by Holton (1992), while the more specialised, and advanced, book by Andrews et al. (1987) is a key reference for an exposition of the dynamics and transport of the middle atmosphere (the stratosphere and mesosphere).

For the chemical and transport processes governing the distribution of trace gases in the atmosphere, Brasseur and Solomon (1986) can be recommended as a comprehensive and detailed treatment. The book by Wayne (2000) is also excellent, and might be the first choice for an introduction to the subject. Both of these books present some observational data on the distribution of trace gases in the atmosphere.

Large-scale transport and mixing, associated the atmospheric circulation, play a very important role, in addition to photochemistry, in determining the distribution and evolution of trace gases in the atmosphere, e.g. ozone. The processes involved in transport and mixing are diverse: they include large-scale, overturning circulations associated with radiative heating and cooling in the atmosphere, as well as mixing due to "wave breaking" and instabilities in the flow. As yet, there are no books devoted specifically to this subject. The influential review article by Holton et al. (1995) does, however, give an illuminating treatment of stratosphere-troposphere exchange and related issues.

Unfortunately, there is a lack of reference books that attempt to gather together the wealth of observational data now available to give a broad view of the structure and

R. Swinbank et al. (eds.), Data Assimilation for the Earth System, 251–252.

evolution of the atmosphere. The book by Peixoto and Oort (1992) is a starting point, at least for the structure of the lower atmosphere, though there are very few time series giving a sense of the atmospheric *evolution*. Satellite data, which afford a global, time-evolving view of the atmosphere are increasingly used in modern textbooks, but perhaps the best current sources of atmospheric images are available at web sites and data centres connected with programmes in earth observation.

Table 1. Web Sites

BADC	http://badc.nerc.ac.uk
EOS	http://eospso.gsfc.nasa.gov/eospso_homepage.html
ESA	http://www.esa.int
UARS	http://umpgal.gsfc.nasa.gov/uars-science.html

Web sites (see Table 1) are frequently appearing as new programmes in earth observation are instituted. NASA has an excellent site on its Earth Observing System (EOS), with links to imagery and NASA data centers. An excellent site for data on atmospheric structure, notably of trace gases is the NASA site for the Upper Atmosphere Research Satellite (UARS), which was launched in 1991, and provided a wealth of data from the upper troposphere to mesosphere. The European Space Agency's (ESA) site is also worth browsing for recent imagery and links to data products from ESA missions (As one of many data centers, the British Atmospheric Data Centre (BADC) is a valuable archive of global meteorological data from the Met Office and ECMWF, as well as data from a variety of ground-based observational campaigns and satellite missions.

References

Andrews, D.G. (2000) *An Introduction to Atmospheric Physics*, Cambridge University Press.
Andrews, D.G., Holton, J.R. and Leovy, C.B. (1987) *Middle Atmosphere Dynamics*, Academic Press.
Brasseur, G. and Solomon, S. (1986) *Aeronomy of the Middle Atmosphere*, Reidel, second edition.
Holton, J.R. (1992) *An Introduction to Dynamic Meteorology*, Academic Press, third edition.
Holton, J.R., Haynes, P.H., McIntyre, M.E., Douglass, A.R., Rood, R.B., Pfister, L. (1995) Stratosphere-troposphere exchange, *Rev. Geophys.*, **33**, 403-39.
Peixoto, J.P. and Oort, A. (1992) *Physics of Climate*, American Institute of Physics, New York.
Salby, M.L. (1996) *Fundamentals of Atmospheric Physics*, Academic Press.
Wayne, R.P. (2000) *Chemistry of Atmospheres*, Oxford University Press, third edition.

INTRODUCTION TO ATMOSPHERIC PHOTOCHEMICAL MODELLING

BORIS KHATTATOV, JEAN FRANCOIS LAMARQUE, GUY
BRASSEUR, GEOFF TYNDALL AND JOHN ORLANDO
*Atmospheric Chemistry Division, National Center for Atmospheric Research,
Boulder, CO, USA*

1. Importance of Atmospheric Chemistry

Atmospheric photochemical processes often occurring at altitudes of tens of kilometers above the Earth's surface can be of paramount importance to the existence of life on Earth. The so-called ozone layer formed by a complex variety of chemical and photodissociation processes at altitudes near 22 km, absorbs harmful ultraviolet radiation before it reaches the ground where it can damage living organisms. The deoxyribose nucleic acid molecules (D.N.As.) of most organisms absorb very strongly at wavelengths around 300 nm. Had this radiation not been prevented from reaching the ground, it would have caused immediate and significant tissue damage and lead to formation of cancer cells and genetic mutations.

Knowing what processes control formation and destruction of ozone molecules in the stratosphere is extremely important for monitoring and predicting changes in ozone abundances. The "ozone hole" phenomenon in the Southern Hemisphere discovered in the mid-eighties clearly demonstrated how human (anthropogenic) activities can destroy the natural chemical balance in the atmosphere and potentially lead to disastrous consequences.

It took years of scientific studies to discover the complete chain of related physical effects and chemical reactions and prove that the dramatic rapid destruction of ozone was originally caused by industrial emissions of chlorofluorocarbons

Chlorine and bromine released from these molecules in the process of photodissociation can act as catalysts in fast ozone loss cycles. This property combined with persistent patterns of atmospheric circulation, lack of continents in the Southern Hemisphere, very cold temperatures in the Antarctic region and absence of solar radiation during the long Antarctic winter eventually causes almost complete destruction of ozone in the lower stratosphere, where most of the ozone resides.

Recently, the so-called mini ozone holes were discovered in the Northern Hemisphere above Western Europe and Russia. The exact reasons for their formation are still debated. The fact that these holes occur above some of the most populated regions on Earth is troubling.

The risk of global warming highlights another strong link between concentrations of atmospheric trace gases and global environmental conditions. Some atmospheric gases, for instance, CO_2 and H_2O, trap infrared radiation emitted by the Earth's surface.

R. Swinbank et al. (eds.), Data Assimilation for the Earth System, 253–262.
© *2003 Kluwer Academic Publishers. Printed in the Netherlands.*

Increased concentrations of these gases are likely to lead to temperature increases in the troposphere since normally this radiation would have escaped to space. While the direct connection between recent increases in atmospheric carbon dioxide and observed global temperature increases is difficult to unmask due to existence of sources of natural temperature variability, there exists enough scientific evidence pointing to this connection beyond a reasonable doubt (http://www.wmo.ch/web/Press/Press670.html).

One of the most common and clear illustrations of the importance of atmospheric chemistry to our well-being has to do with the air pollution in the troposphere. Carbon monoxide, nitrogen and complex organic compounds and small particulate matter contained in car exhaust and by-products of industrial production and incomplete combustion lead, through a complex chain of chemical transformations, to formation of smog, acid rains, and tropospheric ozone. While stratospheric ozone shields us from harmful radiation, recent studies show that increased levels of ozone in the troposphere can lead to complications in patients with cardio-vascular diseases and even increased rates of hospital admissions and mortalities recorded the next day.

Chains of chemical transformations leading to formation of a particular chemical are often very long and complex. This is particularly true for tropospheric chemistry, where one wishes to follow concentrations and interactions between hundreds to thousands of chemical compounds. Needless to say, modelling and understanding these processes can be a daunting task requiring significant resources and extensive observational verification. Data assimilation can facilitate this task by uncovering and making use of relationships between observed and modelled parameters in a mathematically consistent and rigorous fashion.

In the following sections we will give a brief practical overview of elementary atmospheric photochemistry. A much more detailed and well explained presentation can be found, for instance, in an excellent textbook by Brasseur et al. (1999).

2. Elementary Chemical Processes

The Earth's atmosphere can be thought of as a combustion system where the energy of the sun drives a variety of chemical transformations. The exact composition of the atmosphere is determined by a complex chemical mechanism that consists of hundreds to thousands of elementary chemical reactions. For example, the photolysis of O_2 in the stratosphere is responsible for initiating the chemistry involved in the production of ozone, O_3

$$O_2 + h\nu \rightarrow O + O$$
$$O_2 + O + M \rightarrow O_3 + M$$

The process of absorption of a photon by a molecule results in a change in the energy level of the model. In the process the photon disappears. Photons come in different "colours" or frequencies corresponding to different energies. "Blue" photons have more energy than "yellow" and more energy than "red". High energy photons can break-up molecules; this process is called photodissociation or photolysis. Examples of photolysis reactions are shown below:

$$O_2 + h\nu \rightarrow O + O$$
$$O_3 + h\nu \rightarrow O_2 + O$$
$$NO_2 + h\nu \rightarrow NO + O$$

Rates of the photodissociation reactions depend on the amount of sunlight (number of photons) and absorption cross-section of the molecule. These rates are often called photodissociation coefficients or photolysis rates, J. Specifics of calculation of these rates can be found, for instance, in Brasseur et $al.$ (1999). For the purposes of this presentation, it is important to note that the rate of change of a particular chemical due to photodissociation is directly proportional to the corresponding photolysis rate multiplied by the chemical's concentration:

$$\frac{d[O_3]}{dt} = -J_{O_3} \cdot [O_3]$$

The most common reactions between atmospheric chemicals are bimolecular (by number of reagents) reactions of the type

$$AB + C \rightarrow A + BC$$

An example of such reaction is

$$NO + O_3 \rightarrow NO_2 + O_2$$

The rate of disappearance of reagents, equal to the rate of appearance of the products, in this reaction is

$$\frac{d[AB]}{dt} = \frac{d[C]}{dt} = -\frac{d[A]}{dt} = -\frac{d[BC]}{dt} = k \cdot [AB] \cdot [C]$$

k in this equation is called the reaction rate. This coefficient is usually a strong function of temperature:

$$k = A \cdot \exp(-\Delta E / RT)$$

Another common type of atmospheric chemical reaction is the tri-molecular reaction of the type

$$A + B + C \rightarrow products$$

Some examples are

$$NO + NO + O_2 \rightarrow NO_2 + NO_2$$
$$O_2 + O + N_2 \rightarrow O_3 + N_2$$

Similarly to bimolecular reactions, the rate of disappearance of reagents (and appearance of products) is given by

$$\frac{d[A]}{dt} = \frac{d[B]}{dt} = \frac{d[C]}{dt} = -\frac{d[products]}{dt} = k \cdot [A] \cdot [B] \cdot [C]$$

We will now consider the case of ozone photochemistry to illustrate how one constructs mathematical models describing the temporal evolution of atmospheric trace gases.

3. Ozone Photochemistry

The concentration of ozone in the stratosphere is determined by a balance between its production and loss rates. In a purely O_2-N_2 atmosphere, processes controlling ozone concentration are:

$$O_2 + h\nu \rightarrow O + O \qquad (J1)$$
$$O_2 + O + M \rightarrow O_3 + M \qquad (k1)$$
$$O_3 + h\nu \rightarrow O_2 + O \qquad (J2)$$
$$O + O_3 \rightarrow O_2 + O_2 \qquad (k2)$$

In this notation M stands for any inert molecule such as O_2 or N_2. The resulting coupled differential equations controlling the evolution of ozone and atomic oxygen are as follows:

$$\frac{d[O]}{dt} = 2 \cdot J1 \cdot [O_2] - k1 \cdot [O][O_2] + J2[O_3] - k2 \cdot [O][O_3]$$

$$\frac{d[O_3]}{dt} = k1 \cdot [O][O_2] - J2[O_3] - k2 \cdot [O][O_3]$$

Given initial conditions, that is concentrations of ozone and atomic oxygen at time $t=t_0$, one can solve this system of coupled first order differential equations and obtain $[O_3]$ and $[O]$ as a function of time.

In reality, concentration of ozone is directly and indirectly affected by many other chemicals. Particularly important are the so-called catalytic ozone destruction cycles of the following form:

$$X + O_3 \rightarrow XO + O_2$$
$$\underline{O + XO \rightarrow O_2 + X}$$
$$\text{Overall:} \quad O + O_3 \rightarrow O_2 + O_2$$

In this cycle chemical X leads to destruction of one ozone molecule without being destroyed itself. As a result, a single molecule X can destroy a very large number of ozone molecules. Some of the most important catalytic ozone loss cycles in the stratosphere involve NO, Cl, and OH in place of X, for example:

$$NO + O_3 \rightarrow NO_2 + O_2$$
$$\underline{O + NO_2 \rightarrow O_2 + NO}$$

Overall: $O + O_3 \rightarrow O_2 + O_2$

Table 1 lists an example of a minimal (by modern standards) set of photochemical reactions one needs to take into account when modelling stratospheric chemistry.

Table 1. A typical set of stratospheric photochemical reactions.

k001: N2O5 + H2O(a) → 2*HNO3;	k042: HCL + OH → H2O + CL;
k002: O + O3 → 2*O2;	k043: HCL + O → OH+ CL;
k003: O1D + O3 → 2*O2;	k044: HOCL + OH → H2O + CLO;
k004: O1D + N2 → O + N2;	k045: CLONO2 + O → CLO +NO3;
k005: O1D + O2 → O + O2;	k046: CLONO2 + OH → HOCL +NO3;
k006: O1D + H2O → 2*OH;	k047: CLO+NO2+ M → CLONO2 + M;
k007: O1D + H2 → H + OH;	k050: NO2 + O → NO+ O2;
k008: O1D + CH4 → OH+ CH3;	k051: NO+ O3 → NO2 + O2;
k009: O + O2+ M → O3+ M;	k052: NO+ HO2 → NO2 + OH;
k016: OH+ CO → CO2 + H;	k053: NO2 + O3 → NO3 + O2;
k017: CH4 + OH → CH3 + H2O;	k054: HNO3 + OH → NO3 + H2O;
k019: H2 + OH → H2O + H;	k055: HNO4 + OH → H2O + O2 +NO2;
k020: H + O3 → O2 + OH;	k057: NO2 + OH+ M → HNO3 + M;
k021: H + HO2 → 2*OH;	k058: NO2 + HO2 + M → HNO4 + M;
k022: OH+ O → O2 + H;	k059: NO3 + NO2 + M → N2O5 + M;
k023: OH+ O3 → O2 + HO2;	k060: N2O5 + M → NO2 + NO3 + M;
k024: OH+ OH → H2O + O;	k061: HNO4 + M → HO2 + NO2 + M;
k025: OH+ HO2 → H2O + O2;	j001: O2 → 2*O
k026: HO2 + O3 → 2*O2 + OH;	j002: O3 → O2 + O
k027: HO2 + O → O2 + OH;	j003: O3 → O2 + O1D
k028: HO2 + HO2 → H2O2 + O2;	j004: HO2 → O + OH
k029: H2O2 + OH → H2O + HO2;	j005: H2O2 → 2*OH
k030: H + O2+ M → HO2 + M;	j006: NO2 → NO+ O
k031: CL+ O3 → CLO + O2;	j007: NO3 → NO2 + O
k032: CL+ CH4 → HCL + CH3;	j008: NO3 → NO+ O2
k033: CL+ H2 → H + HCL;	j009: N2O5 → NO2 + NO3
k034: CL+ HO2 → O2 + HCL;	j010: HNO3 → OH+ NO2
k035: CL+ HO2 → OH+ CLO;	j011: HNO4 → OH+ NO3
k036: CL+ H2O2 → HO2 + HCL;	j012: HNO4 → HO2 + NO2
k038: CLO + O → CL+ O2;	j016: HOCL → OH+ CL
k039: CLO + NO → CL+ NO2;	j017: CLONO2 → CL+ NO3
k040: CLO + OH → HO2 + CL;	j018: CLONO2 → CL+ NO2 + O
k041: CLO + HO2 → HOCL + O2;	j026: HCL → H + CL

If one takes into account all these reactions in Table 1, the new set of coupled first-order ordinary differential equations (ODEs) becomes significantly more complex. Instead of solving a system of two coupled differential equations, as in the case of pure O_2-N_2 atmosphere, one now has to solve a system of 19 coupled first-order ODEs. Some of these equations are listed in Table 2, several equations have been skipped to save space.

Table 2. A system of coupled ODEs corresponding to reactions in Table 1.

[1:] d[H   ]/dt = j026*HCL + k007*O1D*H2 + k016*OH*CO + k019*H2*OH - k020*H*O3 - k021*H*HO2 + k022*OH*O - k030*H*O2*M + k033*CL*H2

[2:] d[OH   ]/dt = j004*HO2 + 2*j005*H2O2 + j010*HNO3 + j011*HNO4 + j016*HOCL + 2*k006*O1D*H2O + k007*O1D*H2 + k008*O1D*CH4 - k016*OH*CO - k017*CH4*OH - k019*H2*OH + k020*H*O3 + 2*k021*H*HO2 - k022*OH*O - k023*OH*O3 - 2*k024*OH*OH - k025*OH*HO2 + k026*HO2*O3 + k027*HO2*O - k029*H2O2*OH + k035*CL*HO2 - k040*CLO*OH - k042*HCL*OH + k043*HCL*O - k044*HOCL*OH - k046*CLONO2*OH + k052*NO*HO2 - k054*HNO3*O - k055*HNO4*OH - k057*NO2*OH*M

....

....

....

[15:] d[N2O5   ]/dt = - j009*N2O5 - k001*N2O5*H2O(a) + k059*NO3*NO2*M - k060*N2O5*M

[16:] d[O   ]/dt = 2*j001*O2 + j002*O3 + j004*HO2 + j006*NO2+ j007*NO3 + j018*CLONO2 - k002*O*O3 + k004*O1D*N2+ k005*O1D*O2 - k009*O*O2*M - k022*OH*O + k024*OH*OH- k027*HO2*O - k038*CLO*O - k043*HCL*O - k045*CLONO2*O - k050*NO2*O

[17:] d[O1D   ]/dt = j003*O3 - k003*O1D*O3 - k004*O1D*N2 - k005*O1D*O2 - k006*O1D*H2O - k007*O1D*H2 - k008*O1D*CH4

[18:] d[O3   ]/dt = - j002*O3 - j003*O3 - k002*O*O3 - k003*O1D*O3 + k009*O*O2*M - k020*H*O3 - k023*OH*O3 - k026*HO2*O3- k031*CL*O3 - k051*NO*O3 - k053*NO2*O3

---

4. Photochemical Models

Formally, in order to compute concentrations of stratospheric trace gases at all times one has to solve the following system of equations:

$$\frac{d\vec{x}}{dt} = \mathbf{f}(\vec{x}(t))$$

where vector x contains instantaneous values of concentrations of all 19 related chemicals:

$$\vec{X} = \left[H, OH, HO2, H2O2, CL, CLO, ..., HNO3, HNO4, N2O5, O, O1D, O3, H2O(a) \right]^T$$

and non-linear vector function **f** represents the system of equations described in Table 2.

After an appropriate discrete numerical scheme is introduced, time evolution of **x** can be formally written as

$$\vec{x}(t + \Delta t) = \mathbf{M}(\vec{x}(t))$$

where **M** is often called the propagator operator. In practice the propagator is given by a computer program that calculates vector **x** at a later time given a value of **x** at the initial time. A typical solution for several components of vector **x** is shown in Figure 1.

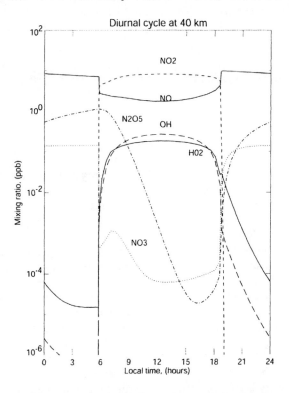

Figure 1. Time evolution of several stratospheric trace gases.

Atmospheric chemistry mechanisms are the most computationally intensive components of photochemical models of the atmosphere. This computational burden is partly due to the fact that atmospheric chemical kinetic systems are very "stiff", *i.e.*, they include reactions ranging from very fast to very slow; this requires the use of elaborate numerical integration schemes ("stiff solvers"). To demonstrate this point, note the very sharp changes in concentrations of several chemicals in Figure 1 at around 6 am and 7 pm. These changes are due to rapid variability of the photodissociation coefficients J at sunrise and sunset. At these times concentrations of many trace gases can increase or decrease by several orders of magnitude on timescales of seconds. On the other hand, concentrations of other chemicals, such as ozone or nitric acid, are fairly constant with time. The development of photochemical models which accurately describe atmospheric chemistry while being computationally efficient is a difficult undertaking and an active area of research.

Computer codes for constructing and solving the described photochemical model can be found at: http://acd.ucar.edu/~boris/research.htm.

In practice chemical solvers must be coupled with dynamical, radiation transfer, land surface and other modules and solved globally in order to properly simulate the chemical composition of the atmosphere. Here we present a very brief overview of global three-dimensional chemistry-transport models (CTM). For a more detailed description the reader is referred to Brasseur *et al.* (1999).

At the heart of any CTM is the continuity or mass conservation equation:

$$\frac{\partial n}{\partial t} + u\frac{\partial n}{\partial x} + v\frac{\partial n}{\partial y} + w\frac{\partial n}{\partial z} = D\left(\frac{\partial^2 n}{\partial x^2} + \frac{\partial^2 n}{\partial y^2} + \frac{\partial^2 n}{\partial z^2}\right) + P - L(n)$$

where

n   - is the concentration of the atmospheric trace gas in question

u,v,w - are the three components of the wind

D   - is the diffusion coefficient describing both molecular and turbulent diffusion

P   - is the source term for the species, including both chemical production and emissions

L   - is the chemical loss term

Generally, this equation has to be solved for each trace gas included in the model formulation. For numerical solution of this equation a 3-D global grid has to be introduced and an appropriate finite difference numerical scheme implemented either in geographical or spectral space. Brasseur *et al.* (1999) show several examples of numerical schemes and discuss their advantages and disadvantages.

Additionally, many physical processes affecting distributions of atmospheric chemicals act on spatial and/or temporal scales much shorter than the finite grid spacing in the numerical model. In these cases one has to use empirical parameterizations in order to account for these effects. Most common parameterizations address processes in the planetary boundary layer, convective mass fluxes, turbulent and molecular diffusion and land surface processes.

Computer programs representing 3-D CTMs are very complex. They consist of thousands lines of code and many modules -- radiation, chemistry, large-scale transport, sub-grid parameterizations (*e.g.*, convection, boundary layer), land surface models, etc. Entire careers have been spent developing foundations of a particular module. Usually these programs are written by many scientists, post-docs, and graduate students with different programming styles and skills and over many years. This can easily result in un-readable and un-documented codes that are prone to errors and make it hard to introduce new modules or to bring in data assimilation. Earth system modelling has become a tight mix of computer sciences, mathematics, physics and chemistry, with computer sciences component becoming crucial for continuing healthy development in this area.

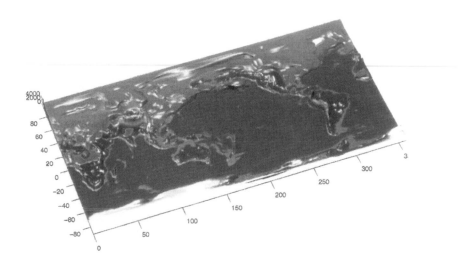

Figure 2. A global iso-surface of CO distribution computed with MOZART 2 model.

As an example of CTM simulations consider results obtained with the NCAR CTM MOZART 2 (Model of OZone And Related Tracers). The model description can be found at http://acd.ucar.edu/models/MOZART/. In the configuration used here the model provides the global distribution of 60 chemical constituents between the Earth's surface and 60 km. The concentration of each species is predicted by individually solving a mass conservation equation taking into account advective, convective, and diffusive transport as well as surface and in-situ emissions, photochemical conversions, and wet and dry surface deposition. Dynamical fields in the model are taken from the ECMWF reanalysis. Figure 2 shows an example of model simulations – global distribution of carbon monoxide in the atmosphere presented as a 180 ppt (parts per trillion) isosurface. An animation of model simulations can be found at http://earthobservatory.nasa.gov/Study/GlobalTraveler/Images/co_transport_sor.mov

5. Summary

Monitoring and understanding changes in atmospheric chemical composition is crucial for maintaining healthy environmental conditions. Modern computer models of atmospheric chemistry take into account a variety of complex chemical and physical processes, acting on very different time and space scales. Some of these processes are well understood while others are not. Systematic comparisons of model results with available observations provide the only means of verifying our understanding of the environment and lead to model improvements.

Methods of data assimilation provide a unified mathematical framework for objective analysis of discrepancies between model results (our theoretical knowledge) and observations (the reality). Such analysis should lead to advances in our ability to model and forecast changes in the environment. Practical applications of data assimilation in studies of atmospheric chemical composition will be discussed in the chapter *Multivariate Chemical Data Assimilation*.

References

Brasseur, G. P., J.J., Orlando, and G. S. Tyndall (eds)., 1999: *Atmospheric Chemistry and Global Change*, Oxford University Press.

Suggested Reading

Madronich, S., 1993: The atmosphere and UV-B radiation at ground level, in: *Environmental UV Photobiology*, A. Young et al., (eds.), Plenum Press New York.

Ramaroson, R., M. Pirre, and D. Cariolle, 1992: A box model for on-line computations of diurnal variations in a 1-D model: Potential for application in multidimensional cases. *Ann. Geophys.*, **10**, 416.

OZONE ASSIMILATION

RICHARD B. ROOD
NASA/Goddard Space Flight Center
Greenbelt, MD, USA

1. Introduction

This chapter discusses the assimilation of atmospheric ozone, or more generally, the assimilation of trace constituents into a model that explicitly represents atmospheric transport and chemistry. The chemical production and loss terms will be parametrized using methods similar to those described in Stolarski and Douglass (1985). To derive this parametrization, the constituent being assimilated is assumed to be in near equilibrium with the environmental conditions. Perturbations from that chemical equilibrium, caused by transport or temperature dependent chemistry, return to equilibrium with calculated time constants. The time constants vary from many months to a few minutes as a function of season, altitude, and latitude. Besides ozone, the techniques discussed here are directly applicable to satellite measurements of, for instance, nitrous oxide, methane, carbon monoxide, the chlorofluoromethanes, and water vapour. Assimilation of reactive constituents whose concentrations are dependent on the concentration of other constituents, *i.e.* "full chemistry," is discussed in the chapters *Introduction to Atmospheric Photochemical Modelling* and *Multivariate Chemical Data Assimilation*. A list of reading material that serves as an introduction to the discipline is given at the end.

The next section will discuss the motivation to investigate the assimilation of ozone as well as some aspects of the ozone problem that require special attention if assimilated ozone data are to be of sufficient quality to add information beyond what is directly available from the observational record. Following that is a discussion of the ozone observing system and a description of the ozone data assimilation system used at NASA's Goddard Space Flight Center. It will be stated as a goal that an assimilation-based ozone product of geophysical interest will require global mapping with vertical resolution comparable to that directly observed by limb-viewing research instruments, order 3 km or better. The sensitivity of the vertical profile to changes in the assimilation system will be discussed. Finally, a summary and identification of near-term challenges will be given.

R. Swinbank et al. (eds.), Data Assimilation for the Earth System, 263–277.

2. Goals and special considerations

Ozone is an important trace gas in the atmosphere. The study of ozone is often broken into two major research areas. The first is stratospheric ozone. In the stratosphere ozone absorbs solar radiation and is also active in the infrared spectrum. Therefore, ozone is important to the stratospheric energy balance. The absorption of solar ultraviolet radiation reduces the amount of ultraviolet that reaches the Earth's surface making it important in biological processes. Further, ozone is highly reactive and central to the photochemistry of many important trace gases. The second primary research area is tropospheric ozone. In the troposphere, ozone is an indicator of polluted air. Because of its reactivity, ozone in the troposphere is hazardous to many life forms, including humans, as well as damaging to many other materials. Most of the ozone, greater than 80%, is contained in the stratosphere. A good summary of atmospheric ozone and its chemistry is found in Dessler (2000) and Hobbs (2000).

There are a number of goals that motivate the assimilation of ozone. These are listed below with brief descriptions of the goal.

1) Mapping: There are spatial and temporal gaps in the ozone observing system. A basic goal of ozone assimilation is to provide vertically resolved global maps of ozone.

2) Short-term ozone forecasting: There is interest in providing operational ozone forecasts in order to predict the fluctuations of ultraviolet radiation at the surface of the earth (*e.g.* Long *et al.*, 1996).

3) Chemical constraints: Ozone is important in many chemical cycles. Assimilation of ozone into a chemistry model provides constraints on other observed constituents and helps to provide estimates of unobserved constituents. (see the chapter, *Multivariate Chemical Data Assimilation*)

4) Unified ozone data sets: There are several sources of ozone data with significant differences in spatial and temporal characteristics as well as their expected error characteristics. Data assimilation provides a potential strategy for combining these data into a unified data set.

5) Tropospheric ozone: Most of the ozone is in the stratosphere, and tropospheric ozone is sensitive to surface emission of pollutants. Therefore, the challenges of obtaining accurate tropospheric ozone measurements from space are significant. The combination of observations with the meteorological information provided by the model offers one of the better approaches available to obtain global estimates of tropospheric ozone.

6) Improvement of wind analysis: The photochemical time scale for ozone is long compared with transport timescales in the lower stratosphere and upper troposphere, which means ozone can be used as a tracer of atmospheric motion. A multivariate assimilation scheme could obtain information about the wind field from the ozone measurements.

7) Radiative transfer: Ozone is important in both longwave and shortwave radiative transfer. Therefore, accurate representation of ozone is important

in the radiative transfer calculations needed to extract (retrieve) information from many satellite instruments. In addition, accurate representation of ozone has the potential to have an impact on the quality of the temperature analysis in multivariate assimilation.

8) Observing system monitoring: Ozone assimilation offers an effective way to characterize instrument performance relative to other sources of ozone observations as well as the stability of measurements over the lifetime of an instrument.

9) Retrieval of ozone: Ozone assimilation offers the possibility of providing more accurate initial guesses than are currently available for ozone retrieval algorithms.

10) Assimilation research: Ozone (constituent) assimilation can be productively approached as a univariate linear problem. Therefore it is a good framework for investigating assimilation science; for example, the impact of flow dependent covariance functions.

11) Model validation: Ozone assimilation provides several approaches to contribute to the validation of models.

These goals, combined with the characteristics of the observing system, establish a set of special considerations for ozone assimilation. As will be discussed in the next section there are many sources of ozone observations. The observations are of two primary types. The first is of total column ozone, which is the amount of ozone between the surface and the top of the atmosphere. The second type of observation is of the vertically resolved profile of ozone. Satellite measurements of both types of observations, suitable for assimilation, have been available on a nearly continuous basis since 1979.

Because researchers have paid high attention to determining ozone trends, the satellite observing system has been designed to be able to detect a trend in total ozone of 1% per decade. A set of high-quality research instruments has been deployed to supplement the total column ozone observations and to facilitate the understanding of the chemical and dynamical mechanisms that are responsible for the trends. The attention of the world's community of ozone scientists has been focused by international studies of ozone trends and their causes. Therefore, directly from ozone observations, the global distribution of ozone is known with better accuracy than most environmental parameters. For an ozone data set obtained by assimilation into a chemistry-transport model to impact the basic knowledge of ozone in the atmosphere, the data set needs to provide global coverage with vertical profile information comparable to that of the research instruments, order 3 km or better. The data set must also convincingly discriminate between tropospheric and stratospheric ozone. This level of accuracy is not currently attainable by assimilating ozone observations into a meteorological model, and therefore, represents an ambitious target for the researcher.

Some of the goals mentioned above can be meaningfully addressed without reaching this level of accuracy. It is straightforward to produce global maps of total column ozone which can be used in, for instance, radiative transfer calculations. Significant errors might still remain in the polar night, where the operational measurement systems do not measure. The use of ozone measurements to provide constraints on other reactive species is an application that has been explored since the 1980's and modern data assimilation techniques could potentially advance this field. Jackman *et al.* (1987) constrained a chemistry transport model with satellite observations in what could be interpreted as an early assimilation experiment. The impact of ozone assimilation on the meteorological analysis of temperature and wind, and hence improvement of the weather forecast, is also possible. The most straightforward impact would be on the temperature analysis in the stratosphere. The improvement of the wind analysis is a more difficult challenge and confounded by the fact that where improvements in the wind analyses are most needed, the tropics, the ozone gradients are relatively weak. The goal of producing unified ozone data sets from several instruments will be difficult until bias can be correctly accommodated in data assimilation.

The use of ozone assimilation to monitor instrument performance and to characterize new observing systems is currently possible and productive. The improvement of retrievals using assimilation techniques to provide ozone first guess fields that are representative of specific environmental conditions is also an active research topic.

3. Ozone observing system

There are some surface-based observations of the total column dating back to the early part of the twentieth century, with a quasi-global network in existence since the International Geophysical Year in 1957-1958. Total column measurements are taken with both Dobson and M-83 spectrophotometers. Profile measurements are also taken with these instruments, but their ability to provide profile information is poor. These measurements are suitable for validation and calibration. Ozonesondes, similar in concept to radiosondes, are launched on balloons. Ozonesondes provide high quality measurements well into the stratosphere. The errors of the ozonesonde observations often increase above, approximately, 25 km. The spatial and temporal coverage of ozonesonde measurements is low, and with regards to data assimilation, they are primarily useful for validation and calibration. Ozonesondes are the most important regular measurements of tropospheric ozone.

Other measurements suitable for validation include aircraft measurements, lidars, and surface measurements taken for monitoring air quality. During aircraft campaigns there is a wealth of high quality observations from many sources. Some measurements from a small number of commercial airliners are available as well. The Network for Detection of Stratospheric Change (NDSC) maintains a network of observing stations to monitor chemical trends and includes a number of ground-based lidars. These data are suitable for characterising the temporal stability of long-term assimilated ozone data sets. Information about the NDSC can be found at http://www.ndsc.ncep.noaa.gov/.

Sources of ozone information suitable for assimilation come from satellite instruments. Since the middle of the 1990's the number of instruments has increased, with launches by many countries and international consortia. Both total column and profile resolving measurements are taken from space. Some of the instruments are self-calibrating, providing accuracies uncommon in satellite observations. Two instruments, the Total Ozone Mapping Spectrophotometer (TOMS) and the Solar Backscatter

Ultraviolet (SBUV), are distinguished by the length of their records, global coverage, and data access in near real time. These instruments are nadir viewing spectrometers and are space-based approaches similar in concept to those used in the Dobson spectrophotometer. Other measurement types are limb sounding instruments and occultation instruments. Below, examples are given to highlight basic measurement characteristics. It is beyond the scope of this article to review all of the sources of observations (see Table 1).

Before discussing the instrument characteristics, the general attributes of the ozone profile will be presented. Figure 1 shows ozone, temperature, and humidity profiles from an ozonesonde launched by the National Weather Service in the United States. The ozone profile is in blue and is repeated in three different units commonly used to represent ozone. Basic attributes of the profile are the stratospheric peak and the rapid

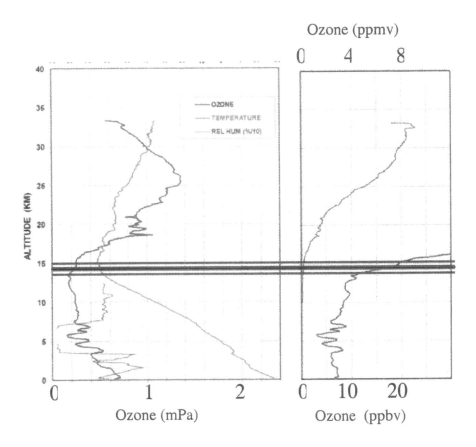

Figure 1. Ozone profiles at Old Hickory, Tennessee, United States on July 12, 1999. These profiles were obtained from the United States Weather Service. The blue lines are the ozone measurements expressed in three units commonly used for ozone. The red line is temperature and the green line relative humidity from the same balloon flight. The horizontal line is an estimate of the range where the tropopause would be specified from meteorological analyses.

increase in ozone at the tropopause. The panel on the left shows partial pressure. The profile has layered structures in both the troposphere and stratosphere which are related to a mix of dynamical and chemical processes. When the ozone is represented in parts per million by volume (ppmv), the unit often used to represent stratospheric ozone, the layered structure is obscured and the small amount of ozone in the troposphere relative to the stratosphere becomes obvious (Right panel, top axis). The profile in parts per billion by volume (ppbv) shows the tropopause increase most clearly (Right panel, bottom axis). The bold horizontal line shows the tropopause plus or minus 1 km, and suggests the sensitivity of the estimation of tropospheric ozone to the definition of the tropopause.

The upper part of the profile, above approximately 35 km, is primarily determined by photochemistry. Below this altitude, both transport and photochemistry are important, with the day-to-day variability of ozone near the tropopause being transport dominated. It is unreasonable to expect an assimilation to reproduce the fine scale layered structure in the ozone profile. To achieve the goal of geophysical usefulness, accurate location of the stratospheric peak and the tropopause, representation of the profile between those two points, and successful partitioning of ozone between the troposphere and stratosphere is needed. Accurate representation near the tropopause is difficult to achieve because of a shortage of resolved information in either the observations or the model forecast.

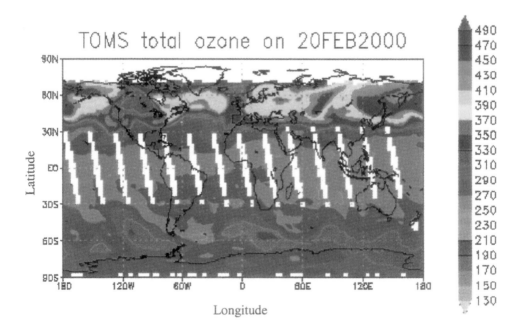

Figure 2. Total ozone observations from the Total Ozone Mapping Spectrophotometer for February 20, 2000. The data void at high northern latitudes is because there is no sunlight to provide the ultraviolet backscatter. The data void in the tropics is the gap between the observational scans. The observations are collected over a 24-hour period. The units are Dobson Units.

Using ultraviolet radiation, TOMS measures total column ozone. TOMS measurements have been highly calibrated to support trend studies (McPeters *et al.*, 1996). There is formally no information from TOMS about the vertical structure of the ozone profile. The TOMS instrument scans across the sub-satellite path and provides nearly global coverage over a 24-hour period. Figure 2 shows the TOMS coverage. There are some gaps between orbits in the tropics. At high latitudes in the winter hemisphere there are no observations because of lack of sunlight to make the observations. TOMS errors are known to be large at high solar zenith angles.

Like the TOMS instrument the SBUV instrument relies on scattered solar ultraviolet radiation (Bhartia *et al.*, 1996). Therefore, there are no measurements in the polar night. The 24-hour SBUV coverage is shown in Figure 3. Since SBUV is a nadir viewing instrument the gaps between the orbits are larger than TOMS. SBUV provides total column measurements as well as vertical profile information. The vertical resolution is about 8 km above the stratospheric peak (see Figure 1). SBUV profiles are reported to the Earth's surface; however, the information below the peak is largely defined by the climatological profile used to initiate the retrieval with modification to match the global column measurement and the resolved profile information at higher altitudes. The

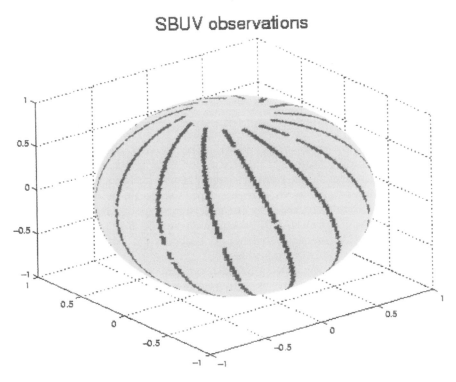

Figure 3. Location of ozone observations from Solar Backscatter Ultraviolet (SBUV) instrument. The data void at high northern latitudes is because there is no sunlight to provide the ultraviolet backscatter. The data void in the tropics is the gap between the orbits. The observations are collected over a 24-hour period.

Global Ozone Monitoring Experiment (GOME) is a newer instrument that provides observations similar to TOMS and SBUV and has been used in assimilation experiments (Eskes *et al.*, 2002).

Limb sounding measurements, historically utilising microwave and infrared emission measurement techniques, provide much better vertical resolution than SBUV. A typical instrument provides 3 km resolution in the stratosphere. Some information is obtained across the tropopause and into the troposphere, though the ability to resolve the tropopause is still limited (see, Figure 1). Measurements in the troposphere are limited by clouds, spectral resolution, and the challenges of extracting information from below the stratospheric ozone peak. The orbital footprint of most limb-scanning instruments is similar to that of SBUV in Figure 3. The infrared and microwave limb-viewing instruments, however, provide both daytime and nighttime measurements, and the polar night data voids of TOMS and SBUV do not exist. The improved vertical resolution comes with less resolution in the horizontal because of the long line of sight through the Earth's limb. The measurements do not extend to the ground, and only partial column information is available. The coverage of many of these instruments is appropriate for assimilation; however, the data are generally not available in near real time and the measurements are taken from research satellites with no guaranteed data continuity.

Occultation measurements are also made by instruments looking through the limb at a source of radiation that rises or sets. These measurements can be made using the sun, the moon, stars, and potentially orbiting manmade sources, and utilise the absorption spectra of constituents in the atmosphere. The characteristics of the profile are similar to those of the limb emission techniques, with 3 km or better profile resolution, and profiles often extending into the troposphere. Instruments, such as the Halogen Occultation Experiment (HALOE) on the Upper Atmosphere Research Satellite (UARS), are generally considered to have exceptional accuracy because of their robust calibration techniques (Bruhl *et al.*, 1996). The viewing patterns of occultation instruments are complex, because of the reliance on rising and setting of astronomical sources. Instruments that use the sun as a source obtain about 14 measurements a day in two latitude bands, with the bands changing depending on the orbital characteristics. Instruments that use the moon and stars can achieve more global coverage. For a solar occultation instrument such as HALOE, assimilation must be done in concert with other instruments to provide adequate coverage to constrain the model. The data are especially valuable for validation and calibration.

Measurements from nadir viewing infrared sounders, which are normally used for temperature and humidity profiling, are also used in ozone assimilation. Estimates of ozone can be obtained, for instance, from the TIROS Operational Vertical Sounder (TOVS). These measurements have been embraced by the operational weather community, but their errors are larger than those from the ultraviolet instruments such as TOMS and SBUV. While the information from these instruments has potential utility in certain applications, the ability for these instruments to contribute to the goal of an ozone data set with global coverage and vertical resolution of 3 km is small. Information about TOVS ozone can be found at http://www.cpc.noaa.gov/products/stratosphere/tovsto/tovsto_info.html.

Currently there are many instruments measuring ozone or scheduled to be launched. Some of these instruments follow in the heritage of the research and operational instruments described above. Others will try to measure tropospheric ozone directly. Table 1 is a list of instruments and more information can be found on the World Wide Web by searching for the instrument name. Some instruments of particular note: GOMOS and SCIAMACHY are on Envisat. GOMOS uses stellar occultation to achieve more spatial coverage than previous generation occultation instruments. SCIAMACHY can take nadir and limb measurements as well as utilising solar occultation. The Troposphere Emission Spectrometer (TES) will fly on the Earth Observing System (EOS) Aura platform and will be the first instrument designed to resolve tropospheric profiles. The Stratospheric Aerosol and Gas Experiment (SAGE) provides the longest occultation-based data record.

Table 1. Satellite Instruments that Measure Ozone
(grouped by measurement type)

Ultraviolet	Total Ozone Mapping Spectrometer	TOMS
	Solar Backscatter Ultraviolet	SBUV
	Global Ozone Monitoring Experiment	GOME
	Scanning Imaging Absorption Spectrometer for Atmospheric Chartography	SCIAMACHY
	Ozone Monitoring Instrument	OMI
Occultation	Stratosphere Aerosol and Gas Experiment	SAGE
	Halogen Occultation Experiment	HALOE
	Polar Ozone and Aerosol Measurement	POAM
	Global Ozone Monitoring by Occultation of Stars	GOMOS
	Atmospheric Chemistry Experiment	ACE
Limb Sounding	Limb Infrared Monitor of the Stratosphere	LIMS
	Cryogenic Limb Array Etalon Spectrometer	CLAES
	Improved Stratospheric and Mesospheric Sounder	ISAMS
	Microwave Limb Sounder	MLS
	Michelson Interferometer for Atmospheric Sounding	MIPAS
	High Resolution Dynamics Limb Sounder	HIRDLS
Weather Sounders (Infrared)	TIROS Operational Vertical Sounder	TOVS
	Atmospheric Infrared Sounder	AIRS
Tropospheric	Tropospheric Emission Spectrometer	TES

4. Ozone Assimilation

The ozone assimilation system used at NASA's Goddard Space Flight Center is shown in Figure 4. The components of the Goddard ozone data assimilation system are representative of the components of other data assimilation systems. The model equation that represents ozone transport and chemistry is:

$$\frac{\partial O_3}{\partial t} = -\mathbf{V} \bullet \nabla O_3 + P - LO_3$$

This equation states that the time rate of change of the ozone mixing ratio is equal to the advection of ozone by the atmospheric winds, **V**, plus the photochemical production, P, minus the loss coefficient, L, times the ozone mixing ratio. The system is detailed in Štajner *et al.* (2001).

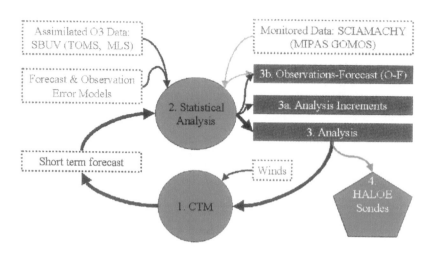

Figure 4. Schematic diagram of Goddard Space Flight Center ozone assimilation system. A short-term forecast is made by the Chemical Transport Model (CTM) using winds from the meteorological analysis (1). Using a specification of the forecast and observational errors SBUV and TOMS data are assimilated into the model using the physical-space statistical analysis system (2). The system can also assimilate Microwave Limb Sounder data. The output products (3) are validated with independent sources of data (4). New data types, e.g. SCIAMACHY are compared with the forecast, statistics are generated, but the data are not yet assimilated.

As currently operated, this system performs a univariate assimilation of ozone that is applied, sequentially, following a meteorological analysis. The meteorological analysis provides an estimate of winds, temperature, and a variety of other parameters, such as cloud mass flux, that might be useful for calculating transport or chemical terms. The model performs a short term forecast, 15 minutes. At the end of the forecast the ozone observations are combined with the model forecast using the physical-space statistical analysis system (Cohn *et al.*, 1998). The quality of the assimilation is assessed using observation minus forecast statistics as well as through direct comparison with other data sources. Some of these data sources, such as HALOE and ozonesondes, are so well characterised that they are used as validation standards. Other data sources are instruments which are being monitored to understand the error characteristics of the new instruments relative to existing TOMS and SBUV data used in the assimilation.

This form of assimilation, where the winds and other parameters from the meteorological analysis are archived then used as input for the chemistry transport model, is called offline assimilation. Current research activities focus on relaxing various simplifications and assumptions of the current system. For example, we have performed experiments with flow dependent covariance models and are currently building an online system that allows the assimilated ozone to be used in the radiative transfer calculation. In the online work the ozone will impact the temperature analysis and, indirectly, the wind analysis. In addition we are performing assimilation experiments with both Microwave Limb Sounder (MLS) and Polar Ozone and Aerosol Measurement (POAM) observations to address problems in the polar night and improve the vertical profile representation near the tropopause.

In the assimilation there are two fundamental sources of information, the observations and the model. The observations were discussed in the previous section. The information provided by the model must also be evaluated. With the assimilation of TOMS and SBUV, there is little observational information about the structure of the ozone profile below the peak. Hence, if there is going to be definition of the lower stratospheric profile and the stratosphere to troposphere transition, then this information will come from the model. The models used in the current experiment have a vertical resolution of about 2 km. As stated earlier, the chemistry determines the shape of the profile above the peak, and for much of the globe and much of the year, the specifics of the chemistry package have little impact on the day-to-day shape of the profile in the lower stratosphere. There are biases in long-term simulations. Hence, to first order the shape of the profile is dependent on the model transport. The model-generated profile is altered by the observations in order to correct the column amount. Most of this correction is in the lower stratosphere and the upper troposphere. Precise ways of altering the profile to match the integrated column measurements are difficult to define.

The chemical aspects of the model require further consideration. The parametrization used here has its biggest impact above the peak. While this is often adequate for providing an estimate of total ozone over much of the globe, this is inadequate for providing profile information. First, there is fast photochemistry in the lower stratosphere during winter dependent upon aerosol surface chemistry. Without this polar night chemistry, only crude approximations to the high latitude vertical profile are possible. Second, chemistry occurs on time scales similar to transport in the troposphere. Because of the relatively small amount of ozone in the troposphere, this is not so important to stratospheric studies, but is important to the goals pursued by current ozone assimilation activities. The chemistry in the troposphere is linked to surface processes such as urban pollution, biomass burning, and convection. Hence, accurate estimates of ozone require many sources of ancillary information, which are currently beyond the scope of the scientific and observational capabilities.

The performance characteristics of the ozone assimilation system are demonstrated in Figure 5. Figure 5 shows ozone profiles at Hohenpeissenberg, Germany, typical of the northern hemisphere middle latitudes. On the figure are three profiles, the red line is from the ozone assimilation, the black is an ozonesonde, and the green is the SBUV profile nearest to the ozonesonde station. Since the SBUV information below the peak is of limited value, only the partial profile is included in the assimilation. The profile estimated by the assimilation captures the basic features of the ozonesonde. The quality of the ozonesonde measurement decreases at 30 hPa due to instrument error. In general, comparison with the ozonesondes shows the assimilation capturing variability of the ozone in the lowermost stratosphere, which is to first order representative of the ability of the meteorological analysis to capture the variability of the tropopause. Using other data sources as validation shows that the assimilation captures the height of the peak in mid-latitudes, but not in the tropics. The tropical peak is more directly influenced by the chemical parametrization.

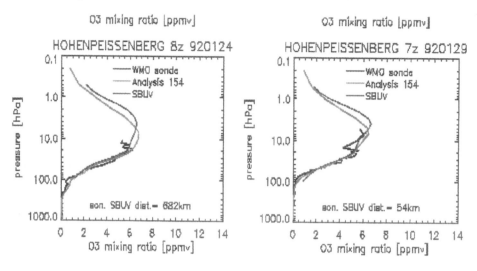

Figure 5. Profiles of ozone at Hohenpeissenberg, Germany. The black line is from an ozonesonde flight. The red line is from the assimilation of TOMS and SBUV data. The green is the SBUV profile closest to the ozonesonde station. These measurements are from January 24 and 29, 1992.

Examination of the profile in Figure 5 also reveals a high bias between the assimilation and the ozonesonde at the tropopause. Seasonal and annual comparisons reveal that this is a consistent bias with a maximum at 150 hPa of 150 ppbv. By examination with the profile in Figure 1, this is an error of nearly 100%. Though this bias is small in comparison to stratospheric amounts and is obscured in the calculation of total column, it is of significant geophysical consequence. With a bias of this size, there is little chance of providing, for instance, credible estimates of tropospheric ozone and stratospheric-tropospheric fluxes would be misrepresented.

The current state of the art in ozone assimilation captures the middle latitude variability quite well. Even in the troposphere, during time periods dominated by transport rather than surface sources, the variability is reasonably well captured. However, during the wintertime at the poles, even if the total column is represented, the assimilation does not place the aerosol-related depletion at the correct altitude. The assimilation might be positively impacted by improvements in any part of the data assimilation system, improved winds, improved chemical models, improved specification of error statistics, and incorporation of new data types. It is the latter of these, incorporation of new data types with resolved vertical information below the peak, that would be expected to have the most robust impact.

Figure 6 shows that changing the error statistics can have a positive impact on the 150 hPa bias. Figure 2 shows the coverage of the scanning TOMS instrument, and Figure 3 shows the coverage of the nadir viewing SBUV instrument. It is reasonable to assume that the high density of the TOMS observations have errors that are correlated with each other and are, hence, not independent. The results shown in Figure 6 show the impact of assuming that the TOMS errors are correlated with a correlation length of 150 km. The figure shows observation minus forecast statistics comparing the assimilation to ozonesondes. There is a systematic reduction of the bias at 150 hPa. This experiment reduces the influence that the TOMS observations have on the assimilation. Other experiments, withholding the TOMS data altogether, reveal the difficulty that the analysis routine faces when trying to distribute the total column ozone across the profile.

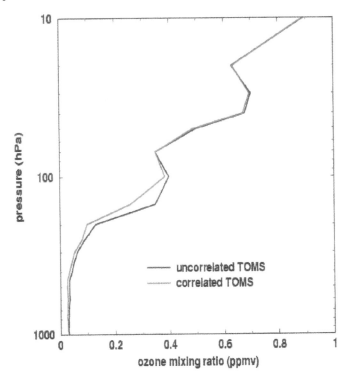

Figure 6. Observation minus forecast increments of ozone mixing ratio from two assimilation experiments using TOMS and SBUV observations. This is the mean for July 1998, and the differences are from all available ozonesondes. The black line is from the control experiment where each TOMS observation is assumed to be an independent observation. The red line is from an experiment where the TOMS measurements are assumed to have correlated errors and, hence, have a smaller impact on the assimilation. The bias near the tropopause has been systematically reduced.

5. Conclusions

There are many sources of ozone observations and several motivations to undertake assimilation experiments. In the near term, ozone assimilation can help define the characteristics and stability of the instruments and improve the retrieval or extraction of ozone information from the observed radiances. In order to add value to the direct knowledge of ozone, beyond that which is obtained directly from the observations, an assimilated ozone data set needs to have global coverage, with vertical resolution comparable to or better than that measured by the occultation instruments. High accuracy is required near the tropopause in order to get the partitioning of ozone between the troposphere and stratosphere correct. The information provided by either the observations or the model is uncertain near the tropopause. The most robust way to improve the analysis near the tropopause is by using new observations that directly resolve the lower stratosphere and the upper troposphere. However, better treatment of the error statistics, improved representation of transport, and improved chemical modelling all stand to benefit the analysis near the tropopause. This offers encouragement that assimilation of ozone might, ultimately, be useful for all of the goals discussed above.

Acknowledgements. I thank Ivanka Štajner and Nathan Winslow for their many implicit and explicit contributions to this article. I thank Anne Douglass, Richard Stolarski, and P. K. Bhartia for many discussions on the role of ozone assimilation in understanding atmospheric transport and chemistry.

References

Bhartia, P.K., R.D. McPeters, C. L. Mateer, L.E. Flynn, and C. Wellemeyer, 1996: Algorithm for the estimation of vertical ozone profile from the backscattered ultraviolet (BUV) technique. *J. Geophys. Res.,* **101**, 18793-18806.

Bruhl, C., S. R. Drayson, J. M. Russell III, P. J. Crutzen, J. M. McInerney, P. N. Purcell, H. Claude, H. Gernandt, T. J. McGee, I. S. McDermid, and M. R. Gunson, 1996: Halogen Occultation Experiment ozone channel validation. *J. Geophys. Res.,* **101**, 10217-10240.

Cohn, S. E., A. M. da Silva, J. Guo, M. Sienkiewicz, and D. Lamich, 1998: Assessing the effects of data selection with the DAO physical-space statistical analysis system. *Mon. Weather Rev.,* **126**, 2913-2926.

Dessler, A. E., 2000: *The Chemistry and Physics of Stratospheric Ozone,* Academic Press.

Eskes, H. J., P. F. J. van Velthoven, and H. M. Kelder, 2002: Global ozone forecasting based on ERS-2 GOME observations, *Atmos. Chem. Phys.,* **2**, 271-278.

Hobbs, P. V., 2000: *Introduction to Atmospheric Chemistry,* Cambridge University Press.

Jackman C. H., P. D. Guthrie, J. A. Kaye, 1987: An intercomparison of nitrogen-containing species in Nimbus 7 LIMS and SAMS data. *J. Geophys. Res.,* **92**, 995-1008.

Long, C. S., A. J. Miller, H. T. Lee, J. D. Wild, R. C. Przywarty, and D. Hufford, 1996: Ultraviolet index forecasts issued by the National Weather Service. *Bull. Amer. Meteorol. Soc.,* **77**, 729-748.

McPeters, R. D., P. K. Bhartia, A. J. Kruger, J. R. Herman, B. M. Schlesinger, C. G. Wellemeyer, C. J. Seftor, G. Jaross, S. L. Taylor, T. Swissler, O. Torres, G. Labow, W. Byerly, and R. P. Cebula, 1996: *Nimbus-7 Total Ozone Mapping Spectrometer (TOMS) data products users guide,* NASA Reference Publication 1384, National Aeronautical and Space Administration, Washington, D.C., U.S.A.

Štajner, I., L. P. Riishøjgaard, and R. B. Rood, 2001: The GEOS ozone data assimilation system: Specification of error statistics. *Q. J. R. Meteorol. Soc.,* **127**, 1069-1094.

Stolarski, R. S., and A. R. Douglass, 1985: Parameterization of the photochemistry of stratospheric ozone including catalytic loss processes. *J. Geophys. Res.,* **90**, 709-718.

Suggested Reading

Austin, J., 1992: Toward the four dimensional assimilation of stratospheric constituents. *J. Geophys. Res.,* **97**, 2569-2588.

Eskes, H. J., P. F. J. van Velthoven, P. J. M. Valks, and H. M. Kelder, 2003: Assimilation of GOME total ozone satellite observations in a three-dimensional tracer transport model. *Q. J. R. Meteorol. Soc.*, **129**, in press.

Fisher, M., and D. J. Lary, 1995: Lagrangian four-dimensional variational assimilation of chemical species. *Q. J. R. Meteorol. Soc.*, **125**, 723-757.

Khattatov, B. V., J. C. Gille, L. V. Lyjak, G. P. Brasseur, V. L. Dvortsov, A. E. Roche, and J. W. Waters, 1999: Assimilation of photochemically active species and a case analysis of UARS data. *J. Geophys. Res.*, **104**, 18715-18737.

Kondratyev, K. Ya, A. A. Buznikov, O. M. Pokrovskii, and Yu. B. Yanushanets, 1993: Methods of assimilating satellite information for analysis and prediction of atmospheric ozone, *Soviet J. Remote Sensing*, **10(3)**, 407-420.

Levelt, P. F., M. A. F. Allaart, and H. M. Kelder, 1996: On assimilation of total ozone satellite data. *Ann. Geophys.*, **14**, 1111-1118.

Levelt, P. F., B. V. Khattatov, J. C. Gille, G. P. Brasseur, X. X. Tie, and J. W. Waters, 1998: Assimilation of MLS ozone measurements in the global three-dimensional chemistry transport model ROSE. *Geophys. Res. Lett.*, **25**, 4493-4496.

Lyster, P. M., S. E. Cohn, R. Ménard, L.-P. Chang, S.-J. Lin, and R. G. Olsen, 1997: Parallel implementation of a Kalman filter for constituent data assimilation. *Mon. Weather Rev.*, **125**, 1674-1686.

Ménard, R. and L.-P. Chang, 2000: Assimilation of stratospheric chemical tracer observations using a Kalman filter. Part II: Validation results and analysis of variance and correlation dynamics. *Mon. Weather Rev.*, **128**, 2672-2686.

Ménard, R., S. E. Cohn, L.-P. Chang, and P. M. Lyster, 2000: Assimilation of stratospheric chemical tracer observations using a Kalman filter. Part I: Formulation. *Mon. Weather Rev.*, **128**, 2654-2671.

Riishøjgaard, L. P., I. Štajner, and G.-P. Lou, 2000: The GEOS ozone data assimilation system. *Adv. Space. Res.*, **25**, 1063-1072.

Struthers, H., R. Brugge, W. A. Lahoz, A. O'Neill, and R. Swinbank, 2002: Assimilation of ozone profiles and total column measurements into a global General Circulation Model. *J. Geophys. Res.*, **107**, 10.1029/2001JD000957.

MULTIVARIATE CHEMICAL DATA ASSIMILATION

BORIS KHATTATOV

Atmospheric Chemistry Division, National Center for Atmospheric Research, Boulder, CO, USA

1. Introduction

We present an overview of the mathematical formalism of data assimilation applied to photochemical atmospheric models. Examples of Kalman filter and variational assimilation analysis are presented along with time-dependent linearization and error covariance matrices for a typical stratospheric chemical system described in the Chapter *Introduction to Atmospheric Photochemical Modelling*.

Let vector **x** represent the state of a time-dependent box photochemical model, *i.e.*, concentrations of all modelled species in a parcel of air at a given instant. The photochemical box model **M** describes the transformation of vector **x** from time t to time $t+\Delta t$. Formally,

$$\vec{x}(t + \Delta t) \; = \; \mathbf{M}(\vec{x}(t)) \tag{1}$$

Let vector **y** contain observations of the state, *i.e.*, observations of the chemical composition of the atmosphere. Usually, the dimension of **y** is significantly less than N, the dimension of the model space. Moreover, the locations of observations in the real physical space can be different from the locations of the grid points at which the model space is defined. The connection between **y** and **x** can be established through the so-called observation operator **H**, which represents mapping of the state variables from the locations of the grid points to the locations of the observations:

$$y = \mathbf{H}(x) \tag{2}$$

Everywhere in this discussion we assume that the interpolation errors associated with operator **H** are negligible. The results are easily extended to the case when this is not true. Combining the above two equations, we get

$$y = \mathbf{H}(\mathbf{M}(x)) \tag{3}$$

R. Swinbank et al. (eds.), Data Assimilation for the Earth System, 279–288.

The data assimilation problem is then to find the "best" value of **x**, which inverts this equation for a given **y** allowing for observation errors and other prior information (Lorenc 1986). In most cases, dimensions of vectors **x** and **y** will be different, and this problem will be either overdetermined or underdetermined. Therefore inversion of (3) should be done in the statistical sense.

"Best" here means that the errors of the final analysis are minimal. An exact value of a physical quantity can rarely be determined. One can only say that this value lies within a certain range with a certain probability, and therefore all estimates of the best value of **x** obtained from the observed **y** are probabilistic in nature. A mathematically robust definition of best or optimal **x** is, for instance, the value corresponding to the maximum of the probability density function (PDF) of **x** given observations **y**. This is the so-called maximum likelihood definition.

The exact shapes of the PDFs in both **x** and **y** spaces are generally unknown. In order to solve the posed analysis problem one needs to establish a relationship between the PDF of **x** and the PDF of **y**. Formal transformation of PDFs by the model from the parameter space **x** to the model space **y** is described by the so-called Fokker-Kolmogorov equation (*e.g.*, Jazwinski 1970), which is impossible to solve in most practical applications. This is one of the reasons why simplifications are needed in order to be able to solve practical problems in data assimilation.

One simplification is that the probability density functions can be approximated by Gaussian functions:

$$PDF(\mathbf{x}) \sim \exp[-0.5(\mathbf{x}-\hat{\mathbf{x}})^T \mathbf{C}^{-1}(\mathbf{x}-\hat{\mathbf{x}})] \qquad (4)$$

where $\hat{\mathbf{x}}$ is the true (unknown) value of **x**, and **C** is the corresponding error covariance matrix. Its diagonal elements are the uncertainties (standard deviations) of $\hat{\mathbf{x}}$ and the off-diagonal elements represent correlation between uncertainties of different elements of vector **x**. The covariance matrix **C** is defined as

$$\mathbf{C} = <(\mathbf{x}-\hat{\mathbf{x}})(\mathbf{x}-\hat{\mathbf{x}})^T> \qquad (5)$$

where angle brackets represent averaging over all available realizations of **x**.

We also assume that there exists a prior, independent estimate of **x**, or $\mathbf{x}_b$, often called the background, with the corresponding background error covariances **B**. The solution minimizing the final analysis errors is given by a minimum of the following functional (Lorenc 1986):

$$J(\mathbf{x}) = [\mathbf{y} - \mathbf{H}(\mathbf{M}(\mathbf{x}))]^T (\mathbf{O} + \mathbf{F})^{-1} [\mathbf{y} - \mathbf{H}(\mathbf{M}(\mathbf{x}))] + [\mathbf{x} - \mathbf{x}_b]^T \mathbf{B}^{-1} [\mathbf{x} - \mathbf{x}_b] \qquad (6)$$

Here **O** is the observational error covariance matrix, **F** is the error covariance corresponding to operators **M** and **H**, and **B** is the background error covariance matrix. They characterize our confidence in the measurements, the model and observation operator, and the a priori background estimate. $J(\mathbf{x})$ is often called the misfit or cost function. Once the optimal estimate of **x** is obtained, (1) can be used to derive the best estimate of the state.

In practical applications one has to find an appropriate way to compute the error covariances and to minimize $J(\mathbf{x})$. In most cases, in order to be able to do this we need to introduce the linear approximation. In the linear approximation we assume that for small perturbations of the parameter vector $\Delta\mathbf{x}$ the following is a good approximation:

$$\mathbf{M}(\mathbf{x}+\Delta\mathbf{x}) = \mathbf{M}(\mathbf{x}) + \mathbf{L}\Delta\mathbf{x} \qquad (7)$$

Note that in this expression **L** is a matrix, while **M** is, in general, a nonlinear operator. Formally, **L** is a derivative of **M** with respect to **x**:

$$\mathbf{L} = \frac{d\mathbf{M}}{d\mathbf{x}} \qquad (8)$$

The linearization **L** of the original model **M** will be used in two ways. First, minimization of $J(\mathbf{x})$ often requires knowledge of the derivative of $J(\mathbf{x})$ with respect to **x**. This, in turn, requires knowledge of d**M**/d**x**. In this discussion we assume that observational operator **H** is simply a linear interpolation from times/locations of the model grid points to times/locations of the observations.

Second, for small variations of **x** one can show that transformation of error covariance matrix $\mathbf{C_x}$ in the parameter space to the error covariance matrix $\mathbf{C_y}$ in the model space is as follows:

$$\mathbf{C_y} = \mathbf{L}\,\mathbf{C_x}\,\mathbf{L}^\mathrm{T} \qquad (9)$$

This, in turn, allows one to establish correspondence between the PDF of **x** in the parameter space and the PDF of **y** in the model space. If **M** represents the original non-linear model, matrix **L** is said to be the tangent-linear model and its transpose, $\mathbf{L}^\mathrm{T}$, is said to be the adjoint of **M**. The linearization matrix describes time evolution of small perturbations of the model state:

$$\delta\mathbf{x}(t+\Delta t) = \mathbf{L}\delta\mathbf{x}(t)$$

$$\begin{bmatrix} H \\ OH \\ \ldots\ldots \\ \ldots\ldots \\ O3 \\ H2O(a) \end{bmatrix}_{t+\Delta t} = \begin{bmatrix} L_{11} & L_{12} & L_{13} & \cdots & \cdots \\ L_{21} & L_{22} & L_{23} & \cdots & \cdots \\ \cdots & \cdots & \cdots & \cdots & \cdots \\ \cdots & \cdots & \cdots & \cdots & \cdots \\ \cdots & \cdots & \cdots & \cdots & L_{NN} \end{bmatrix} \begin{bmatrix} H \\ OH \\ \ldots\ldots \\ \ldots\ldots \\ O3 \\ H2O(a) \end{bmatrix}_t$$

This is for the example introduced in the chapter *Introduction to Atmospheric Photochemical Modelling* (Table 1).

2. Linearization Matrix

The linearization matrix L is in general a function of the time interval Δt. This is easy to understand if we take the extreme case of $\Delta t = 0$. In this case the final perturbation is the same as the initial perturbation and L is the identity matrix. As time interval increases, the linearization matrix changes its structure (Figure 1). After a few hours of integration a pattern emerges in the distribution of the nonzero elements, with only a few columns containing most of the nonzero values. This demonstrates that a relatively small number of species or families, determine concentrations of all constituents in the model at later times.

We found that for the example from the chapter *Introduction to Atmospheric Photochemical Modelling* (table 1) the matrix L is not invertible for Δt of a few hours or longer. The reason for this is that the rank of L, *i.e.*, number of linearly independent rows or columns, quickly decreases with time. The rank is shown in Figure 1 on top of each plot. For this example, after just 6 hours the rank decreases from 19 to 11 and becomes 9 after 4 days of integration. This means that, in general, only nine linear combinations of initial species concentrations completely define concentrations of all 19 constituents after 4 days. Formally, on day 4, matrix L represents a transformation from 19-dimensional space to 9-dimensional space. This is a multidimensional equivalent of

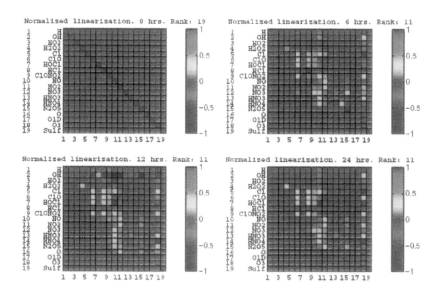

Figure 1. Time evolution of linearization matrix L for the example discussed in the Chapter "Introduction to Atmospheric Photochemical Modelling", Table 1. (Khattatov et al 1999).

multiplication by zero along some of the dimensions. No matter how large some concentrations were initially, in a few hours or a few days their impact might be completely negligible. This behaviour is due to a strong diurnal cycle and the short lifetime of some species in the model.

An interesting consequence of this result is that the past state of the modelled stratospheric chemical system can never be determined from present observations of the system since L cannot be inverted. On the other hand, it means that one does not have to know concentrations of all species to predict the state at some later time. For this example, provided that the model is fairly realistic, only nine linear combinations of species concentrations spanning the orthogonal space of matrix L need to be known in order to predict concentrations of all 19 model constituents 4 days later.

Computer codes for constructing and solving the described photochemical model as well as for computing the linearization and covariance matrices can be found at: http://acd.ucar.edu/~boris/research.htm.

3. Covariance Matrices

One can think of the photochemical model M as a transformation from the N-dimensional space (19-dimensional for the example discussed in the chapter *Introduction to Atmospheric Photochemical Modelling*) of constituent concentrations at present time to some future time, as schematically illustrated in Figure 2.

In most practical cases, the value of x at the initial time is not known precisely; instead one can specify a region of probable values of x. Instead of a point-to-point transformation, we now have a region-to-region transformation (Figure 3). The "shapes" of these regions are described by probability density functions.

Evolution of probability density functions is very hard to compute in practice due to the high dimensionality of the model space and the high computational requirements of the model propagators. However, the Gaussian assumption and the linearization approximation allow one to use (9) to compute evolution of the Gaussian error covariance matrices. An example of temporal evolution of error covariance matrices computed this way is shown in Figure 4.

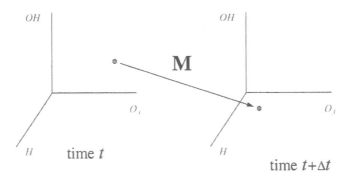

Figure 2. Action of photochemical model.

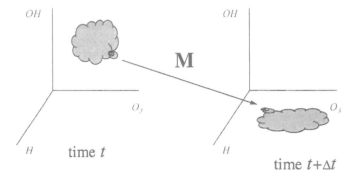

Figure 3. Transformation of PDFs.

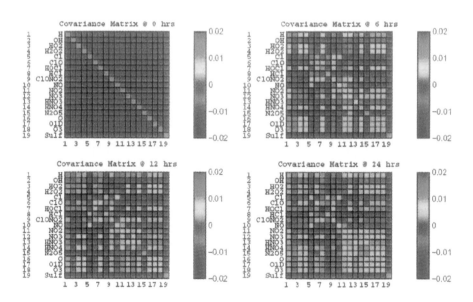

Figure 4. Time evolution of error covariance matrices (Khattatov et al 1999).

4. Variational Technique

In the variational method a minimization algorithm is used to find model initial conditions **x** that minimize a misfit between model results and observations for the whole analysis period. The analysis period is usually much longer than the model time step Δt.

Usually, there is more than one observation available inside the analysis interval, and (6) becomes

$$J(\mathbf{x}) = \sum_i [\mathbf{y}^i - \mathbf{H}^i(\mathbf{M}^i(\mathbf{x}))]^T (\mathbf{O}^i + \mathbf{F}^i)^{-1} [\mathbf{y}^i - \mathbf{H}^i(\mathbf{M}^i(\mathbf{x}))] + [\mathbf{x} - \mathbf{x}_b]^T \mathbf{B}^{-1} [\mathbf{x} - \mathbf{x}_b] \qquad (10)$$

Here index i corresponds to a particular time t_i, $\mathbf{y}^i$ represents an observation at this time, $\mathbf{H}^i(\mathbf{M}^i(\mathbf{x}))$ is the model estimate of the state interpolated, if necessary, to the time/location of $\mathbf{y}^i$, and $\mathbf{O}^i$ and $\mathbf{F}^i$ are the error covariance matrices corresponding to i.

The variational data assimilation technique can be thought of as a constrained least squares fit to a set of observations distributed over some period of time. The constraints are given by the model equations. The choice of the analysis period is somewhat arbitrary and is dictated by the frequency of the observations and the characteristic timescales of the modelled system. The solution inside the analysis interval is smooth, while usually there is a discontinuity at the ends of adjacent analysis periods.

Since $\mathbf{M}$ is almost always nonlinear, the explicit solution is rarely possible, and a minimization algorithm has to be used for finding the optimal $\mathbf{x}$. Most minimization algorithms require knowledge of the gradient of the cost function $J(\mathbf{x})$ with respect to $\mathbf{x}$. The so-called adjoint method (*e.g.*, Talagrand and Courtier, 1987; Fisher and Lary, 1995) is often used to compute $dJ(\mathbf{x})/d\mathbf{x}$. The adjoint method relies on the linearization of the model. This imposes additional requirements on the length of the analysis period, which should be short enough for the linear approximation to be valid.

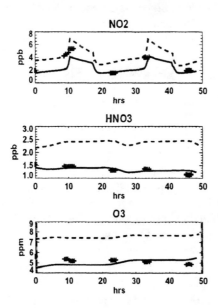

Figure 5. An example of variational approach application (Khattatov et al 1999). Dotted lines represent first model simulation, crosses correspond to actual measurements and solid lines represent assimilation analysis.

Khattatov *et al.* (1999) used the variational approach in conjunction with a photochemical box model and a trajectory model for simultaneous assimilation of several photochemically active gases. They used data from two instruments onboard the Upper Atmosphere Research Satellite (UARS). Figure 5 presents results of the assimilation procedure for one trajectory.

5. Sequential Technique (Kalman Filter)

In the sequential approach each observation is processed separately, and the analysis length is the time between two consecutive observations. Model forecast at the end of the analysis interval is considered to be the a priori background estimate x_b. The optimal analysis equation is used to obtain the best estimate of x from the forecast value and a corresponding observation. The result is used as the model initial condition for the next analysis period (*e.g.*, Lorenc 1986). Therefore $x_t = x_b$ is the model forecast value of x at time t; y is the observation at time t; and x is the optimal estimate of model state at time t obtained from x_t and y.

Equation (6) thus becomes

$$J(\mathbf{x}) = [\mathbf{y}\text{-}\mathbf{H}(\mathbf{x})]^T \mathbf{O}^{-1}[\mathbf{y}\text{-}\mathbf{H}(\mathbf{x})] + [\mathbf{x}\text{-}\mathbf{x}_t]^T \mathbf{B}_t^{-1}[\mathbf{x}\text{-}\mathbf{x}_t] \qquad (11)$$

Here $\mathbf{B}_t$ is the forecast error covariance at time t. Note that operator $\mathbf{M}$ is not explicitly included in the cost function. Since $\mathbf{M}$ is not included in the equation, an explicit solution is possible provided that $\mathbf{H}$ is linear. To minimize (11), one has to solve $dJ(\mathbf{x})/d\mathbf{x} = 0$. The solution is

$$\mathbf{x} = \mathbf{x}_t + \mathbf{K}(\mathbf{y} - \mathbf{H}\mathbf{x}_t) \qquad (12)$$

$$\mathbf{K} = \mathbf{B}_t \mathbf{H}^T (\mathbf{H}\mathbf{B}_t\mathbf{H}^T + \mathbf{O})^{-1} \qquad (13)$$

Matrix $\mathbf{K}$ is called the Kalman gain matrix.

At the end of each analysis period the model value ($\mathbf{x}_t$) and the corresponding observation ($\mathbf{y}$) are "mixed" with weights inversely proportional to their respective errors according to (12) and (13) to produce $\mathbf{x}$. Then the model is integrated forward in time starting from the derived $\mathbf{x}$. Once an observation has been incorporated in the model, the analysis error covariance should be updated to reflect this. It can be shown that the new analysis error covariance can be expressed as (Lorenc 1986):

$$\mathbf{B} = \mathbf{B}_t - \mathbf{B}_t \mathbf{H}^T (\mathbf{H}\mathbf{B}_t\mathbf{H}^T + \mathbf{O})^{-1} \mathbf{H}\mathbf{B}_t \qquad (14)$$

In the absence of observations, the model state is updated using (1), while evolution of the error covariance is obtained from the linearized model equations as in (9):

$$\mathbf{B}_{t+\Delta t} = \mathbf{L}\,\mathbf{B}_t\,\mathbf{L}^T \tag{15}$$

Equations (12)-(15) are often referred to as the extended Kalman filter. In the case of a linear model, $\mathbf{L}=\mathbf{M}$ and (12)-(15) will become the Kalman filter equations.

In reality, the forecast model is usually not perfect, and therefore these equations should include additional terms due to the forecast model error. Everywhere in this chapter we assume that the forecast model error is negligible relative to the observational error and is therefore omitted from the above equations.

Sequential analysis has a discontinuity whenever a new observation is encountered during forward model integration. If $\mathbf{x}$ is a function of both time and space, the four-dimensional analysis, as in the case of the variational technique when the analysis is done simultaneously in space and time, is replaced by a sequence of three-dimensional analyses performed at different times. Figure 6 presents results of Kalman filter assimilation of UARS data.

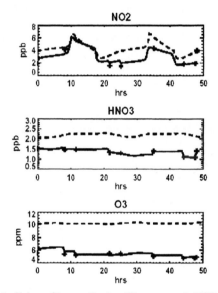

Figure 6. An example of the Kalman filter application (Khattatov et al. 1999). Dotted lines represent first model simulation, crosses correspond to actual measurements and solid lines represent assimilation analysis.

6. Summary

Both variational assimilation and the extended Kalman filter technique can be used for assimilation of a number of chemical species using relatively complex numerical photochemical models. Depending on the application, either method can be used for as long as estimates of the analysis errors are provided. The extended Kalman filter analysis, generally speaking, does not satisfy the model equations. Clearly, discontinuities seen in Figure 6 are unrealistic and can never be reproduced by solving the model photochemical equations. This, however, does not make the sequential approach less valuable than the variational method, where a smooth solution might appear as a result of adherence to possibly incorrect model dynamics. The extended Kalman filter method might be more advantageous when the goal is to predict the state of the photochemical system in between two consecutive observations or if it is needed to predict the state at some later time, when no observations are available.

We also have presented a formal framework for computing the time evolution of uncertainties in a chemical system and applied it to a case for typical stratospheric conditions. Uncertainties, or variances, can be considered to be a quantitative measure of the amount of useful information about the chemical system under consideration. The proposed framework allows one to assess how this "information" changes with time. The described case of a typical stratospheric system is largely academic and its results confirm quantitatively what is already known, *e.g.*, that concentrations of several key species or linear combinations of species (families) control future evolution of the system. These results are encouraging and we believe that this framework will be practically most useful when applied to complex and poorly studied chemical systems involving a hundred or more chemicals. Tropospheric chemistry in general and boundary layer chemistry in particular are examples of systems where this methodology can provide quantitative guidance and help to establish measurement priorities.

References

Fisher, M., and Lary, D., J., 1995: Lagrangian four dimensional variational data assimilation of chemical species. *Q. J. R. Meteorol. Soc.*, **121**, 1681-1704.

Khattatov, B. V., Gille, J. C., Lyjak, L. V., Brasseur, G. P., Dvortsov, V. L., Roche, A. E., and Waters, J., 1999: Assimilation of photochemically active species and a case analysis of UARS data. *J. Geophys. Res.*, **104**, 18,715-18,737.

Jazwinski, A. H., 1970: *Stochastic Processes and Filtering Theory*, Academic Press.

Lorenc, A. C., 1986: Analysis methods for numerical weather prediction. *Q. J. R. Meteorol. Soc.*, **112**, 1177-1194.

Talagrand, O., and Courtier, P., 1987: Variational assimilation of meteorological observations with the adjoint vorticity equation. I: Theory. *Q. J. R. Meteorol. Soc.*, **113**, 1311-1328.

USES OF OCEAN DATA ASSIMILATION AND OCEAN STATE ESTIMATION

KEITH HAINES

Environmental Systems Science Centre, Reading University
3 Earley Gate, Reading RG6 6AL, UK

1. Introduction

Many of the methods used for ocean data assimilation are of course generic, in particular, techniques for estimation and error propagation, and these methods are adequately discussed elsewhere in this book. Instead we focus on some of the more specific oceanographic aspects. We start with some examples of the practical uses of ocean data assimilation and then look in more detail at the ocean inverse problem for recovering the time mean steady state ocean circulation. The physical constraints and assumptions are very explicit for this problem and similar ideas later turn out to be useful in the time dependent ocean assimilation context.

In the next chapter, *Altimeter covariances and errors treatment*, we look at the characteristics of some ocean data sets from ships, buoys and satellites, which have been used in ocean assimilation studies. Satellite altimeter sea level anomaly data, and profiles of temperature in the upper ocean (the two most commonly assimilated ocean data sets) are considered in more detail focussing on the relationships between these quantities and the rest of the state vector. The main problem with altimeter sea level data is the covariances with subsurface ocean conditions that can be represented in different ways. The adaptive treatment of model sea level errors will also be considered. For temperature profile data the main problems are that observations are usually of limited depth and that salinity is not measured at the same time so that the ocean density field is not known. The treatment of model bias will be introduced at this point and we look at some recent ideas about how to deal with a biased model.

Finally, in the chapter *Assimilation of hydrographic data and analysis of model bias*, we continue with the important theme of model bias and consider how we might use the assimilation procedure to develop diagnostics that quantify errors in model processes and thus may help us to build better models in the future. Although these ideas are expounded in the context of the ocean assimilation problem it is hoped that some of them may have applications to other areas.

2. Some uses of data assimilation in oceanography

The two important areas of ocean applications associated with surface waves and with coastal sea level are mentioned first only very briefly for the sake of completeness. These data assimilation applications have the following characteristics:

R. Swinbank et al. (eds.), Data Assimilation for the Earth System, 289–296.
© 2003 *Kluwer Academic Publishers. Printed in the Netherlands.*

1. Ocean Wave Forecasting: This is run as an operational service for ships and oil platforms and relies very strongly on meteorological conditions. Models of surface wave spectra and propagation directions contain representations of 'Wind Sea' and 'Swell'. Satellite altimeters can give wave height measurements for assimilation but the spatial/temporal coverage is relatively poor on meteorological timescales. A multi-altimeter mission might improve this. A good overview of surface wave modelling and assimilation is given in Komen *et al.* (1994)

2. Tidal/Storm Surge Forecasting: Operational for coastlines, estuaries, lagoons and tidal rivers. Strong Meteorological dependency with wind driven Ekman and Inverse Barometer (-1mb atmospheric pressure =+1cm sea level) effects building up the sea level during storms. Data from tide gauges along coasts can be assimilated (surges propagate anti-clockwise around basins in the Northern hemisphere). Examples of forecasting systems: the Adriatic (Venice Lagoon), and for the North Sea (the Thames Flood barrier).

Both of the above examples are mature areas with well-understood dynamical models that are essentially 2-dimensional, so that optimal methods of error treatment, such as the Kalman filter, are tractable. However as for many environmental forecasting systems, the real challenges lie in predicting extreme (dangerous) events, which is much more difficult. We shall not consider these applications again and they are intimately bound up with the problems of meteorological forecasting. The following three areas that will be the focus of chapters dealing with ocean data assimilation, involve deep-sea oceanography where the problems are not so intimately linked with the meteorological prediction.

3. Seasonal Weather Forecasting: Data assimilation is used to initialise the ocean component of ocean-atmosphere coupled models, with the most important data coming from the tropical Pacific that allows the forecasting of El Niño events. Tropical Atmosphere-Ocean (TAO) buoys, www.pmel.noaa.gov/tao/, which are moored to the seabed, provide continuous temperature profiles in the upper ocean (450m) via satellite, and are the main data assimilated. Typically coupled model forecasts run for 6 months, *e.g.*, the 1997 El Niño was successfully forecast from December 1996. The European Centre for Medium Range Weather Forecasts (ECMWF) run Ensemble seasonal forecasts, some of the results from which can presently be found at www.ecmwf.int/products/forecasts/d/charts/seasonal/.

4. Mesoscale Upper Ocean Forecasting: The Global Ocean Data Assimilation Experiment (GODAE), will run from 2003-2005 with real time upper ocean mesoscale forecasting experiments in several countries using eddy resolving ocean models. Near surface currents, coastal upwelling, sea temperatures, and ecosystem parameters will be the main focus. For example Met Office GODAE analyses for the North Atlantic at 1/9° resolution can be found at www.nerc-essc.ac.uk/las, and the French Mercator system analyses may be found at www.mercator.com.fr. Forecasting timescales will be from a few days to a few months. Assimilated data will come from altimeters, from voluntary

observing ships and from ARGO profiling floats (see chapter *Altimeter covariances and errors treatment*, section 1). Uses for GODAE products will be in the areas of, *e.g.*, pollution monitoring, fishing and tourism. See Koblinsky and Smith (2001), www.bom.gov.au/bmrc/ocean/GODAE.

5. Ocean State Estimation: The focus of state estimation is first and foremost to determine the time mean currents and ocean transports of mass, heat and freshwater. These transports are integrated quantities that cannot be measured directly and yet are critical to understanding how the ocean functions as part of the climate system. Full depth observations of temperature T and salinity S, at high accuracy, are needed for the calculations and the World Ocean Circulation Experiment (WOCE), 1990-2002, sampled many regions of the deep ocean for the first time. Figure 1 shows the set of WOCE sections from the Atlantic. Data assimilation is used to combine these data into a self-consistent set of transports across all the available sections. We will study the basis of this procedure in the section below.

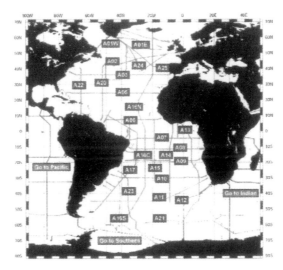

Figure 1. Location of Atlantic cruises where high quality top to bottom hydrographic data were gathered during the World Ocean Circulation Experiment (WOCE) 1990-2002.

3. The Ocean Inverse Problem

Figure 2 shows a North-South (N-S) vertical section of temperature, salinity and silicate (a conservative tracer) measured during WOCE cruise A16. Two kinds of information are implicit within these data. The water properties at the surface give information on how the ocean is forced, and the spreading of these properties through the deep ocean also contain indirect information on the circulation. For example in this section warm shallow thermocline waters reflect the Equator-Pole gradients in surface temperature and salinity (higher salinities in the tropics due to evaporation). Three water masses can be seen spreading down from the surface into the deep ocean with different salinity and silicate properties marking their cores. North Atlantic deep water (NADW), Antarctic intermediate water (AIW), and Antarctic bottom waters (ABW), all show up clearly. Ocean tracers and their conservation on long timescales and in advection over large distances have given a great deal of information about the circulation of the deeper waters of the ocean where the currents are too weak and noisy to observe.

However the combination of temperature and salinity giving the water density, also allows for a more direct calculation of ocean currents as explained below. Away from the Equator the circulation is governed by geostrophic and hydrostatic balance;

$$\rho f v = \frac{\partial p}{\partial x}; \qquad \rho f u = -\frac{\partial p}{\partial y}; \qquad \frac{\partial p}{\partial z} = -g\rho, \tag{1}$$

and thus by the thermal wind relations:

$$\frac{\partial \rho v}{\partial z} = -\frac{g}{f}\frac{\partial \rho}{\partial x}; \qquad \frac{\partial \rho u}{\partial z} = \frac{g}{f}\frac{\partial \rho}{\partial y}. \tag{2}$$

The zonal and meridional currents are u, v, and p is pressure, ρ density, g is gravity and f the Coriolis parameter and the vertical z coordinate is positive upwards. Integrating up from a level $-z_0$ gives:

$$\rho u(x, y, z, t) = \frac{g}{f} \int_{-z_0}^{z} \frac{\partial \rho}{\partial y} dz + \rho u_0(x, y, -z_0, t) = \rho(u_R + u_0), \tag{3}$$

$$\rho v(x, y, z, t) = -\frac{g}{f} \int_{-z_0}^{z} \frac{\partial \rho}{\partial x} dz + \rho v_0(x, y, -z_0, t) = \rho(v_R + v_0). \tag{4}$$

The u_R, v_R are the relative velocities determined entirely by knowledge of the density field, and u_0, v_0 are the reference velocities at level $-z_0$. Ships are able to measure profiles of the density field in the ocean $\rho(z)$, (although density depends on salinity as well as temperature, and salinity is difficult to measure accurately (by conductivity) so many older measurements are suspect). From sections of ship measurements, such as those from WOCE in Figure 1, the u_R/v_R component of velocity perpendicular to the sections can therefore be calculated. However the reference velocities u_0/v_0 are needed to completely determine the currents. This has always been the main barrier to accurate quantification of the ocean circulation.

A useful approximation has been to assume a deep *level of no motion* at say $-z_0$ =2000m where we assume u_0, $v_0 = 0$. Currents in the deep ocean tend to be much smaller than near surface currents and therefore near surface currents can be determined to reasonable accuracy in this way. However this fails if we want to calculate important quantities such as heat or mass transport across sections of the ocean. The problem is that, although deep velocities are small, the volume of water moving may be very large, as shown by the large depth of cold waters of almost uniform properties in Figure 2.

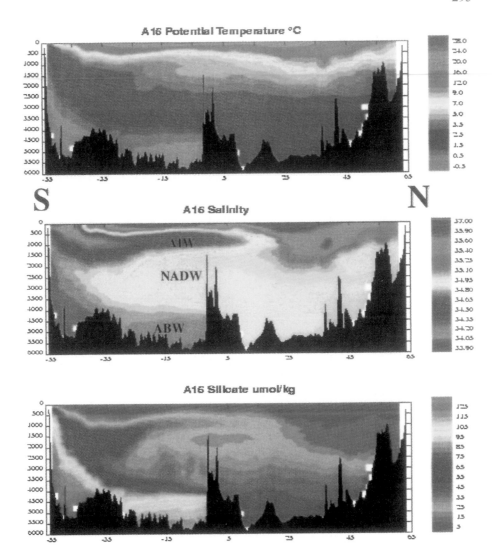

Figure 2: Temperature, Salinity and Silicate, measured during WOCE cruise A16 running approximately N-S through the Atlantic. Property values indicate the spreading of waters through the deep ocean. Antarctic Intermediate water (AIW), North Atlantic Deep water (NADW) and Antarctic Bottom water (ABW) are clearly distinguished.

Classical ocean inverse theory now relies on two assumptions to progress further.

(1) It is assumed that we are seeking a steady state solution in which water properties and circulation are time independent,

(2) Additional non-kinematic equations can be included to express the concepts of mass, heat and salt conservation, as well as conservation of other measured tracers.

Consider for example Figure 2 as representing the steady N-S water property distribution in the Atlantic Ocean and then trying to determine the N-S mean currents which would be consistent with the observed distributions of NADW, AIW and ABW using East-West (E-W) sections of density information running perpendicular to Figure 2 at various latitudes.

Hidaka (1940) was the first to try to use conservation constraints to try to solve for the missing reference velocities. Consider a closed volume of ocean such as the latitude band delimited by two observation sections shown in Figure 3 (from Wunsch; 1996). The transports can be calculated between each pair of depth profiles except for the missing reference velocities that are different for each station pair. However we can assume that there is no net transport of water into the region (one additional constraint), and the steady state assumption will allow us to assume no transport of salt into the region either (if we are to include heat we must also include the air-sea heat flux).

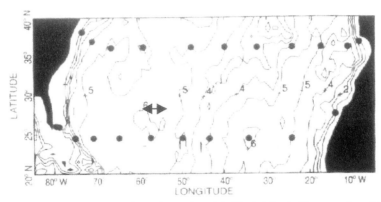

Figure 3. Example of station data enclosing a close region of the North Atlantic. The separation of the jth and $j+1^{th}$ station is labelled δ^j. From Wunsch (1996).

The general equations for the volume conservation can be written

$$\sum_j \rho \int_{-H}^{0} (u_R^j(z) + u_0^j)\delta^j dz = 0,$$

$$(5)$$

and for salt conservation

$$\sum_j \rho \int_{-H}^{0} S^j(z)(u_R^j(z) + u_0^j)\delta^j dz = 0,$$

$$(6)$$

where j represents consecutive station pairs in Figure 3, δ^j is the station separation and now **u** are velocities perpendicular to the track. One could make similar conservation statements about other tracers such as the silicate shown in the bottom panel of Figure 2. In practice for realistic ocean volumes, a whole set of simultaneous equations must be solved with careful consideration for the errors in the measurements and the physical constraints to be imposed. Generically the problem is often written as;

$$Ex + n = y \qquad (7)$$

where x is the vector of unknown reference velocities, n is the noise vector and y is the vector of resolved transports. The conditioning of this problem depends critically on how many physical constraints are imposed, but for real situations the problem is usually under-constrained. Many further constraints can be added by requiring no changes in the water masses within the control volume. This would mean that the volume of waters *with particular T,S, and other properties* which flow into the region is the same as the volume which flows out. This would imply no mixing of water within the region. For Figure 2 this means that the implied flow would be parallel to contours of constant water properties. Although this no mixing case is unrealistic and would be inconsistent with all the observations we can seek a solution that requires the minimum amount of mixing, and this is often sufficient to obtain a unique solution to the reference velocities and transports. We will not describe mathematically how to obtain these solutions, but Wunsch (1996) provides a full overview of the mathematics involved and the ocean applications. Figure 4 shows an ocean inverse solution by Ganachaud and Wunsch (2000) using hydrographic sections from around the globe, many of which were taken during WOCE. The transports of water across the different sections in different 'neutral' density (similar to potential density) classes, along with error estimates, are shown.

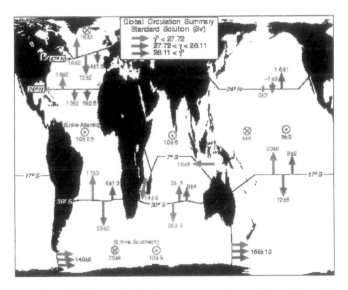

Figure 4. Transports of water in different density classes across various sections of the world ocean and transformations between water classes within regions derived from a global inverse calculation. Units are in Sv where $1\mathrm{Sv}=10^{6}\mathrm{m}^{3}\mathrm{s}^{-1}$. From Ganachaud and Wunsch (2000).

The major drawback of the ocean inverse method is that it assumes a steady state situation whereas in reality this is not the case and the different ocean section data are taken at widely varying times. There is a consortium project (Estimating the Climate and Circulation of the Ocean, ECCO) underway to perform state estimation in a time evolving sense for the period 1992-1997 using a 4-D variational approach, www.ecco.ucsd.edu/OSE/OSE.html. The complexity of this problem is very great and compromises must be made over the amount of data to include and the resolution of the ocean model. However the same kind of careful handling of the conservation laws used to provide closure constraints in the steady state estimation problem above can give insights into how to develop simpler and yet still very useful assimilation methods for the best available ocean models.

References

Ganachaud, A., and C. Wunsch, 2000: Oceanic meridional overturning circulation; mixing, bottom water formation rates and heat transport. *Nature* **408**, 453-456.

Hidaka K., 1940: Practical evaluation of ocean currents. *Proc. Imp. Acad. Tokyo*, **16**, 394-397.

Koblinsky, C.J., and N.R. Smith (Eds.), 2001: *Observing the Oceans in the 21st Century*. GODAE Project Office, Bureau of Meteorology, Melbourne, Australia, 285-306. ISBN 0642 70618 2.

G.J Komen, L. Cavaleri, M. Donelan, K. Hasselmann, and P.A.E.M. Janssen, 1994: *Dynamics and Modelling of Ocean Waves*. Cambridge University Press, UK.

Wunsch C., 1996: *The ocean circulation inverse problem*. Cambridge University Press.

ALTIMETER COVARIANCES AND ERRORS TREATMENT

KEITH HAINES

Environmental Systems Science Centre, Reading University
3 Earley Gate, Reading RG6 6AL, UK

1. Introduction: Ocean Data Sets

There are now a large number of data sources which are being or soon will be used in ocean data assimilation systems for determining the ocean circulation. The following is not a complete list but indicates some of the most important data sources.

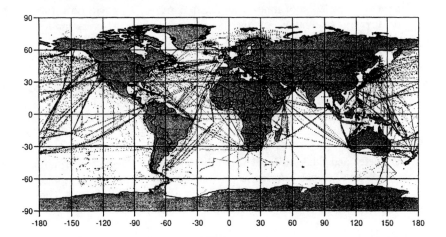

Figure 1. Locations of all the temperature profile data available for 1993. A total of 69980 profiles were used.

1. Temperature profiles: Mostly from Expendable bathythermographs (XBTs) from Voluntary Observing Ships. From 1950-1970 only the top 200-450m were normally sampled, but more recently this has increased to 800m. A typical annual global distribution (1993) is shown in Figure 1 (which includes Tropical Atmosphere-Ocean, TAO, buoy array data from the tropical Pacific)
2. Conductivity Temperature Depth (CTD) instruments measure Temperature and Salinity (and hence density), sometimes to full ocean depth. Until recently confined to research vessels, *e.g.*, Figure 2 in chapter *Uses of Ocean Data Assimilation and Ocean State Estimation*. Historically there are a much lower number of observations, mainly because salinity is hard to measure accurately.

R. Swinbank et al. (eds.), Data Assimilation for the Earth System, 297–308.
© 2003 *Kluwer Academic Publishers. Printed in the Netherlands.*

3. The ARGO neutrally buoyant float program is beginning to provide temperature (T) and salinity (S) profiles in the top 2000m. Floats drift at 2000m and surface every 2 weeks measuring T, S profiles and transmit data via satellite. They have a nominal 5-year deployment lifetime. At the time of writing there are around 800 deployed (the aim is for 3000 during The Global Ocean Data Assimilation Experiment, GODAE). Real time ARGO data are available from www.ifremer.com.fr/coriolis.

4. Satellite Sea Surface Temperatures (SST): These are measured by passive Infra-red (IR) (AVHRR, ATSR) or microwave (SSMI, TRMM) instruments. The data have been little used for assimilation so far due to problems with rapid diurnal variations in surface temperature and skin temperature effects which are not well represented in ocean models. Examples of SST analysis projects include www.nodc.noaa.gov/dsdt/oisst/index.html

5. New satellites measuring the Earths Gravity field, Gravity Recovery and Climate Experiment, GRACE (launched 17 March 2002) and the Gravity and Ocean Circulation Explorer, GOCE (to be launched 2005 or later) missions will greatly improve knowledge of the Earth's geoid. The reason these are mentioned here is because these data can be assimilated with altimeter data to improve knowledge of time-mean ocean currents. This will be very important for inverse calculation of the ocean circulation. Also assumptions must presently be made about the geoid in order to assimilate altimeter data today, as will be discussed below. More information can be found at www.csr.utexas.edu/grace/, and www.esa.int/export/esaLP/goce.html

6. Satellite Altimeters have provided continuous global coverage since 1992 (TOPEX/POSEIDON and ERS-1/2). JASON, GEOSAT Follow On, and Envisat (see chapter *Research Satellites*) all carry altimeters. The altimeter instrument is a microwave radar measuring sea level relative to the satellite accurate to 2-3cm. Corrections for atmospheric signal delays, inverse barometer and tidal sea level variability are usually accounted for. Thereafter any sea level slopes *relative to the geoid* imply surface geostrophic currents. Data are available along tracks or as maps, usually every 10 days. These can be assimilated to provide mesoscale upper ocean currents.

In the following sections we focus on altimeter data in particular, describing how it may be assimilated into an ocean model and looking in some detail at the causes of covariances between sea level and other quantities that can be used to recover subsurface information on the circulation. The great accuracy of satellite altimeter data from the TOPEX/POSEIDON mission onwards has meant that it is these data above all others that has received most attention for data assimilation. The two major problems to be tackled are: (1) the unknown geoid, and (2) the unknown projection of the sea level information with depth. These are explained in more detail below.

2. Altimeter data: the geoid problem

If sea level variations can be measured relative to the surface of constant gravitational potential called the geoid, which we will identify as **z=0**, then they are equivalent to pressure variations which are related geostrophically to the surface flow. Mean sea level is used in the meteorological community as if it were a geopotential surface, however it is not. The true geopotential surface deviates from mean sea level by up to 1m (irrelevant in meteorology but very important for determining ocean currents). Figure 2 left-panel, shows one of the best available global geoids, EGM96, as variations from a reference ellipsoidal Earth. Figure 2 mid-panel, shows the mean sea level determined by altimeter data relative to the same ellipsoidal Earth (Hernandez and Shaeffer, 2000). Figure 2 right-panel, shows the difference, which should be a streamfunction for the mean surface geostrophic flow. Although the large-scale features are reasonably consistent with this, the small mesoscale features are completely unrealistic due to inaccuracies in the geoid data at these scales.

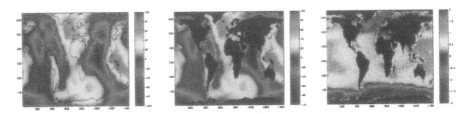

Figure 2. The left panel shows the EGM96 geoid height relative to a reference ellipsoidal earth. The middle panel shows the mean sea level determined by altimeter data relative to the same ellipsoid as EGM96. This mean sea level was produced by Hernandez and Shaeffer (2000). The right panel shows the difference (sea level – geoid). All units are in m. In the left and middle panels, red indicates values larger than the reference ellipsoid, blue indicates values lower than the reference ellipsoid. In the right panel, red indicates positive differences, blue indicates negative differences.

Although the true geoid is not known accurately enough at the mesoscale, the time varying component of the altimeter signal, or the sea level anomaly, can still be assimilated with mesoscale accuracy. This must be compared with an equivalent 'anomaly' sea level from the model, and to define this from the full sea level we need a separate definition of mean sea level. Several ways have been used to define this mean sea level.

1. A previous run of the ocean model without data assimilation is often used to determine a mean sea level over some period (this period should really be the same as that used to define the altimeter sea level anomalies, *e.g.*, 1993-1995). Disadvantages are that it is often known that this model mean sea level is biased in some areas and this bias will be preserved.

2. An independent sea level anomaly can be determined from climatological hydrographic data using dynamic height calculations as a proxy for mean sea level, *e.g.*, Fox *et al.* (2000b), Killworth *et al.* (2001). Disadvantages are that the climatological data will be appropriate to a different time period to the

altimeter data and that the dynamic height calculation does not give a full description of sea level variations (see section 3 in this chapter).

3. A previous run of the ocean model can be performed without altimeter data assimilation but with assimilation of hydrographic data. The resulting sea level can be used for a subsequent assimilation run over the same period with both hydrographic and altimeter data assimilated, Fox and Haines (2003). Both of the disadvantages to (2) listed above are overcome in this way.

4. An independent geoid model with more accurate small-scale information would allow the total altimeter sea level, relative to this geoid, to be assimilated. It is possible to calculate local geoids with the necessary small-scale accuracy but these have not been used up to now in assimilation studies.

Of the above methods, (3) is perhaps the best so far attempted for mesoscale studies, although it will not work where there are very few hydrographic data available (*e.g.* southern ocean). Also sufficient salinity is almost never available to properly constrain the density field but this may be overcome using T/S relationships (see chapter *Assimilation of Hydrographic Data and Analysis of Model Bias*, section 2 on assimilating hydrographic data). We now look at the problem of recovering subsurface information from altimeter data.

3. Altimeter data: the vertical mode problem

Knowledge of the covariance relationships between sea level anomalies and anomalies in other quantities (temperature, salinity, currents at depth) is needed in some form for any altimeter assimilation method. What we will try to emphasise in this section are that there are different ways of representing this information, and some ways are more succinct and easier to verify empirically than others. This is related to the physical relationships between variables.

Sea level or pressure variations on the geoid $p(z=0)$ can be broken down into pressure variations at the sea floor ($z=-H$) and hydrographic variations from the water column density (assuming hydrostatic balance);

$$p(0) = p(-H) - g \int_{-H}^{0} \rho(z) dz.$$

(1)

Large-scale rapid variations in $p(0)$ tend to be barotropic and have variations at $p(-H)$ similar in magnitude and well correlated with $p(0)$ variations. Smaller scale more persistent variations in $p(0)$ tend to be strongly baroclinic with only weak correlations with $p(-H)$. It is necessary to be able to make this distinction if observations of sea level $p(0)$ variations are to be correctly assimilated into a model. In what follows it will also be useful to define;

$$D(x, y) = -g \int_{-z_0}^{0} \rho(x, y, z) dz,$$

(2)

the dynamic height at the sea surface relative to some level $-z_0$, which is determined entirely from hydrographic data. Provided that all horizontal pressure variations at level $-z_0$ are negligible (*i.e.*, the level of no motion assumption) then $p(0)=D(x,y)$.

Methods for obtaining subsurface quantities by projection of sea level anomalies below the surface can be broadly broken into two classes, empirical projection and dynamical projection. Much of this discussion is based on that given in Haines (1994).

3.1. EMPIRICAL PROJECTION

This should be based on concurrent observations of local hydrographic or current meter data and sea level over a long period of time. But usually this criterion cannot be met. The usual way of developing relationships is with Empirical Orthogonal Functions (EOFs). To illustrate the method we look at some early results from De Mey and Robinson (1987). They used one year of data from the POLYMODE current meter array in the North-West (NW) Atlantic to develop EOFs of the vertical pressure variability, shown in Figure 3. The first two modes represent 81.5% and 16.7% of the pressure variance, respectively. They reasoned that if only sea surface height data is available it makes sense to project it onto the surface enhanced first EOF mode and thereby to recover the pressure variations at depth. De Mey and Robinson (1987) used this method to assimilate the surface data alone from POLYMODE and managed to partly recover deeper pressures and currents.

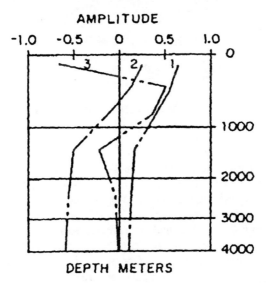

*. 6. The first three vertical empirical modes of eddy
pressure anomaly p' in the Mark 2 domain.

Figure 3. First three Empirical Orthogonal functions of pressure variations in the POLYMODE experiment
from De Mey and Robinson (1987).

The problems with this method are that the vertical modes are probably very variable spatially and possibly also in time, depending on the vertical thermocline structure. Hurlburt *et al.* (1990) developed a much wider set of correlation functions to relate sea level variability at one location with 3-D pressure variations. The problem here is that the only way to develop these full covariances is by using model output data, which may well be strongly biased. Nevertheless, this is still the system used in the US navy's ocean forecasting system www7320.nrlssc.navy.mil/html/dart-home.html.

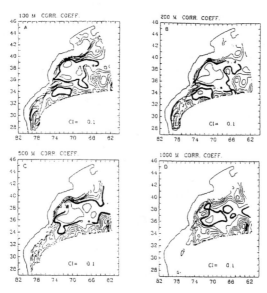

Figure 4. Correlations between surface height variance and subsurface density variance at different levels below the surface. Contour intervals are 0.1 apart and the 0.8 contour is bold. From Mellor and Ezer (1991).

Variations on the EOF theme exist where sea level is correlated directly with hydrographic water properties, temperature and salinity, Mellor and Ezer (1991) and Ezer and Mellor (1994). Figure 4 shows correlations of sea surface height variations and density variations at several depths within Mellor and Ezer's limited area high resolution model of the Gulf stream along the US east coast. The correlations are high even down 1000m below the surface making it feasible to use these correlations for assimilation of sea level data. One big advantage is that the temperature, salinity and density are the appropriate state variables for full Primitive Equation (PE) ocean models, while the pressure field correlations are really only suitable for quasi-geostrophic models. However the correlations still require to be derived from a model and are not therefore necessarily realistic.

A further study by Oschlies and Willebrand (1996) used vertical current correlations obtained from a PE model of the NW Atlantic. This is rather similar to calculating pressure correlations, however they also calculated consistent changes in density (via the thermal wind relation) with the temperature-salinity characteristics preserved in the model. This makes the method similar in some respects to the methods discussed below under dynamical projection. There does not obviously seem to be much to distinguish between these empirical methods, however in the next section we show that there are preferable ways to project altimeter data.

3.2 DYNAMICAL PROJECTION

These ideas were originally developed around quasi-geostrophic theory and it is still useful to review results in this framework. The key result is that a change of state vector will lead to new coordinates in which knowledge about the ocean variability at depth can be more succinctly expressed in terms of the sea level.

Haines (1991) used an idealized 4-layer quasi-geostrophic ocean gyre model to illustrate the assimilation of sea level data. Figures 5a,b show the streamfunction ψ and the potential vorticity q in each layer at some instant. The flow broadly represents subtropical and subpolar gyres with a strong current between them penetrating an ocean basin and going unstable (*cf.* the Gulf stream). Altimeter sea level data is equivalent to observations of ψ_1, the top layer streamfunction. The fields are related by;

$$q_1 = \nabla^2 \psi_1 + \beta y - \gamma_{1,2}^2 (\psi_1 - \psi_2),$$
$$q_2 = \nabla^2 \psi_2 + \beta y - \gamma_{2,1}^2 (\psi_2 - \psi_1) - \gamma_{2,3}^2 (\psi_2 - \psi_3),$$
$$q_3 = \nabla^2 \psi_3 + \beta y - \gamma_{3,2}^2 (\psi_3 - \psi_2) - \gamma_{3,4}^2 (\psi_3 - \psi_4),$$
$$q_4 = \nabla^2 \psi_4 + \beta y - \gamma_{4,3} (\psi_4 - \psi_3).$$

$$(3a\text{-}d)$$

where β is the northward coriolis gradient and γ are the Rossby deformation radii between layers. The point of showing these fields is that q_i $-i=1\text{-}4$ provides an alternative state vector for describing the system and yet the correlations in the vertical are completely different for ψ and q. In particular q is virtually uncorrelated vertically, due largely to its Lagrangian properties and the fact that q gradients have been mixed away below the surface. Haines (1991) suggested a mixed state vector representation for data assimilation purposes using ψ_1 from observations and q_2, q_3, q_4 from the a priori model. This mixed description is complete in the sense that all the other fields may be found, and it removes the need to use vertical correlations in ψ.

Figures 6a,b illustrate the convergence of an identical twin assimilation experiment. Note particularly that although q_2, q_3, q_4 are not changed at all during each assimilation, their errors decrease over time as the model runs forward so that deep property fields (q in this case) are recovered over time despite the lack of correlation with surface streamfunction. This is a powerful and attractive idea when considering the importance of deep tracer fields such as those shown in Figure 2 in chapter *Uses of Ocean Data Assimilation and Ocean State Estimation*.

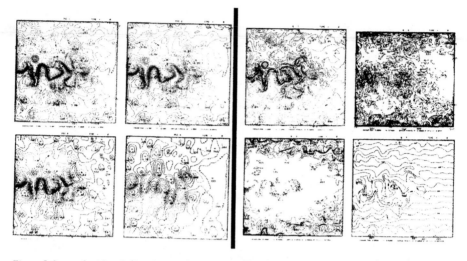

Figure 5. Streamfunction (left) and potential vorticity (right) fields from a 4-layer quasi-geostrophic ocean box model. The surface layer field is in the top left of each group, second layer to the right, third layer below and the fourth layer in the bottom right. Note the very different covariance relations between anomalies in the different fields. From Haines (1991).

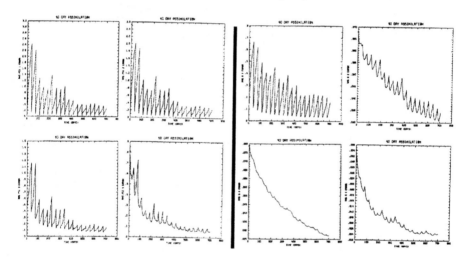

Figure 6. Convergence of the Streamfunction (left) and potential vorticity (right) root-mean-square (RMS) errors during a data assimilation twin experiment in which surface streamfunction data are assimilated (representing surface altimeter data) every 40 days. The layers are as described in Figure 5. From Haines (1991).

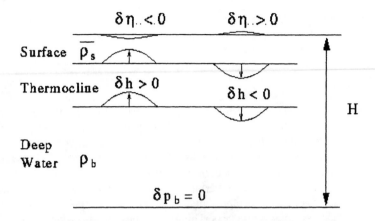

Figure 7. Schematic altimeter assimilation by vertical displacement of the thermocline. The vertical displacements δh are calculated to cancel the sea level change leading to no change in pressure at the ocean floor. From Cooper and Haines (1996).

Cooper and Haines (1996) extended this idea to a primitive equation framework for the oceans in which potential vorticity is given by;

$$q = \frac{f}{\rho_0} \frac{\partial \rho}{\partial z}.$$
(4)

Now using the a-priori model q for the deep oceans means keeping the stratification $\delta\rho/\delta z$ constant *as a function of* ρ. The only way to do this while still changing the density field is to vertically displace the water column, see Figure 7. To close the problem it was assumed that the sea level anomalies are essentially baroclinic, so a constraint of no change to the deep pressure field was imposed;

$$\Delta p(0) = g \int_{-H}^{0} \Delta\rho dz, \quad \text{where} \quad \Delta\rho = \frac{\partial \rho}{\partial z}\Delta h,$$
(5)

for small vertical displacements Δh. It should be noted that the constraint of no change to the deep pressure is different from the solution found in Haines (1991) in the quasi-geostrophic framework. This constraint is equivalent to taking ψ_1 from observations, and q_2, q_3 and ψ_4 from the model a-priori in the quasi-geostrophic framework.

The dynamical approach to the vertical projection of altimeter data, described above, makes a virtue out of not changing certain quantities in the model at all and thus dispensing with multivariate covariance functions completely. The idea can be compared to the constraints imposed during ocean state estimation where minimal changes to water properties are sought. However the necessary recasting of the state vector can also be used to develop a covariance approach. Gavart and De Mey (1997) studied the empirical covariances of sea level and the depth of isopycnals. Since the depth of all density (or temperature) surfaces contains precisely the same information as the density (or temperature) as a function of depth, the two descriptions are obviously equivalent. However the covariance information looks quite different for the two descriptions. Empirical orthogonal functions were calculated from hydrographic profiles measured around the Azores current (SEMAPHORE project) using different descriptions of the data (see Figure 8). Note particularly that the first EOF of vertical displacement is very uniform with depth, providing support for the Cooper and Haines (1996) method. When the hydrographic data were projected only onto the first mode z-level EOF, and dynamic heights calculated from the result, 8.4 cm^2 of sea level variance is left as a residual. When the projection onto the first mode isopycnal EOF is made only 3.6 cm^2 of residual sea level variance remains. Thus the isopycnal coordinate system provides a more compact representation of the hydrographic variance, which can be used in the data assimilation process. This system has now become part of the French Mercator forecasting system.

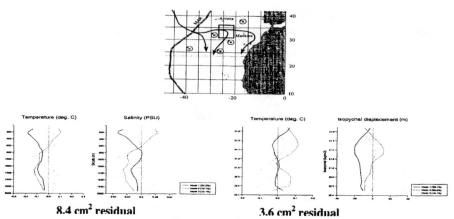

Figure 8. Empirical Orthogonal functions (EOF) describing the vertical variance of temperature and salinity in the Azores box (top panel). The bottom left two panels show the first 3 EOFs as a function of depth. The bottom right two panels show the first 3 EOFs of temperature on an isopycnal and depth of an isopycnal. The residuals below show the dynamic height (sea level) variance NOT explained by the 3 EOFs above. From Gavart and De Mey (1997).

4. Altimeter data: Adaptive treatment of sea level errors

In the section above the multivariate connections between sea level data and the subsurface hydrography have been emphasised. However in a real assimilation problem it is still necessary to perform an error analysis on the sea level information and this can only be done with statistical methods. If we consider assimilating maps of observed altimeter anomalies, the error variance, σ_o^2, that goes with each map is also available. Provided we have an estimate of the model sea level error variance σ_m^2 we can calculate a realistic sea level analysis η_a, from the model sea level η_m, and the altimeter observed sea level η_o.

$$\eta_a = \eta_m + W (\eta_o - \eta_m) \qquad \text{where } W = \sigma_m^2/(\sigma_m^2 + \sigma_o^2) \qquad (6)$$

However calculating a realistic σ_m^2 is very difficult, normally requiring some kind of Kalman filter treatment.

Dee (1995) suggested that for systems with a large number of simultaneous observations (satellite data often falls into this category) it is possible to tune a small number of model covariance parameters 'on-the-fly' without needing to forecast them from the previous assimilation time. This is an important use of the innovation vector information (see chapters on data assimilation theory, Talagrand). There essentially arises a consistency condition between the observation errors and the innovations (which we might assume to be known), and the model errors (which are not known).

$$< (\eta_m - \eta_o)^2 > = <\sigma_m^2> + <\sigma_o^2> - 2C(<\sigma_m^2> <\sigma_o^2>)^{\frac{1}{2}} \qquad (7)$$

where C accounts for any expected correlations between the model sea level and the observed sea level. Dee (1995) writes this relationship much more generally in terms of

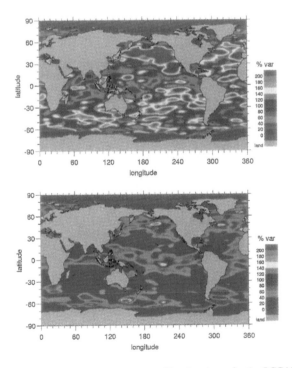

Figure 9. Errors in mean sea level as a percentage of local variance for the OCCAM model with altimeter assimilation. Top panel shows the errors for the first altimeter assimilation and bottom panel show the errors after 6 months. The errors were determined using Equation 7. From Fox *et al.* (2000a). Red indicates largest errors (> 200%), blue/purple indicates errors close to 0%. Grey indicates land.

unknown parameters within the innovation error covariance matrix, which is then compared with the actual measured innovations. In the equation above the <> operator represents some kind of averaging procedure which allows a large number of independent innovations $(\eta_m - \eta_o)$ to give 'low resolution' information on the model error variance σ_m^2. Once the model errors are known the appropriate sea level analysis can be made since W is determined (see Equation 6).

Fox et al. (2000b) used this method to calculate model sea level errors adaptively during a global ocean data assimilation experiment. The ensemble averaging procedure used is just a spatial smoothing of innovations. Figure 9 shows an example of the errors σ_m^2 calculated at the beginning of an assimilation run, and some way into the run. Clearly the model errors after the start are greatly reduced and the weighting W of the new altimeter observations takes this into account. This adaptive calculation of errors is very cheap compared with any Kalman filter approach although it is only suitable for use with large observational data sets. This method is also applicable more widely than simply in ocean data assimilation. The idea is being used at the US Data Assimilation Office (DAO) to look at improved ways of assimilating satellite water vapour data into atmospheric models, Dee (2002).

References

Cooper, M.C., and K. Haines, 1996: Data assimilation with water property conservation. J. Geophys. Res 101, C1, 1059-1077.

De Mey, P., and A. Robinson, 1987: Assimilation of altimeter eddy fields in a limited area quasi-geoostrophic model. J. Phys. Oceanogr. 17, 2280-2293.

Dee D..P., 1995: On-line estimation of error covariance parameters for atmospheric data assimilation. Mon. Weath. Rev. 123, 1128-1145.

Dee D.P., 2002: An adaptive scheme for cycling background error variances during data assimilation. ECMWF/GEWEX workshop on Humidity Analysis, Reading, July 2002..

Ezer, T., and G.L. Mellor, 1994: Continuous assimilation of GEOSAT altimeter data into a three-dimensional primitive equation Gulf Stream model. J. Phys. Oceanogr., 24, 832-847.

Fox, A.D., K. Haines, B. De Cuevas, and D.J. Webb, 2000a: Altimeter assimilation in the OCCAM global model, Part I: A twin experiment. J. Marine Sys. 26, 303-322.

Fox, A.D., K. Haines, B. De Cuevas, and D.J. Webb, 2000b: Altimeter assimilation in the OCCAM global model, Part II: TOPEX/POSEIDON and ERS1 data. J. Marine Sys. 26, 323-347.

Fox. A.D., and K. Haines, 2003: Interpretation of Water Mass Transformations diagnosed from Data Assimilation. J. Phys Oceanogr. 33, 485-498.

Gavart, M., and P. De Mey, 1997: Isopycnal EOFs in the Azores current region: A statistical tool for dynamical analysis and data assimilation. J. Phys. Oceanogr. 27, 2146-2157.

Haines, K., 1991: A direct method for assimilating sea surface height data into ocean models with adjustments to the deep circulation. J. Phys. Oceanogr., 21, 843-868.

Haines, K., 1994: Dynamics and data assimilation in oceanography. NATO Series I, Vol 19, 1-32.

Hernandez, F., and P. Shaeffer, 2000: Altimetric mean sea surfaces and gravity anomaly maps intercomparisons. AVI-NT-011-5242-CLS 48pp, CLS, Ramonville St. Agnes.

Hurlburt, H.E., D.N. Fox, and E.J. Metzger, 1990: Statistical inference of weakly correlated subthermocline fields from satellite altimeter data. J Geophys. Res., 95, C7, 11375-11409.

Killworth, P.D., D.E. Dietrich, Ch. Le Provost, A. Oschlies, and J. Willebrand, 2001: Assimilation of altimetric data into an eddy-permitting model of the North Atlantic. Prog. Oceanogr. 48, 313-335.

Mellor G.L., and T. Ezer, 1991: A Gulf stream model and an altimetry assimilation scheme. J. Geophys. Res., 96, C5, 8779-8795.

Oschlies, A., and J. Willebrand, 1996: Assimilation of Geosat altimeter data into an eddy-resolving primitive equation model of the North Atlantic Ocean. J. Geophys. Res. 101, 14175-14190.

ASSIMILATION OF HYDROGRAPHIC DATA AND ANALYSIS OF MODEL BIAS

KEITH HAINES

Environmental Systems Science Centre, Reading University
3 Earley Gate, Reading RG6 6AL, UK

1. Introduction: Temperature and Salinity in the upper water column.

In this chapter we look at the assimilation of subsurface temperature profile data. Particular attention will be paid to covariances with salinity, and to the analysis of model bias in these fields. Up to now most subsurface data consists of temperature (T) profiles only without coincident salinity, although in the near future the ARGO float program will provide regular salinity measurements and the algorithms described here will need to be augmented. As discussed earlier in chapter *Altimeter Covariances and Errors Treatment*, section 1, the vast majority of T profile data from Expendable bathythermographs (XBTs) or from moorings tend to be of limited depth. These data are the main resource for ocean assimilation for seasonal forecasting activities and we shall illustrate the methods used by reference to results from the European Centre for Medium-range Weather Forecasts (ECMWF) seasonal forecasting system.

2. Assimilation of T profile data

The starting point for T profile assimilation at ECMWF is an Optimal Interpolation method. Observed $T_o(z)$ profiles are compared with model $T_m(z)$ profiles. The misfits $(T_o - T_m)(z)$ are then spread out over some influence radius with some weighting and the calculated innovations are added to the model fields, slowly over a period of days to reduce the assimilation shocks. The details are not important but it is important to note that early schemes (a) made no change to the temperature below the deepest observed $T_o(z)$ (only 450m for Tropical Atmosphere-Ocean, TAO data) and (b) made no updates to the salinity field. The consequences of these omissions have been shown to be quite severe. Of course if sufficient observations existed, covariance matrices could be used to update the salinity and deeper temperature fields but these data are not available.

Troccoli and Haines (1999) offered an alternative method of updating the deeper temperatures in a model, as well as a way of updating the salinity, that has now been adopted at ECMWF. Figure 1 illustrates the problem clearly. Unstable density profiles can easily be created at the base of the observation T profile, or even within the range of the observation T profile, by the standard assimilation method. The solution suggested is to vertically displace the model water column to ensure a temperature match at the

309

R. Swinbank et al. (eds.), Data Assimilation for the Earth System, 309–319.
© 2003 *Kluwer Academic Publishers. Printed in the Netherlands.*

deepest level of the temperature analysis. In addition within the upper ocean it is suggested that the salinity be modified to preserved the T/S relationships present in the model water column. These two constraints ensure that the final analysed water column is continuous in T and S at all levels and is also guaranteed to be statically stable.

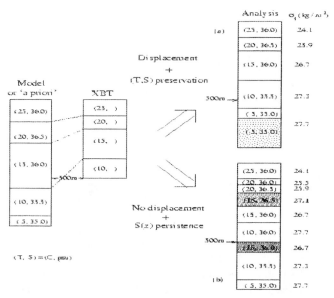

Figure 1. Schematic of assimilation of an observed temperature profile into an a priori model, shown on the left. Bracketed terms show (T,S) of the waters. Lower right shows the simplest assimilation, which directly imports the T data. The values to the right of the water columns show potential density. Upper right shows the analysis of the water-mass conserving scheme of Troccoli and Haines (1999).

This scheme was incorporated into the ECMWF seasonal forecasting model and run to produce ocean analyses over a 10 year period. The impacts on the T and S fields in the tropical Pacific are illustrated in Figure 2 from Troccoli *et al.* (2002). The new method appears to reduce the strong mixing of the temperature and salinity fields that otherwise occur around the Equator. There is also some evidence that this allows some improvements in the ability of the coupled model to make El Niño forecasts Segschneider *et al.* (2001).

The relationship between salinity and temperature can of course conventionally be represented in terms of a T/S covariance function which itself would be a function of depth. However the method described above can also be seen as another example of the use of alternative state vectors to represent the variability in observations. Troccoli and Haines (1999) illustrated this in a study of variances from an intensive conductivity-temperature-depth (CTD) campaign undertaken in a small region of the Tropical Pacific. Figure 3 shows data from 104 T and S profiles taken over a period of 10 days. Clearly there is considerable variability in both the T and S profiles as a function of depth. However the S(T) relationship shows less variance and if the salinity variance is measured as a function of temperature and projected back to depth levels using the average T(z) profile then the effective reduction of variance is very clearly seen.

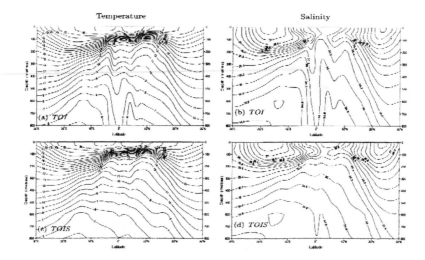

Figure 2: Meridional sections at 30W of 9 year mean temperature (left) and salinity (right) from two data assimilation experiments with the ECMWF seasonal forecasting ocean model. Top panels show conventional assimilation, equivalent to the bottom right of Figure 1. Bottom panels show the result of the water mass conserving scheme. From Troccoli *et al.*, (2002).

The result of the above assimilation method is to preserve the T/S relation during assimilation of T profiles. This can be regarded as having a similar justification to the preservation of the deeper potential vorticity fields during altimeter assimilation in the chapter *Altimeter Covariances and Errors Treatment*. Both T and S are Lagrangian conserved properties of the water and are therefore likely to vary together whenever the cause of variation is associated with advection. Most wave processes in the ocean, whether internal waves or Rossby waves, produce variability of this type. T/S properties are therefore preserved over long timescales and only change in the ocean interior due to mixing processes. This also applies to an ocean model in which assimilation is performed using constraints in the assimilation schemes. We will return to this in the next sections when considering diagnostics of bias from ocean data assimilation experiments.

3. Corrections for bias during data assimilation

Models of the 3-D ocean circulation that we have today contain large biases. Some of these are well known, *e.g.*, the Gulf Stream often flows too far north along the Eastern coastline of the US leading to large biases in sea surface temperatures, or (b) Tropical Pacific models produce too little upwelling in the eastern equatorial Pacific leading to a warm bias at the surface. If such biases were traceable to a few explicit parameters in the model then assimilation could be used to tune these, using a cost function and a 4D-var approach. However this will not usually be the case.

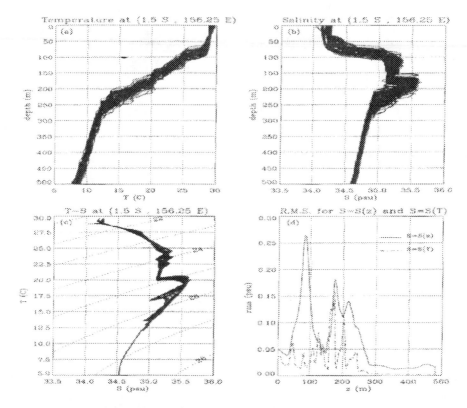

Figure 3. Comparison of 104 CTD profiles from e region of the western tropical Pacific taken over 10 days in 1992. Top left (a) temperature profiles,; top right (b) salinity profiles; bottom left (c) T-S diagram; bottom right (d) S(z) and S(T) rms variance comparison for all profiles. Note that to plot S(T) variance as a function of z in (d) the mean T profile from (a) has been used. From Troccoli and Haines (1999).

In an assimilation context we should consider that;

(1) Much of optimal assimilation theory assumes that the models to be used are unbiased. If this is not the case there seems little point in trying to solve the optimal error equations exactly, when a much cheaper approximate approach is more appropriate,

(2) Detection of bias is easy: if the innovations do not average to zero in the long term then either the model or the observations (or both) are biased,

(3) If we assume the model is biased then one of the main effects of data assimilation is to try to counteract bias or model drift,

(4) Prevention of drift can be successful but if the drift is too rapid this can cause an unrealistic model response every time data is assimilated, which is damaging to the results,

(5) Theoretical methods exist to detect and separate out model bias during the data assimilation procedure but they have rarely been used in an ocean or atmosphere assimilation context. Unfortunately these methods usually require a 'bias model', which is normally a rather *ad hoc* affair,

(6) We should aspire to decode the model bias revealed by data assimilation in order to learn something useful about models failings. But how?

A good recent discussion of generic ways of treating model bias is contained in Dee and Da Silva (1998). In this section we will focus on ways of separating out model bias (see (4) above) at the time of assimilation to prevent unwanted model responses noted in (3) above. In the following section we will look more at the interpretation of bias in a situation where bias has been identified at the end of the assimilation procedure. We will present a bias approach for dealing with model errors in the tropical Pacific. All the results in this section come from Bell *et al.* (2002).

Ocean models used as part of coupled models for El Niño prediction are often unable to maintain a tight enough thermocline at the Equator in the eastern Pacific ocean. Figure 4a shows the time mean temperature innovation during assimilation of TAO data into the Met Office model over several years (this should be zero for an unbiased system). Unfortunately the large transient model response to the assimilation corrections in this region leads to unrealistic upwelling and other undesirable effects, which reduce the value of the model for seasonal forecasting.

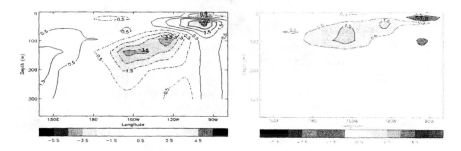

Figure 4. Mean annual temperature increments (°C month^{-1}) along the Equator due to temperature profile assimilation in the Met Office model. Left panel (a) shows the results without a bias model and right panel (b) shows the result after a pressure correction bias model has been applied. From Bell *et al.*, (2002).

Following Bell *et al.* (2002), let the deterministic model equations be;

$$x_{k+1} = M (x_k , u_k) \qquad (1)$$

where M is the model time evolution operator, u_k are model parameters (*e.g.* forcing), and k is the time subscript. We must assume that these are biased so we can write;

$$x_{k+1} = M^t (x_k , u_k) + T(b_k) \qquad (2)$$

where $\mathbf{M}^t$ is the true model and $\mathbf{b}_k$ is the bias vector; $\mathbf{T}$ allows the bias to have smaller dimensions than $\mathbf{x}_k$. Effectively we have extended the model state vector to include bias variables. To complete the problem we need to define a model for bias evolution;

$$\mathbf{b}_{k+1} = \mathbf{W}(\mathbf{b}_k, \mathbf{x}_k) + \zeta_k, \tag{3}$$

where ζ_k, allows for white noise error growth. Sequential data assimilation can be carried out on this system, updating $\mathbf{x}_k$ and $\mathbf{b}_k$, at assimilation times and it can be shown that, at least for the linear problem, $\mathbf{x}_k$ will converge to the true solution *provided we have a true model for the bias* $\mathbf{W}$. For further details of the theory of bias modelling see Nichols, *Treating model errors in 3-d and 4-d data assimilation*, this volume.

Bell *et al.* (2002) presumed that the bias in their tropical Pacific ocean model was caused by a *constant incorrect wind stress* but they argued that it could be compensated for with a pressure field bias $\mathbf{p}^b$ in the model, giving effective equations of motion of the form;

$$\frac{\partial \mathbf{u}^m}{\partial t} + \Gamma(\mathbf{u}^m) - f\mathbf{v}^m = -\frac{\partial(\mathbf{p}^m + \mathbf{p}^b)}{\partial x} + \frac{\partial \tau_{xz}}{\partial z},$$

$$\frac{\partial \mathbf{v}^m}{\partial t} + \Gamma(\mathbf{v}^m) + f\mathbf{u}^m = -\frac{\partial(\mathbf{p}^m + \mathbf{p}^b)}{\partial y} + \frac{\partial \tau_{yz}}{\partial z}, \tag{4a,b}$$

where superscripts $\mathbf{m}$ refer to the model variables, $\mathbf{b}$, the bias variables, and $\Gamma(.)$ are the non-linear terms. The $\mathbf{p}^b$ is assumed not to evolve between assimilation times.

It may seem odd to represent, what are thought to be errors in the surface wind stress, by a 3D pressure bias $\mathbf{p}^b$. However this 'pressure correction' method does simplify the analysis and it appears to have good convergence properties. This is perhaps because it generally allows a level of decoupling between the ocean mass field (relating temperature and salinity to the density and hence the model pressure $\mathbf{p}^m$) and the equations for momentum (now controlled by the tuneable $\mathbf{p}^b$ as well as $\mathbf{p}^m$).

Figure 4b shows the greatly reduced average innovations in the eastern equatorial Pacific when the pressure correction method is applied. It allows a good ocean thermal structure to be maintained by assimilation (important for air-sea interaction in coupled modelling of El Niño) without inducing unrealistic dynamical responses (*e.g.* upwelling, waves) that would contaminate the model. For practical purposes of running the model without unphysical transients the method has worked. However the solution has not lead to any insights into whether the bias really is due to wind stress error? Could the 'true' wind stress be recovered as well as biases in the downward mixing of stress? This question presently remains unanswered.

4. Interpretation of bias after a data assimilation experiment

A different approach to studying bias in ocean data assimilation is presented in Fox and Haines (2003). Here the aim is not to alter the data assimilation in any way but simply to diagnose from the average innovations some useful information about model physics. Therefore unlike in section 3 above, the full innovations are applied to the model variables on the assumption that the subsequent model drift away from the data does not cause a problem.

The assimilation experiment used OCCAM (Ocean Circulation and Climate Array Model), a global ¼ degree, 36 level, z-level ocean model, (Webb *et al.* 1998). This model was forced by 6-hourly ECMWF wind stress and relaxed at the sea surface to monthly evolving analyses of sea surface temperature, (Reynolds and Smith; 1994), and a climatological surface salinity, (Levitus and Boyer 1994), over a 5-year period 1992-1996. T profile data was assimilated once per month using the methods of section 2 in this chapter and altimeter data was assimilated as mapped surface height anomalies every 10 days using the methods of chapter *Altimeter and Covariances and Errors Treatment*, section 3.2. For further details see Fox and Haines (2003).

Figure 5 shows four sections of temperature in the upper ocean following the track of a WOCE cruise N-S across the Pacific ocean in July-August 1993, bottom panel. The top panel is a simulation with realistic forcing but no assimilation of data. Clearly there is a consistent bias compared with the World Ocean Circulation Experiment (WOCE) section, in that the thermocline is very diffuse. The second panel shows the result of assimilating T profile data alone, run for the purposes described in chapter *Altimeter and Covariances and Errors Treatment*, section 2, point (3). Note the much tighter and more realistic thermocline at all latitudes. The third panel shows the best assimilation with both T profile and altimeter data included. The assimilation looks to be very successful but how do we quantify and interpret the correction of the thermocline bias physically?

We develop a heat budget analysis for the assimilation experiment, with an explicit term for data assimilation, and examine the balance of terms. Figure 6 shows all the terms in the local vertically integrated heat budget over the whole globe for 1993-96;

Trend = Advective Convergence +Assimilation +Surface Heat flux +Mixing (5)

All terms are expressed as equivalent surface heat flux (Wm^{-2}). The mixing is very weak in this vertically integrated picture because horizontal diffusive mixing of heat is small. The trend is generally small and positive (which later will be shown to be due to warming of deep waters where observations are not available). Surface heat fluxes are also small in comparison to the dominant balance between advective convergence and assimilation. The results in Figure 6 show that locally the heat budget is dominated by what might be called the *'dynamical bias'*, *i.e.*, the advection of heat to the wrong places, while assimilation redistributes that heat to compensate. Two regions may be readily identified: (1) advection tends to warm the ocean north of Cape Hatteras on the US east coast (Gulf stream overshoot), and (2) advection tends to warm the eastern equatorial Pacific (collapsing thermocline leads to surface convergence of heat and is a similar error to that in Figure 4a).

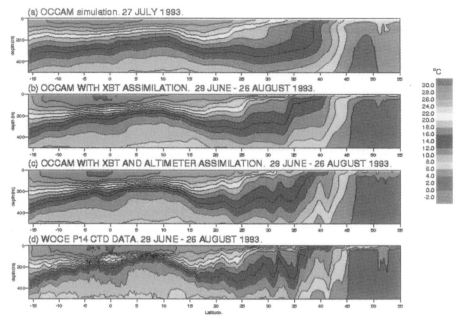

Figure 5: Temperature sections at 179E in the Pacific ocean in July 1993. (a) OCCAM simulation, (b) OCCAM with XBT assimilation, (c) OCCAM with XBT and altimeter assimilation, (d) WOCE section P14 along the same line made from 29 June- 26 Aug 1993. Units are C.

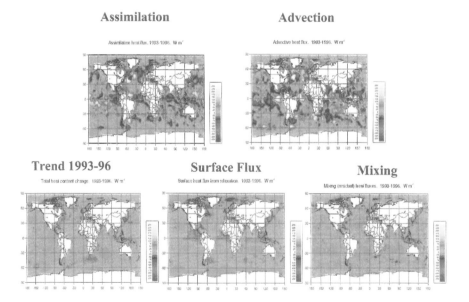

Figure 6: Vertically integrated local heat budget over 1993-96 for the OCCAM assimilation experiment. Assimilation represents both XBT and altimeter assimilation and Advection represents convergence of heat. These are the two largest terms in the balance. All units are in Wm^{-2} of equivalent surface heat flux.

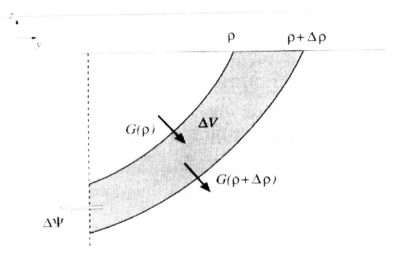

Figure 7. Schematic of diapycnal water transformations G and inflow/outflow ΔΨ, controlling the volume of water in each potential density class within a region of the ocean. From Nurser *et al.*, (1999), after Walin (1982).

Other diagnostics can be formulated to look at the *'thermodynamical bias'* by breaking the heat budget down in the vertical but now integrating over a horizontal area. Walin (1982) recast the heat budget as a volume budget over an area of the ocean for waters of each potential temperature class (see Figure 7). This is very similar to the constraints discussed for the ocean inverse problem calculations in chapter *Uses of Ocean Data Assimilation and Ocean State Estimation*. The volume of water in each potential temperature class in a region in determined by inflow/outflow at the boundaries and the processes of air-sea heat flux and interior mixing which transform waters from one potential temperature to another. A volume budget can therefore be constructed;

$$\textbf{dV(T)/dt = Boundary advection(T) + Surface flux convergence(T)}$$
$$\textbf{+ Mixing convergence(T)} \qquad (6)$$

Units of these quantities are Sv (10^6 m^{3}s^{-1}). Figure 8 from Speer (1997), shows how this budget looks from COADS (the Comprehensive Ocean Atmosphere Data Set) heat fluxes over the North Atlantic from 11S-80N.

In our data assimilation experiment the assimilation is also involved in converting waters from one temperature class to another so;

$$\textbf{dV(T)/dt = Boundary advection(T) + Surface flux convergence(T)}$$
$$\textbf{+ Mixing convergence(T) + Assimilation convergence(T)} \qquad (7)$$

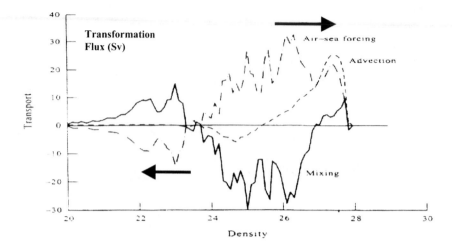

Figure 8. Diapycnal volume transformation fluxes in Sv for the Atlantic ocean 11S-80N. 'Air-sea forcing' is based on climatological flux data sets, and 'Advection' is from inverse modelling of a hydrographic section at 11S, plus Ekman. The residual 'Mixing' assumes water volumes are in a steady state. Positive means transformation towards higher density. From Speer (1997).

In an unbiased model the assimilation would not need to convert waters in a net sense but in a biased model the assimilation must compensate for conversion errors in surface fluxes and mixing in order to maintain the correct volumes of water in each thermal class.

Figures 9a,b show the results of this analysis for the data assimilation experiment for 1993-96 with the OCCAM model over the whole North Atlantic, bounded by the Equator and the Bering straits. Points to note are:

• The trend in water volumes is only significant for the coldest waters,

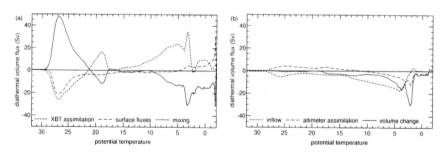

Figure 9. Diathermal volume transformation fluxes, in Sv, for the North Atlantic for 1993-96 inclusive from the OCCAM data assimilation experiment. (a) Shows the XBT assimilation, the surface heat flux and the model mixing contributions. (b) Shows the altimeter assimilation, the advective inflow and the total volume change over 4 years. From Fox and Haines (2002).

- 17Sv advection of water warmer than 4C into the North Atlantic and 17Sv of colder waters advected out.
- Altimeter assimilation does not transform waters (assimilation is by vertical displacement of isopycnals)
- T profile assimilation separates waters producing both warm and cold waters at the expense of water of intermediate temperature. This is usually the signature of an air-sea heat flux.
- The surface heat flux induced by surface temperature relaxation is very weak for T< 21C
- Mixing transformations are strong in the model (peaking at 45Sv this is probably too strong because assimilation is maintaining much sharper temperature gradients than the free running model).
- Assimilation maintains the volume of 18C mode waters with a formation rate of 16Sv. This is consistent with other estimates based directly on air-heat fluxes (e.g. Speer 1997, Marshall et al.1999).

The conclusion to be drawn from section 4 in this chapter is that diagnostics based on data assimilation innovations can be developed specifically to identify bias in model processes such as air-sea heat flux, interior mixing and advection. This can be done by writing budgets in conservative properties and identifying the role of data assimilation in maintaining the correct property distributions. This method of diagnostics can be contrasted with the process of tuning model parameters with a 4D-var assimilation scheme. Here we do not need to specify the parameterisation of processes a priori. Instead we diagnose the biases in the innovations directly as physical processes. It is then left to the scientist to decide how to improve the model in order to reduce the bias. In the example above it would appear that the best approach would be to repeat the experiment with more accurate surface fluxes, which should mean that the assimilation does not need to cause so much water transformation to maintain water mass volumes in the North Atlantic. Although this method is applied to an ocean assimilation experiment there is no reason why similar ideas may not be carried across into other applications of data assimilation, for example for chemical tracers or water vapour in the atmosphere.

5. Ocean data assimilation concluding remarks

If there is one message that unites the chapters on data assimilation applied to deep sea circulation models, it is that the models are not currently good enough and we should be very careful about building into our assimilation schemes (e.g. through covariances) anything which may turn out to be a bias of a particular model. This is largely a result of not having good enough data until very recently in order to improve our models. Thus we should avoid calculating covariances from models and we should not assume our models are unbiased for the purposes of 'optimally assimilating' our ocean data. It is possible to rely on knowledge of ocean physics and dynamics directly in building our covariance models. It is also important to consider carefully the information content of ocean data sets and use them carefully in combination to get the best ocean analyses. Finally it should be possible to learn about the bias errors in our ocean models through the process of diagnosing the innovations from data assimilation experiments and this should be an interesting area of future research. Many of the above remarks are likely to carry across and apply to data assimilation in other domains.

References

Bell, M.J., M.J. Martin, and N.K. Nichols, 2002: Assimilation of data into an ocean model with systematic errors near the equator. *The Met. Office, Ocean Applications Tech. Note.No. 27, March 2001, 27 pp and Submitted to Q. J. R. Meteorol. Soc.*

Dee, D., and A. da Silva, 1998: Data assimilation in the presence of forecast bias. *Q. J. R. Meteorol. Soc.,* 124, 269-295.

Fox, A.D., and K. Haines, 2003: Interpretation of Water Mass Transformations diagnosed from Data Assimilation. *J. Phys Oceanogr.* **33**, 485-498.

Levitus, S., and T.P. Boyer , 1994: World Ocean Atlas 1994. Technical Report, US Dept. of Commerce, NOAA.

Marshall, J., D. Jamous, and J. Nilsson, 1999: Reconciling 'thermodynamic' and 'dynamic' methods of computation of water-mass transformation rates. *Deep Sea Res. I,* **46**, 545-572.

Nurser, A.J.G., R. Marsh, and R.G. Williams, 1999: Diagnosing water mass formation from air-sea fluxes and surface mixing. *J. Phys. Oceanogr.,* **29**, 1468-1487.

Reynolds, R. W., and T. M. Smith, 1994: Improved global sea surface temperature analyses using optimum interpolation. *J. Climate,* **7**, 929-948.

Segschneider J., D.L.T. Anderson, J. Vialard, M. Balmaseda, T.N. Stockdale, A. Troccoli, and K. Haines, 2001: Initialization of seasonal forecasts assimilating sea level and temperature observations. *J. Climate,* **14**, 4292-4307.

Speer, K.G., 1997: A note on average cross-isopycnal mixing in the North Atlantic ocean. *Deep-Sea Res. I,* **44**, 1981-1990.

Troccoli, A., and K. Haines, 1999: Use of the Temperature-Salinity relation in a data assimilation context. *J. Atmos. Ocean Tech.,* **16**, 2011-2025.

Troccoli A., M. Balmaseda, J. Segschneider, J. Vialard, D.L.T. Anderson, K. Haines, T. Stockdale, F. Vitart, and Fox A.D., 2002: Salinity adjustments in the presence of temperature data assimilation. *Mon. Weather Rev.,* **130**, 89-102.

Walin G., 1982: On the relation between sea-surface heat flow and thermal circulation in the ocean. *Tellus* **34**, 187-195.

Webb, D.J., A.C. Coward, B.A. de Cuevas, and C.S. Gwilliam, 1998: A multiprocessor ocean general circulation model using message passing. *J Atmos Ocean Tech.* **14**, 175-183.

LAND SURFACE PROCESSES

PAUL R. HOUSER
NASA Goddard Space Flight Center,
Greenbelt, Maryland 20771 USA

1. Introduction

Through their regulation of water and energy transfer between the land and atmosphere, the dynamics of terrestrial water stores are an important boundary condition for the global water cycle at weather and climate timescales. The basis for a concerted integrated research effort is now provided by breakthroughs in techniques to observe: (1) global and regional precipitation, (2) surface soil-moisture, (3) snow, (4) surface soil freezing and thawing, (5) surface inundation, (6) river flow, and (7) total terrestrial water-storage changes, combined with better estimates of evaporation. As the primary input of water to the land surface, precipitation defines the terrestrial water cycle. The partitioning of this precipitation between infiltration (and subsequently evapotranspiration) and runoff is determined by surface physics, vegetation, snow and soil-moisture conditions, and soil-moisture dynamics.

Accurate initialization of land surface moisture and energy stores in fully-coupled climate system models is critical for seasonal-to-interannual, climatological and hydrological prediction because of their regulation of surface water and energy fluxes between the surface and atmosphere over a variety of time scales. Subsurface moisture and temperature stores exhibit persistence on seasonal-to-interannual time scales; together with external forcing and internal land surface dynamics, this persistence has important implications for the extended prediction of climatic and hydrologic extremes. Because these are integrated states, errors in land surface forcing and parametrization accumulate in subsurface moisture and temperature stores, which lead to incorrect surface water and energy partitioning. However, many innovative new land surface observations are becoming available that may provide additional information necessary to refine and constrain the physical parametrizations and initialization of land surface states critical for seasonal-to-interannual prediction. These constraints can be imposed in three ways. First, by forcing the land surface primarily by observations (such as precipitation and radiation), the often severe atmospheric numerical weather prediction land surface forcing biases can be avoided. Second, by employing innovative land surface data assimilation techniques, observations of land surface storages (such as soil temperature and moisture) can be used to constrain unrealistic simulated storages. Third, the land surface physical parametrizations themselves are improved through the use of observed parameters, and through the data assimilation process where model states are constantly being evaluated against observations.

R. Swinbank et al. (eds.), Data Assimilation for the Earth System, 321–329.

2. Land surface modelling

Figure 1 shows recent advances in understanding soil-water dynamics, plant physiology, micrometeorology, and the hydrology, all of which control biosphere-atmosphere interactions. These advances have spurred the development of land surface models, whose aim is to represent in a simple, yet realistic way, the transfer of mass, energy, and momentum between a vegetated surface and the atmosphere (Dickinson *et al.*, 1993; Sellers *et al.*, 1986). Land surface model predictions are regular in time and space, but these predictions are influenced by model structure, errors in input variables and model parameters, and inadequate treatment of sub-grid scale spatial variability. Consequently, land surface model predictions of land surface hydrology and land surface states are much improved by the assimilation of land surface observations.

Three recent land surface models warrant further discussion. These are: (1) the Mosaic land model of Koster and Suarez (1992) and Koster *et al.* (1998), (2) the National Centers for Environmental Prediction, Oregon State University, United States Air Force, and Office of Hydrology, land surface model, called Noah, and (3) the Community Land Model.

The Mosaic land model addresses the problem of subgrid heterogeneity by subdividing each grid cell into a user-specified mosaic of tiles (after Avissar and Pielke, 1989), each tile having its own vegetation type and hence water and energy balance. Surface flux calculations for each tile are similar to those described by Sellers *et al.* (1986). Tiles do not directly interact with each other, but influence each other indirectly, by their collective influence on the overlying atmosphere. Like the plethora of land surface models that have been developed over the past decade (*e.g.* the Project for Intercomparison of Landsurface Parameterization Schemes, PILPS, participants, Henderson-Sellers *et al.* 1993), the Mosaic land model is well suited to modelling the vertical exchange of mass, energy and momentum with the overlying atmosphere, but includes a poor representation of lateral moisture movement, which significantly controls variations in soil water, surface energy fluxes and runoff. Recognizing this weakness, Koster *et al.* (2000) developed a new, catchment-based land surface model that includes a more realistic representation of hydrological processes, including the lateral transport of soil water through the subsurface. The catchment-based model, which relies heavily on concepts originally put forth by Famiglietti and Wood (1991) and Famiglietti and Wood (1994) (*i.e.* the Topmodel based Land- Atmsophere-Transfer-Scheme, TOPLATS) will represent a major advance in land surface models for the following two reasons. First, the Topmodel (Beven and Kirkby, 1979) topographically-based framework will result in improved runoff prediction, and consequently, more realistic catchment-scale water balance. Second, the downslope movement of moisture within the watershed will yield sub-catchment-scale variations of surface and unsaturated-zone moisture content, which will result in more realistic prediction of within-catchment variations in surface fluxes. Improved simulation of runoff will ultimately result in a more realistic flux of continental streamflow from the land to the oceans in the coupled model, and similarly, the within-catchment variations in surface fluxes result in more representative catchment-average exchanges with the atmosphere.

The Noah land model simulates soil moisture (both liquid and frozen), soil temperature, skin temperature, snow water equivalent, snow density, canopy water content, and the traditional energy flux and water flux terms of the surface energy balance and surface water balance. This model has been used in: a) the National Oceanic and Atmospheric Administration (NOAA) Office of Hydrology submission to the Project for Intercomparison of Landsurface Parameterization Schemes phase 2d tests for the Valdai, Russia site, b) the real-time, United States domain, Land Data Assimilation System, c) the coupled National Centers for Environmental Prediction (NCEP) mesoscale Eta model (Chen *et al*, 1997) and the Eta model's companion 4-dimensional Data Assimilation System, as well as in d) the coupled National Centers for Environmental Prediction global Medium-Range Forecast model and its companion 4-dimensional Global Data Assimilation System.

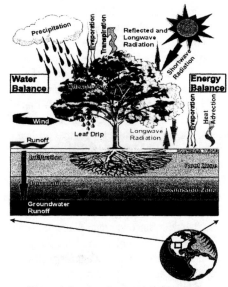

Figure 1: Land surface modelled processes.

The Community Land Model is being developed by a *grass-roots* collaboration of scientists who have an interest in making a general land model available for public use. By *grass roots*, we mean that the project is not being controlled by any single organization or scientist, but rather, the scientific steering is judged by the community. However, the project began at a sub-group meeting at the 1998 National Center for Atmospheric Research (NCAR) Climate System Model meeting, and it was implemented in the National Center for Atmospheric Research Climate System Model in early 2000. The Community Land Model development philosophy is that only proven and well-tested physical parametrizations and numerical schemes shall be used. The current version of the Community Land Model includes the best components from each of three contributing models: the Land Surface Model (Bonan, 1996), the Biosphere Atmosphere Transfer Scheme (Dickinson *et al.*, 1986) and Chinese Institute of Atmospheric Physics land surface model (Dai and Zeng, 1997). The Community Land Model code management is similar to *open source* in that, use of the model implies that any scientific gain will be included in future versions of the model. Also, the land model has been run for a suite of test cases including many of the Project for Intercomparison of Landsurface Parameterization Schemes case studies. These include the First International Satellite Land Surface Climatology Project Field Experiment (Kansas, USA), Cabauw (Netherlands), Valdai (Russia), Hydrological and Atmospheric Pilot Experiment in the Sahel (France), the Amazon Region Micrometeorology Experiment, and the Anglo-Brazilian Amazonian Climate Observational Study. These cases have not been rigorously compared with observations, but will be thoroughly evaluated in the Project for Intercomparison of Landsurface Parameterization Schemes framework.

3. Land surface modelling considerations

There are strong justifications for studying a land surface model uncoupled from atmospheric and ocean models. Coupling the land surface model to an atmospheric model allows the study of the interaction and feedbacks between the atmosphere and land surface. However, in coupled models the atmospheric model can impose strong land surface forcing biases on the land surface model. For example, biases in precipitation and radiation can overwhelm the behaviour of land surface model physics. In fact, several numerical weather prediction centres must "correctively nudge" their land surface model soil moisture toward climatological values to eliminate its drift. By using an uncoupled land surface model, we can better specify land surface forcing using observations, use less computational resources, and address most data assimilation development questions. The physical understanding and modelling insights gained from implementing distributed, uncoupled land-surface schemes with observation-based forcing has been vividly demonstrated in recent Global Energy and Water Cycle Experiment retrospective off-line land surface modelling projects known as the Project for Intercomparison of Landsurface Parameterization Schemes phase 2c and the Global Soil Wetness Project (Koster and Milly, 1997).

Runoff-routing schemes allow model validation and assimilation using ground-based and remote streamflow observations (Lohmann, *et al.*, 1996). Graham *et al.* (1999) and Olivera *et al.* (2000) describe the development of river transport methods for the National Center for Atmospheric Research Climate System Model. Since current plans for runoff routing models include a parametrization of lakes and wetlands, coupling the land surface models with the runoff routing transport scheme also may allow for assimilation of altimetry-derived water heights.

The interaction of the land surface with the atmosphere is strongly dependent on the scale at which it is modelled. It is well established that sub-grid variability can profoundly impact land-surface predictability. Therefore, the dependence of land surface predictions on spatial resolution and parameter aggregation can be assessed with multi-scale sensitivity experiments. Land surface model performance can vary widely over a range of spatial scales (typically land surface models are implemented on scales ranging from 1 to 200 kilometres). Parameters and forcing can be transferred between spatial scales using standard aggregation algorithms, along with simple interpolation. Scaling studies help to identify: (1) the sensitivity of land surface model predictions to spatial resolution, (2) mechanisms for relating and transferring results between spatial scales, and (3) the importance of sub-grid scale heterogeneity on land-surface storage estimation.

4. Remote sensing of the land surface

The emphasis of land surface data assimilation research is to assimilate remotely-sensed observations of the land surface that previous research suggests will provide memory to the land-atmosphere interaction. Remote observations of interest include: (1) temperature, (2) soil moisture (surface moisture content, surface saturation, total water storage), (3) other surface water bodies (lakes, wetlands, and large rivers) and (4) snow (areal extent, snow water equivalent). The remote sensing potential and availability of each of these quantities is described in more detail below.

Remote sensing of surface temperature is a relatively mature technology. The land surface emits thermal infrared radiation at an intensity directly related to its emissivity and temperature. The absorption of this radiation by atmospheric constituents is smallest in the 3 to 5 and 8 to 14 micrometre wavelength ranges, making them the best atmospheric windows for sensing land surface temperature. Some errors due to atmospheric absorption and improperly specified surface emissivity are possible, and the presence of clouds can obscure the signal. Generally, surface temperature remote sensing can be considered an operational technology, with many spaceborne sensors making regular observations (for example, the Landsat Thematic Mapper, Advance Very High Resolution Radiometer, AVHRR, the Moderate Resolution Imaging Spectroradiometer, MODIS, and the Advanced Spaceborne Thermal Emission and Reflection Radiometer, ASTER) (Lillesand and Kiefer, 1994). The evolution of land surface temperature is linked to all other land surface processes through physical relationships. These land surface process interconnections can be exploited in a data assimilation framework to constrain all of the predicted land surface states.

Remote sensing of soil moisture content is a developing technology, although the theory and methods are well established (Eley, 1992). Long-wave passive microwave remote sensing is ideal for soil moisture observation, but there are technical challenges involved in correcting for the effects of vegetation and roughness. Soil moisture remote sensing has previously been limited to aircraft campaigns (*e.g.* Jackson, 1997a), or analysis of the Defense Meteorological Satellite Program (DMSP) Special Sensor Microwave Imager (SSM/I) (Engman, 1995; Jackson, 1997b). The Special Sensor Microwave Imager has also been successfully employed to monitor surface saturation/inundation (Achutuni and Scofield, 1997; Basist and Grody, 1997). The Earth Observing System Advanced Microwave Sounding Unit (AMSR) will provide additional C-band microwave observations that may be useful for soil moisture determination. The Tropical Rainfall Measuring Mission's Microwave Imager (TRMM-TMI), which is very similar to the Advanced Microwave Sounding Unit, is much better suited to soil moisture measurement (because of its 10 MHz channels) than Special Sensor Microwave Imager. All of these sensors have adequate spatial resolution for land surface applications, but have a very limited quantitative measurement capability, especially over dense vegetation. However, Sipple *et al.*, (1994) demonstrated that it is possible to determine saturated areas through dense vegetation using the Scanning Multichannel Microwave Radiometer (SMMR), which can greatly aid land surface predictions. Because of the large error in remotely-sensed microwave observations of soil moisture, there is a real need to maximize its information by using algorithms that can account for its error and that extend its information in time and space.

An important and emerging technology with respect to land surface observation is the potential to monitor variations in total water storage (ground water, soil water, surface waters (*e.g.* lakes, wetlands, rivers), water stored in vegetation, snow and ice) using satellite observations of the time variable gravity field. The Gravity Recovery and Climate Experiment (GRACE), an Earth System Science Pathfinder mission launched in 2002, will provide highly accurate estimates of changes in terrestrial water storage in large watersheds. Wahr *et al.* (1998) note that the Gravity Recovery and Climate Experiment will provide estimates of variations in water storage to within 5 millimetres on a monthly basis. Rodell and Famiglietti (1999) have demonstrated the potential utility of these data for hydrologic applications, including their application in large (>150,000 km^2) watersheds; and they further discuss the potential power of Gravity Recovery and Climate Experiment observations for constraining modelled water storage in land surface models when combined with surface soil moisture and altimetry observations. Birkett (1995, 1998) demonstrated the potential of satellite radar altimeters to monitor height variations over inland waters, including climatically-sensitive lakes and large rivers and wetlands. Such altimeters are currently operational on the European Space Agency Remote Sensing Satellite 2 (ERS-2), the Topex Poseidon satellite, the European Space

Agency Environment Satellite (Envisat), and the Jason 1 satellite (see chapter *Altimeter covariances and errors treatment*).

Key snow variables of interest to land surface understanding include area coverage and snow water equivalent. While the estimation of snow water equivalent by satellite is currently in research mode, snow areal extent can be routinely monitored by many operational platforms, including The Advanced Very High Resolution Radiometer, the Geostationary Operational Environmental Satellite (GOES) and the Special Sensor Microwave Imager. Recent algorithm developments even permit the determination of the fraction of snow cover within Landsat Thematic Mapper pixels (Rosenthal and Dozier, 1996). Cline *et al.* (1998), describe an approach to retrieve snow water equivalent from the joint use of remote sensing and energy balance modelling.

Precipitation is the most important forcing for the land surface. Since precipitation is not predicted in an uncoupled land surface model, it cannot be formally assimilated. However, we can perform sensitivity experiments to understand how errors in precipitation affect our ability to quantify the variability of terrestrial water.

Precipitation is generally poorly predicted by numerical weather prediction models because we have not mastered the complex prediction of cloud physics and dynamics, which can lead to gross errors in land surface simulations. Therefore, we generally replace these fields by observational products, when available. Unfortunately most high-quality long-term global land surface observations have been processed on monthly time scales for use in climate variability studies, and therefore lack the high temporal resolution required by land surface modelling efforts. These low temporal resolution observations can still be used to improve global land surface predictions by correcting the longer-term land surface forcing biases. Essentially, we use the numerical weather prediction surface fields as high-resolution temporal weights on the longer-term observation averages when high-resolution observed forcing is unavailable. It is recognized that the timing of forcing is also of particular importance in land surface prediction, and therefore state-of-the-art temporal and spatial downscaling techniques are explored to mitigate these effects. As our understanding of the various sources of precipitation information has matured, we have recognized that these sources have disparate and often complementary information. We are therefore exploring the use of a data assimilation analysis scheme to optimally merge the various precipitation observations prior to their use as land forcing. Assimilation of precipitation information to constrain an atmospheric model is also an area of active research that shows great promise for further reducing the errors in land surface forcing, but such precipitation assimilation is outside the scope of this book.

Two general categories of satellite-derived precipitation exist, each with severe limitations. However, it is generally acknowledged that they have less bias, and better location and timing of precipitation when compared to model estimates. The first category of satellite-based precipitation observation is Geostationary Operational Environmental Satellite Precipitation Index estimates (Arkin and Meisner, 1987). This very simple method uses the cloud infrared brightness to directly estimate precipitation using a lookup table. This method can provide hourly precipitation estimates, but is limited to convective precipitation structures in the 40 degrees North to 40 degrees South latitude band. The second category uses shortwave passive microwave, as available with the Special Sensor Microwave Imager instrument, the Tropical Rainfall Measurement Mission Microwave Imager, and the Advanced Microwave Scanning Radiometer, which are sensitive to cloud water vapour quantities and raindrops, and can therefore provide better estimates of precipitation. Because these satellites are not geostationary, their temporal coverage is limited. Many research groups have investigated the derivation of precipitation from this data using methods ranging from simple empirical systems to neural network techniques. The quality of precipitation estimates is expected to be highest from microwave sensors, moderate from Geostationary Operational Environmental Satellite Precipitation Index estimates, and lowest from the numerical model predictions. Generally, the best available precipitation observations are used, or the available observations are optimally merged following Houser *et al.* (1999). Additional corrections for each data type based on rain gauge and climatological information further increase the accuracy of remotely observed precipitation.

5. Land surface validation

Data that are assimilated are generally not useful for subsequent validation because they are not independent from the prediction. Therefore, we also want to compare against independent data sets, either *in-situ* or remote, as available. Some useful land surface validation observations are listed here:

(a) Regionally-averaged *in-situ* validation: Various *in-situ* land observations are available for direct validation of land surface predictions (*e.g.* the First International Satellite Land Surface Climatology Project Field Experiment, the Boreal Ecosystem-Atmosphere Study, the Southern Great Plains Experiments, the Cooperative Atmosphere Surface Exchange Study, the Mackenzie Global Energy and Water Cycle Experiment Study, and the Global Energy and Water Cycle Experiment Continental-Scale International Project).

(b) Streamflow validation: A streamflow routing algorithm (*e.g.* Lohmann *et al.*, 1996) can be used to facilitate validation of predicted states and fluxes via comparison with widely available streamflow observations.

(c) Surface temperature validation: Simulated surface temperature is dependent on model forcing, land surface characteristics, soil water storage, and internal model physics. Thus, surface temperature can provide an integrated assessment of land surface predictive quality. High quality land surface temperature observations are available from a large number of infrared instruments, such as the Advanced Very High Resolution Radiometer (Dubayah *et al.*, 1997).

(d) Snow extent and water equivalent validation: Snow cover information derived from Advanced Very High Resolution Radiometer, the Geostationary Operational Environmental Satellite and the Special Sensor Microwave Imager is available from various operational centres (Robinson *et al.*, 1993), and can be useful for evaluating the snow accumulation and melt processes. Further, validation data for model predictions of snow water equivalent are compiled by the National Weather Service National Operational Hydrologic Remote Sensing Center. These data include *in-situ* snow course measurements and airborne snow water equivalent measurements compiled in a gridded map format for the United States and Canada.

(e) Long-term budget partitioning validation: Over long time periods, the land surface model should estimate the correct partitioning of available surface energy into sensible, latent, and ground heat fluxes, and of precipitation into evaporation, runoff, and groundwater recharge. Relatively reliable estimates of these partitionings have been established for various sub-regions and watersheds, and will be a valuable check on performance.

(f) Model cross-validation: Comparing results between models that have different forcing, parametrizations, or resolution can provide additional validation.

328

References

Achutuni, R., and R. A. Scofield, 1997: The spatial and temporal variability of the DMSP SSM/I global soil wetness index. *AMS Annual Meeting, Proceedings of the 13th Conference on Hydrology*, 188-189.

Arkin, P.A., and B.N. Meisner, 1987: The relationship between large-scale convective rainfall and cold cloud over the western hemisphere during 1982-84. *Mon. Weather Rev.*, 115, 51-74.

Avissar, R., and R. Pielke, 1989: A parameterization of heterogeneous land surfaces for atmospheric numerical models and its impact on regional meteorology. *Mon. Weather Rev.*, 117, 2113-2136.

Basist, A., and N. Grody, 1997: Surface wetness and snow cover. *AMS Annual Meeting, Proceedings of the 13th Conference on Hydrology*, 190-193.

Beven, K., and M. Kirkby, 1979: A physically-based variable contributing area model of basin hydrology. *Hydrol. Sci. J.*, 24, 43-69.

Birkett, C. M., 1995: The contribution of the TOPEX/POSEIDON to the global monitoring of climatically sensitive lakes. *J. Geophys. Res.*, 100, 25179-25204.

Birkett, C. M., 1998: Contribution of the TOPEX NASA radar altimeter to the global monitoring of large rivers and wetlands. *Wat. Resour. Res.*, 34, 1223-1239.

Bonan, G.B., 1996: A land surface model (LSM version 1.0) for Ecological, Hydrological, and Atmospheric Studies: Technical description and user's guide. *NCAR Technical Note* NCAR/TN-417+STR, National Center for Atmospheric Research, Boulder, Colorado, 150pp.

Chen, F., K. Mitchell, J. Schaake, Y. Xue, H. Pan, V. Koren, Y. Duan, M. Ek, and A. Betts, 1996: Modeling of land-surface evaporation by four schemes and comparison with FIFE observations. *J. Geophys. Res.* 101, 7251-7268.

Cline, D. W., R. C. Bales and J. Dozier, 1998: Estimating the spatial distribution of snow in mountain basins using remote sensing and energy balance modeling. *Wat. Resour. Res.*, 34, 1275-1285.

Dai Y., and Q.-C. Zeng, 1997: A land surface model (IAP94) for climate studies, Part I: formulation and validation in off-line experiments. *Advance Atmospheric Sciences* 14, 433-460.

Dickinson, R.E., A. Henderson-Sellers, P.J. Kennedy, and M.F. Wilson, 1986: Biosphere-Atmosphere Transfer Scheme (BATS) for the NCAR Community Climate Model. *NCAR Technical Note: NCAR/TN-275+STR*, p. 69.

Dickinson, R. E., A. Henderson-Sellers, and P. J. Kennedy, 1993: Biosphere-Atmosphere Transfer Scheme (BATS) Version 1e as Coupled to the NCAR Community Climate Model. *NCAR Technical Note* 387+STR.

Dubayah, R., D. P. Lettenmaier, K. Czajkowski, and G. O'Donnell, 1997: The Use of Remote Sensing in Land Surface Modeling. *Presented at the American Geophysical Union Spring Meeting, Baltimore*.

Eley, J., 1992: Summary of Workshop, Soil Moisture Modeling. Proceedings of the NHRC Workshop held March 9-10, 1992, *NHRI Symposium Proceedings* 9.

Engman, E. T., 1995: Recent Advances in Remote Sensing in Hydrology. *Reviews of Geophysics*, Supplement, 967-975.

Famiglietti, J. S., and E. F. Wood, 1991: Evapotranspiration and Runoff from Large Land Areas: Land Surface Hydrology for Atmospheric General Circulation Models. *Surveys in Geophysics*, 12, 179-204.

Famiglietti, J. S. and E. F. Wood, 1994: Multi-Scale Modeling of Spatially-Variable Water and Energy Balance Processes. *Wat. Resour. Res.*, 30, 3061-3078.

Georgakakos, K.P., and Smith, G.F., 1990: On improved hydrologic forecasting - Result from a WMO real time forecasting experiment. *J. Hydrol.*, 114, 17-45.

Graham , S. T., J. S. Famiglietti, and D. R. Maidment, 1999: 5-Minute, 1/2 Degree and 1-Degree Data Sets of Continental Watersheds and River Networks for Use in Regional and Global Hydrologic and Climate System Modeling Studies. *Wat. Resour. Res.*, 35, 583-587.

Henderson-Sellers, A., Z.-L. Yang, and R. E. Dickinson, 1993: The Project for Intercomparison of Land-surface Parameterization Schemes, *Bull. Amer. Meteorol.. Soc.*, 74, 1335-1349.

Houser, P., E. Douglass, R. Yang, and A. Silva, 1999: Merging Precipitation Observations with Predictions to Develop a Spatially & Temporally Continuous 3-hour Global Product. *GEWEX Conference, Beijing, China*.

Jackson, T. J., 1997a: Southern Great Plains 1997 (SGP97) Hydrology Experiment Plan, http://hydrolab.arsusda.gov/~tjackson.

Jackson, T. J., 1997b: Soil moisture estimation using special satellite microwave/imager satellite data over a grassland region. *Wat. Resour. Res.*, 33, 1475-1484.

Koster, R. D., and M. J. Suarez, 1992: Modeling the land surface boundary in climate models as a composite of independent vegetation stands. *J. Geophys. Res.*, 97, 2697-2715.

Koster, R. D., M. J. Suarez, A. Ducharne, M. Stieglitz, and P. Kumar, 2000: A catchment-based approach to modeling land surface processes in a GCM, Part 1, Model Structure, *J. Geophys. Res.*, 105, 24809-24822.

Koster, R. D., and P. C. D. Milly, 1997: The interplay between transpiration and runoff formulations in land surface schemes used with atmospheric models. *J. Climate*, 10, 1578-1591.

Olivera, F., J. S. Famiglietti, and K. Asante, 2000: Global-Scale Flow Routing Using a Source-to-Sink Algorithm. *Wat. Resour. Res.*, 36, 2197-2207.

Robinson, D., Bevins, R.E., and G. Rowbotham, 1993: The characterization of mafic phyllosilicates in low-grade metabasalts from eastern North Greenland. *American Mineralogist* ,**78**, 377-390.

Rodell, M., and J. S. Famiglietti, 1999: Detectability of variations in continental water storage from satellite observations of the time dependent gravity field. *Wat. Resour. Res.*, **35,** 2705-2723.

Rosenthal, C. W., and J. Dozier, 1996: Automated mapping of montane snow cover at subpixel resolution from the Landsat Thematic Mapper, *Wat. Resour. Res.*, **32** , 115-130.

Sellers, P. J., Y. Mintz, Y. C. Sud, and A. Dalcher, 1986: A simple biosphere model (SiB) for use with general circulation models. *J. Atmos. Sci.*, **43,** 505-531.

Sipple, S., S. Hamilton, J. Melak, and B. Choudhury, 1994: Determination of inundation area in the Amazon river flood plain using SMMR 37 GHz polarization difference. *Remote Sensing Environment*, **48**, 70-76.

Wahr, J.;, M. Molenaar, and F. Bryan, 1998: Time variablity of the Earth's gravity field: Hydrological and oceanic effects and their possible detection using GRACE. *J. Geophys. Res.*, **103** , 30205-30229.

ASSIMILATION OF LAND SURFACE DATA

PAUL R. HOUSER
NASA Goddard Space Flight Center,
Greenbelt, Maryland 20771 USA

1. Introduction

Charney *et al.* (1969) first suggested combining current and past data in an explicit dynamical model, using the model's prognostic equations to provide time continuity and dynamic coupling amongst the fields (Figure 1). This concept has evolved into a family of techniques known as *four-dimensional data assimilation.* "Assimilation is the process of finding the model representation which is most consistent with the observations" (Lorenc, 1995). In essence, data assimilation merges a range of diverse data fields with a model prediction to provide that model with the best estimate of the current state of the natural environment so that it can then make more accurate predictions. The application of data assimilation in hydrology has been limited to a few one-dimensional, largely theoretical studies (*i.e.* Entekhabi *et al.*, 1994; Milly, 1986), primarily due to the lack of sufficient spatially-distributed hydrologic observations (McLaughlin, 1995). However, the feasibility of synthesizing distributed fields of soil moisture by the novel application of four-dimensional data assimilation applied in a hydrological model was demonstrated by Houser *et al.* (1998). Six Push Broom Microwave Radiometer images gathered over the United States Department of Agriculture, Agricultural Research Service Walnut Gulch Experimental Watershed in southeast Arizona were assimilated into a land surface model using several alternative assimilation procedures. Modification of traditional assimilation methods was required to use these high-density Push Broom Microwave Radiometer observations. The images were found to contain horizontal correlations with length scales of several tens of kilometres, thus allowing information to be advected beyond the area of the image. Information on surface soil moisture was also assimilated into the subsurface using knowledge of the surface-subsurface correlation. Newtonian nudging assimilation procedures were found to be preferable to other techniques because they nearly preserve the observed patterns within the sampled region, but also yield plausible patterns in unmeasured regions, and allow information to be advected in time.

The feasibility of land surface data assimilation methods has been recently tested in research projects conducted at Goddard Space Flight Center, Massachusetts Institute of Technology, and several other institutions. This research focuses on: (1) the use of a one-dimensional Kalman filtering based land assimilation strategy that expands upon the one-dimensional, theoretical assimilation algorithms developed by Entekhabi *et al.* (1994) and Milly, (1986), and (2) the four-dimensional data assimilation strategies developed by Houser *et al.* (1998) and Walker *et al.* (1999).

R. Swinbank et al. (eds.), Data Assimilation for the Earth System, 331–343.

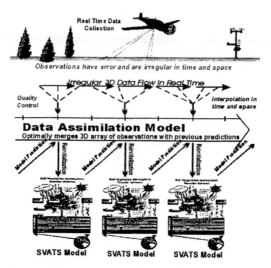

Figure 1: The land surface data assimilation process.

2. The Kalman filter

The Kalman filter attempts to obtain an optimal estimate of the land surface state by combining observations of that state with a land surface model forecast of that state. The Kalman filter has been extensively utilized in data assimilation research (Ghil *et al.*, 1981; Cohn, 1982). The Kalman filter assimilation scheme is a linearized statistical approach that provides a statistically optimal update of the system states, based on the relative magnitudes of the covariances of both the model system state estimate and the observations. The principal advantage of this approach is that the Kalman filter provides a framework within which the entire system is updated with covariances representing the reliability of the observations and model prediction.

The Kalman filter algorithm (Kalman, 1960) tracks the conditional mean of a statistically optimal estimate of a state vector X through a series of forecasting and update steps. To apply the Kalman filter, the equations for evolving the system states must be written in the linear state space formulation of Equation (1). When these equations are non-linear, the Kalman filter is called the extended Kalman filter, and is an approximation of the non-linear system that is based on first-order linearization. Walker and Houser (2001) have implemented a one-dimensional version of the extended Kalman filter with the simplifying assumption that errors in different catchments are uncorrelated.

The ensemble Kalman filter is an alternative to the extended Kalman filter for non-linear problems (Evensen, 1994; Houtekamer and Mitchell, 1998). The ensemble Kalman filter is based on the propagation of an ensemble of states from which the required covariance information is obtained at the time of the update. This approach has successfully been introduced into ocean assimilation at the National Aeronautics and Space Administration Seasonal-to-Interannual Prediction Project (NSIPP) by Keppenne and Rienecker (2000). Reichle *et al.* (2001) applied the ensemble Kalman filter to the soil moisture estimation problem and found it performs well, with the distinct advantage that the error covariance propagation is better behaved in the presence of large model nonlinearities. In general form, the nonlinear land surface model is expressed as:

$$X_k = f_k(x_k) + w_k \tag{1}$$

where X_k is the state vector at time k, and w_k is the model error with covariance $Q = E[w\,w^T]$.

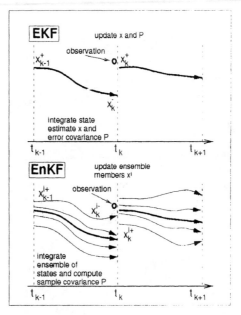

Figure 2: Schematic of the extended Kalman filter (EKF) and the ensemble Kalman filter (EnKF). The extended Kalman filter approximates the error covariance propagation by linearising the land surface model. The ensemble Kalman filter nonlinearly propagates an ensemble from which sample covariances are derived at the update time.

Both the extended Kalman filter and the ensemble Kalman filter work sequentially from one measurement time to the next, applying in turn a forecast step and an update step. Figure 2 illustrates the difference between the ensemble Kalman filter and the extended Kalman filter. During the forecast step, the extended Kalman filter propagates a single estimate of the state vector (from x_{k-1}^+ to x_k^-), and integrates the uncertainty (error covariance) (from P_{k-1}^+ to P_k^-) of that state, which will be used in the model forecast at the update step. The ensemble Kalman filter simultaneously propagates an ensemble of state vectors each state vector representing a particular realization of the possible model trajectories, and the error covariance is computed from the distribution of the model states in the ensemble.

Using superscripts – and + to refer to the state estimates, individual ensemble members or covariances before and after the update step, respectively, the state variables, covariances during forecast and update steps are expressed as follows:

State forecast:

extended Kalman filter: $X_{k+1}^{-} = f_k(x_k^{+})$

(2)

ensemble Kalman filter: $X_{k+1}^{i-} = f_k(x_k^{i+}) + w_k^{i}$

(3)

Covariance forecast:
extended Kalman filter:

$$P_{k+1}^{-} = F_k P_k^{+} F_k^{T} + Q_k$$

(4)

$$[F_k]_{mn} = \frac{\partial f_m}{\partial x_n}\Big|_{x_k^{-}}$$

(5)

ensemble Kalman filter

$$P_{k+1}^{-} = \frac{1}{N-1} D_{k+1} D_{k+1}^{T}$$

(6)

$$D_{k+1} = [x_{k+1}^{1-} - x_{k+1}^{-}, \ldots\ldots, x_{k+1}^{N-} - x_{k+1}^{-}]$$

(7)

$$x_{k+1}^{=} = \frac{1}{N \sum_{i=1}^{N} x_{k+1}^{i-}}$$

(8)

During the update step, the observation vector Y is linearly related to the system state vector X and the state independent terms XS_0,

$$y_k = H_k x_k + x s_{0k} + v_k$$

(9)

where v_k is the measurement error with covariance $R=E[v\, v^T]$.

During the update step, the extended Kalman filter revises its estimate of the state (from x_k^{-} to x_k^{+}) using the observation and the prognostic state error covariance P_k^{-} and the state error covariance is also updated (from P_k^{-} to P_k^{+}). On the other hand, the ensemble Kalman filter updates each ensemble members separately, using the observations and the diagnosed state error covariance P_k^{-}.

During the update step the Kalman gain weights the observations against the model forecast. Its weighting is determined by the relative magnitude of model uncertainty with respect to the observation covariance. The Kalman gain is given by

$$K_k = P_k^- H_k^T [H_k P_k^- H_k^T + R_k]^{-1}$$

(10)

If no observations are available at time k we set K_k=0. Next, the state estimate by the extended Kalman filter or each ensemble member by ensemble Kalman filter is updated using a linear combination of forecast model states and the observations:

Update:

extended Kalman filter: $x_k^+ = x_k^- + K_k[y_k - H_k x_k^- - xk_k]$

(11)

$$P_K^+ = P_k^- - K_K H_k P_k^-$$

(12)

ensemble Kalman filter: $x_k^{i+} = x_k^{i-} + K_k[y_k - H_k x_k^{i-} - xs_k + v_k^i]$

(13)

It is also important to continuously reevaluate the assumptions and problems that exist in the modelling and assimilation algorithms, so as to not attribute errors to the incorrect error source. A few considerations are outlined below.

As illustrated by Hollingsworth *et al.* (1986), assumptions on the bias and horizontal correlation structure of the model and observations can have a significant impact on error estimation. In practice, data assimilation is often implemented with the assumption that observations and predictions are unbiased and uncorrelated in space. These assumptions work reasonably well for *in situ* observations, but satellite observations are usually biased by use of inaccurate algorithms, and their errors are usually horizontally correlated because the same sensor is making all the observations.

In cooperation with the National Aeronautics and Space Administration's Seasonal-to-Interannual Prediction Project, we have evaluated the benefits and drawbacks of using a 1-dimensional (vertical) or a 3-dimensional (vertical and horizontal) Kalman filter. At the horizontal scale of the Advanced Microwave Sounding Unit (60 kilometres), soil moisture and snow exhibit little spatial correlation. Further, land surface models have little or no explicit treatment of the physical processes associated with these horizontal scales. 3-dimensional Kalman filtering is also much more computationally intensive than 1-dimensional Kalman filtering. However, there is potential benefit for 3-dimensional assimilation to: (1) extend observations into data-sparse regions, and (2) be used with higher resolution observations that exhibit more spatial correlation.

The potential benefits and drawbacks of using direct radiance assimilation are also being explored. It is possible to design a data assimilation system that assimilates radiances directly, rather than derived quantities such as retrievals, by including a forward model in the assimilation. Such a system is under development in our effort to derive soil moisture from the Tropical Rainfall Measuring Mission's Microwave Imager (TRMM-TMI), and may allow for more precise identification of calibration problems in the Advanced Microwave Sounding Unit.

3. Streamflow assimilation

While altimetry data could be used to provide additional observation of streamflow where the rivers are sufficiently wide (greater than 250 metres) and a stage discharge relationship is available, altimetry data is most likely to be used to update lake/reservoir storage. In this situation we can assimilate the lake/reservoir level and hence storage. This approach requires the existence of volume-elevation relationships depending on the bathymetry of the water body.

The first key to the implementation of a successful streamflow assimilation scheme for correction of the land surface states with the Kalman filter is the adequate specification of the model land surface state error covariance matrix. The important factor is that the model system state error covariance matrix correctly identifies the cross correlation between the soil moisture and snow prognostic variables. Likewise, successful runoff assimilation with the Kalman filter for correction of the runoff forecasts upstream of the streamflow gauging station will depend on adequate specification of the model runoff error covariance matrix. However, the model runoff error covariance matrix is not required for updating the model land surface states.

The variance of the observations can be identified reliably in most cases, since it depends on the characteristics of the measuring device (Georgakakos and Smith, 1990). Observation error estimates are not available with the streamflow data, but with knowledge of the gauging station instrumentation, these error estimates may be readily obtained.

The second key to implementation of a successful streamflow assimilation scheme for the correction of the land surface states with the Kalman filter is the adequate specification of the observation equation. Since runoff is a land surface model flux, the relationship between this flux and the model states, such as soil moisture and snow, must be established. The difficulty with this approach is the time lag between runoff from the source and runoff observed at the gauging station. Hence, this will be the single most difficult component of the runoff assimilation scheme. To ensure that observations are related back to the correct time for each land surface states may require continuous or near-continuous assimilation of runoff data and identification of the time lags associated with the runoff from each catchment.

An effective evaluation of any large-scale modelling endeavour is the most difficult and yet most important aspect. We evaluate the assimilation of runoff data through comparison of model derived surface soil moisture and snow states with remotely sensed surface soil moisture and snow observations. The experimental design involves synthetic studies where model output of soil moisture, snow, and runoff is used for the observation and evaluation data, followed by experiments where these fields are replaced by actual observations. The initial studies are for an individual catchment where runoff at the catchment outlet will be assimilated to correct soil moisture and snow forecasts, evaluating the within-catchment flow routing component. This is followed by the assimilation of runoff for a small number of catchments without any reservoir storage to evaluate the assimilation scheme when the between-catchment routing is included. The final stage is to evaluate the assimilation when there is a reservoir (or reservoirs) included in the runoff network.

4. Soil moisture assimilation

A Kalman filter soil moisture assimilation strategy has been developed (Walker and Houser, 2001). The principal advantage of this approach is that the Kalman filter provides a framework within which the entire system is modified, with covariances representing the reliability of the observations and model prediction. We have used a one-dimensional Kalman filter for updating the soil moisture prognostic variables of the Koster *et al.* (2000) catchment-based land surface model. A one-dimensional Kalman-filter was used because of its computational efficiency and the fact that horizontal correlations between soil moisture prognostic variables of adjacent catchments at the scales of interest to climate modelling are likely only through the large-scale correlation of atmospheric forcing. Moreover, all calculations for soil moisture in the catchment-based land surface model are performed independent of the soil moisture in adjacent catchments.

Forecasting of the soil moisture covariance matrix using Kalman filter theory requires a linear forecast model. However, forecasting of the soil moisture prognostic variables (surface excess, root zone excess and catchment deficit) in the catchment-based land surface model is non-linear. Hence, forecasting of the soil moisture prognostic variables covariance matrix was achieved through linearization of the soil moisture forecasting equations. The linearization was performed by a first order Taylor series expansion of the non-linear forecasting equations.

In order to perform an update of the soil moisture prognostic variables with the Kalman filter, the observations (near-surface soil moisture) must be linearly related to the soil moisture prognostic variables. In the catchment-based land surface model, the soil moisture prognostic variables are the surface excess, root zone excess, and catchment deficit, which are related to the observed volumetric soil moisture of the surface layer through a complicated non-linear function.

A set of numerical experiments have been undertaken for North America to illustrate the effectiveness of the Kalman filter assimilation scheme in providing a more accurate estimate of the soil moisture storage throughout the entire soil profile (Figure 3). Moreover, the corresponding positive influence on the water balance components, namely evapotranspiration and runoff, has been investigated. In this experiment, atmospheric forcing data and soil and vegetation properties from the first International Satellite Land-Surface Climatology Project (Sellers et al., 1996) have been used as model input for the year 1987.

Using the land surface model of Koster et al. (2000), the initial conditions from spin-up, and the model input data described above, the temporal and spatial variation of soil moisture across North America was forecast for 1987. The forecasts of near-surface soil moisture were output every 3 days to represent the near-surface soil moisture measurements from remote sensors. In addition to soil moisture, the land surface model provided estimates of evapotranspiration and runoff for each of the catchments. This simulation provided the "true" soil moisture and water balance data for comparison with degraded simulations. Moreover, it allowed evaluation of the effectiveness of assimilating near-surface soil moisture data for improving the land surface model forecast of soil moisture and water budget components, when initialized with poor soil moisture initial conditions. In the degraded simulation, the initial conditions for the soil moisture prognostic variables from the spin-up were set to arbitrarily wet values uniformly across all of North America. The land surface model was then forced with the same atmospheric data as in the "truth" simulation. The wet initial condition causes over-estimation of evapotranspiration and runoff. The final simulation was to assimilate the near-surface "observations" from the "truth" simulation into the degraded simulation every 3 days. The effect of assimilation on the soil moisture forecasts can be seen in Figure 3. These results show that after only 1 month of assimilation, the "true" soil moisture has been retrieved for the majority of North America.

338

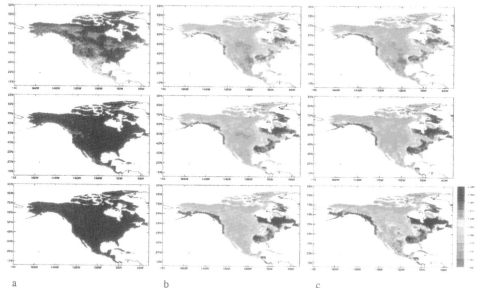

Figure 3: Comparison of gravimetric soil moisture on 30 January 1987 in near-surface (top row), root zone (middle row) and entire profile (bottom row) from: (a; left column) simulation with degraded initial conditions for soil moisture; (b; middle column) simulation with spin-up initial conditions ("truth"); and (c; right column) degraded simulation with assimilation of near-surface soil moisture from the "truth" simulation once every 3 days.

5. Snow assimilation

Snow plays an important role in governing both the global energy and water budgets, due to its high albedo, thermal properties, and being a medium-term water store. However, the problem of accurately forecasting snow in regional and global atmospheric and hydrologic models is difficult, as a result of subgrid-scale variability of snow, and errors in the model forcing data. Hence, any land surface model snow initialization based on model spin-up will be affected by these errors. By assimilating snow observation products into the land surface model, a best estimate of snow states may be obtained and model bias can be corrected. We are implementing a snow assimilation scheme that optimally merges remotely-sensed snow observations with the catchment-based land surface model forecast. As a first step, identical twin experiments have been performed to test and validate a snow data assimilation scheme. Synthetic observations of snow water equivalent are assimilated and other snow states are subsequently reanalyzed using the updated snow water equivalent. Preliminary results show good agreement between the assimilation and simulated truth. Figure 4 shows snapshots of truth, assimilation and forecast results of 24 continuous catchments in North America on March 16 1987. The assimilation starts from January 1, 1987, with a poor initial condition that assumes no snow is present anywhere. It produces satisfactory estimates of snow water equivalent, snow depth and snow temperature, while the model forecast with the same poor initial condition produces very different states. In the near future, snow water equivalent estimated from the Scanning Multichannel Microwave Radiometer (SMMR) and the Special Sensor Microwave/Imager (SSM/I) will be attempted..

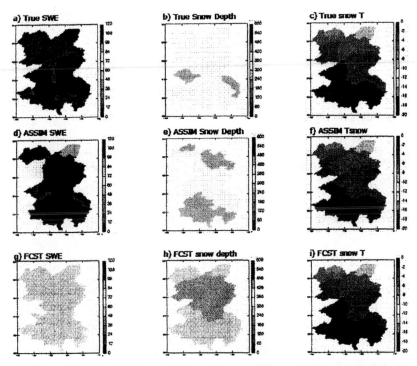

Figure 4: Snapshot of truth, assimilation and forecast (from poor initial condition) on 16 March 1987 from 3-month assimilation starting from 1 January 1987. Top-row: a) to c). Middle row: d) to f). Bottom row: g) to i). Left column: snow water equivalent (millimetres); Middle column: snow depth (millimetres); Right column: temperature (C).

6. Skin temperature assimilation

The land surface skin temperature state is a principle control on land-atmosphere fluxes of water and energy, is closely related to soil water states, and is easily observable from space and aircraft infrared sensors in cloud-free conditions. The usefulness of skin temperature in land data assimilation studies is limited by its very short memory (on the order of minutes) due to the very small heat storage it represents. We used the Physical-space Statistical Analysis System (PSAS, Cohn *et al.*, 1998) in a 2.5 degrees longitude by 2.0 degrees latitude global land surface model to assimilate surface skin temperature observations from International Satellite Cloud Climatology Project (ISCCP). The Physical-space Statistical Analysis System algorithm obtains the best estimate of the state of the system by combining observations with the forecast model first guess. The analysis equation, which encapsulates the Physical-space Statistical Analysis System scheme, is

$$w^a = w^f + K\left(w^o - Hw^f\right) \tag{14}$$

where w^a denotes the analyzed field, w^f represents the model forecast first guess field, w^o is the observational field, K are the weights of the analysis, and H is the interpolation operator which maps model variables into observables. The observed skin temperature minus the forecast first guess skin temperature values are input to the Physical-space Statistical Analysis System. The Physical-space Statistical Analysis System retrieves a grid space average analysis increment ($\delta w^a = K(w^o - H\,w^f)$), that is mapped into the land surface tile space. The analyzed field is then obtained by adding the tile space analysis increment to the first guess skin temperature field.

Results showed that simply correcting the land surface modeled skin temperature with the analysis increment from Equation (14) every 3 hours was insufficient. Since w^f is biased, the traditional analysis equation such as (14) produces a biased w^a (Dee and da Silva, 1998). Therefore, a variant of the Dee and da Silva (1998) bias correction scheme was implemented where,

$$\delta w^a = K\left(w^o - Hw^f + b^f\right) \tag{15}$$

$$w^a = w^f - b_{k-1}^{f} + \delta w^a \tag{16}$$

$$b_k^{f} = b_{k-1}^{f} - \gamma \cdot \delta w^a \tag{17}$$

b_k^f is the updated bias estimate, b_{k-1}^f is the bias estimate based on the previous analysis increment (δw_{k-1}^a) and δw^a is the analysis increment at time t_k. This scheme was inadequate because the surface skin temperature bias acted very quickly. As a result, an incremental bias correction scheme was introduced, where a bias correction term is added to the skin temperature tendency equation at every time step to counteract the subsequent forcing of the analyzed skin temperature back to the initial state. For this scheme, b^f is computed as in Equation (17) and the bias correction term is calculated as,

$$f_b = \frac{b^f}{\tau} \tag{18}$$

where $\tau = 3$ hours is the frequency of the International Satellite Cloud Climatology Project dataset, *i.e.*, the frequency of assimilation. This scheme effectively removed the time mean bias, but did not remove the bias in the mean diurnal cycle. To account for this deficiency, we modelled the time-dependent bias as,

$$b^f(t) = \sum_j \left(a_j \cos \omega_j t + b_j \sin \omega_j t\right) \tag{19}$$

and estimated the Fourier coefficients

$$a_j = a_j - \gamma \delta w^a \cos \omega_j t \tag{20}$$

$$b_j = b_j - \gamma \delta w^a \sin \omega_j t \tag{21}$$

We determined that to adequately account for the diurnal bias changes it is necessary to keep diurnal ($\omega_1 = 2\pi/24h$) and semi-diurnal ($\omega_2 = 2\pi/12h$) harmonics.

The results presented here are based on the evaluation of the discussed techniques for a July 1992 International Satellite Cloud Climatology Project surface skin temperature dataset. The test runs (Table 1) include the simulation without assimilation or bias correction (Model), with Physical-space Statistical Analysis System temperature assimilation (Assimilation I; Equation 14), with bias correction (Assimilation II; Equations 15-17), with incremental bias correction (Assimilation III; Equation 18), with diurnal bias correction (Assimilation IV; Equations 19-21) and with semi-diurnal bias correction (Assimilation V). Figure 5 shows the July 1992 global monthly mean standard deviations of surface skin temperature between the experiments and the observations. The standard deviation decreases gradually with each successive improvement to the methodology, and therefore substantiates the techniques developed. However, the monthly mean standard deviation does not reveal the more visible impact of the diurnal bias correction on the monthly mean diurnal cycle.

Table 1: Description of experiments.

Experiment	Description
Model	No assimilation
Assimilation I	Physical-space Statistical Analysis System assimilation
Assimilation II	Physical-space Statistical Analysis System with bias correction every 3 hours
Assimilation III	Physical-space Statistical Analysis System with incremental bias correction
Assimilation IV	Physical-space Statistical Analysis System with diurnal bias correction
Assimilation V	Physical-space Statistical Analysis System with semi-diurnal bias correction

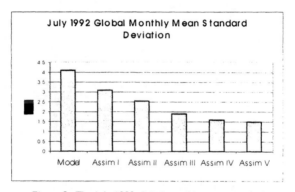

Figure 5: The July 1992 global monthly mean standard deviations of surface skin temperature between the assimilation experiments and the International Satellite Cloud Climatology Project observations (see Table 1).

The July 1992 monthly mean diurnal cycle of surface skin temperature over North America for the International Satellite Cloud Climatology Project observations, Model, Assimilation IV, and Assimilation V, are presented at the top of Figure 6. The effectiveness of implementing semi-diurnal bias correction is shown by how closely Assimilation V matches the observations. Two-metre temperature (middle) and specific humidity (bottom), also displayed in Figure 6, reveal that the inclusion of the bias correction scheme also impacts the surface meteorology fields. Thus, for a decrease in surface skin temperature, due to the bias correction, there is a corresponding decrease in the 2 metre temperature and specific humidity. Figure 6 allows only for model intercomparison, and we are in the process of obtaining a verification dataset.

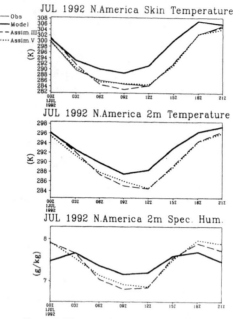

Figure 6. The July 1992 monthly mean diurnal
cycle of skin temperature (top), 2 m temperature
(middle) and 2 m specific humidity (bottom)
over North America for the observations (light
solid), Model (heavy solid), Assimilation III
(dashed) and Assimilation V (dotted).

Similarly, the same corrective effect is visible in the Western Europe surface skin temperature
for Assimilation V. The sensible heat flux and latent heat flux also show that the bias correction
technique has a substantial impact on the energy budget, where the reduction in skin temperature
causes a decrease in the sensible and latent heat flux.

In this study, the Mosaic land model (Koster and Suarez, 1992, 1996) has been forced with near
surface atmospheric conditions derived by the Goddard Earth Observing System Data Assimilation
System (GEOS-DAS). The Physical-space Statistical Analysis System was used with the Mosaic
land model in order to assimilate three hourly International Satellite Cloud Climatology Project
surface skin temperature data. Bias correction techniques were developed, since traditional analysis
with Physical-space Statistical Analysis System of a biased forecast lead to a biased analysis. The
bias correction algorithms that were evaluated included bias correction every 3 hours, incremental
bias correction every time step, and bias correction to the mean diurnal and mean semi-diurnal cycle.
The results for a July 1992 test case have shown that the semi-diurnal bias correction was most

effective. The monthly mean diurnal cycle from the semi-diurnal bias correction experiment closely matched the diurnal cycle from the observations. Also, the semi-diurnal bias correction results show the lowest standard deviation for the global monthly mean between the experiment and the observations.

References

Charney, J. G., M. Halem, and R. Jastrow, 1969: Use of incomplete historical data to infer the present state of the atmosphere. *J. Atmos. Sci.,* **26,** 1160-1163.

Cohn, S., 1982: Methods of Sequential Estimation for Determining Initial Data in Numerical Weather Prediction. Ph.D. thesis, Courant Institute of Mathematical Sciences, New York University, New York, NY, 183 pp.

Cohn, S. E., A. da Silva, J. Guo, M. Sienkiewicz, and D. Lamich, 1998: Assessing the effects of data selection with the DAO Physical-space Statistical Analysis System. *Mon. Weather Rev.,* **126,** 2913-2926.

Dee, D. P., and A. da Silva, 1998: Data assimilation in the presence of forecast bias. *Q. J. R. Meteorol. Soc.* **124,** 269-295.

Entekhabi, D., H. Nakamura, and E. G. Njoku, 1994: Solving the inverse problem for soil moisture and temperature profiles by sequential assimilation of multifrequency remotely sensed observations. *IEEE Trans. Geosci. Remote Sensing,* **32,** 438-448.

Evensen, G., 1994: Sequential data assimilation with a nonlinear quasi-geostrophic model using Monte Carlo methods to forecast error statisitics. *J. Geophys. Res.,* **99,** 10143-10162.

Georgakakos, K.P., and Smith, G.F., 1990: On improved hydrologic forecasting - Result from a WMO real time forecasting experiment. *J. Hydrol.,* **114,** 17-45.

Ghil, M., S. Cohn, J. Tavantzis, K. Bube, and E. Isaacson, 1981: *Application of estimation theory to numerical weather prediction. Dynamic Meteorology: Data Assimilation Methods,* Springer-Verlag, New York, NY, 139-224.

Hollingsworth, A., D. Shaw, P. Lonnberg, L. Illari, K. Arpe, and A. J. Simmons, 1986: Monitoring of Observation and Analysis Quality by a Data Assimilation System. *Mon. Weather Rev.,* **114,** 861-879.

Houser, P. R., W. J. Shuttleworth, H. V. Gupta, J. S. Famiglietti, K. H. Syed, and D. C. Goodrich, 1998: Integration of Soil Moisture Remote Sensing and Hydrologic Modeling using Data Assimilation. *Wat. Resour. Res.,* **34,** 3405-3420.

Houtekamer, P. L. and H. L, Mitchell, 1998: Data assimilation using a Ensemble Kalman filter techniques. *Mon. Weather Rev.,* **126,** 796-811.

Kalman, R.E., 1960: A new approach to linear filtering and prediction problems. *Trans. ASME, Ser. D, J. Basic Eng.* **82,** 35-45.

Keppenne, C.L., and M.M. Rienecker, 2001: Development and initial testing of a parallel Ensemble Kalman filter for the Poseidon isopycnal ocean general circulation model. *Mon. Wea. Rev.,* submitted.

Koster, R. D., M. J. Suarez, and M. Heiser, 2000: Variance and predictability of precipitation at seasonal-to-interannual timescales. *J. of Hydrometeorology,* **1,** 26-46.

Koster, R. D., and M. J. Suarez, 1992: Modeling the land surface boundary in climate models as a composite of independent vegetation stands. *J. Geophys. Res.,* **97,** 2697-2715.

Koster, R.D., M.J. Suarez, A. Ducharne, M. Stiglitz, and P. Kumar, 2000: A catchment-based approach to modeling land surface processes in a GCM. Part 1: Model structure. *J. Geophys. Res.,* **105,** 24809-24822.

Lorenc, A. C., 1995: Atmospheric Data Assimilation. *Meteorological Office Forecasting Research Div.,* **34,** The Met Office, UK.

McLaughlin, D., 1995: Recent developments in hydrologic data assimilation. *Reviews of Geophysics,* 977-984.

Milly, P. C. D., 1986: Integrated remote sensing modeling of soil moisture: sampling frequency, response time, and accuracy of estimates. *Integrated Design of Hydrological Networks - Proceedings of the Budapest Symposium,* **158,** 201-211.

Reichle, R.H., J.P. Walker, and R.D. Koster, and P.R. Houser, 2002: Extended versus Ensemble Kalman Filtering for Land Data Assimilation. J. *Hydromet.,* **3,** 728-740.

Sellers, P. J., B. W. Meeson, J. Closs, J. Collatz, F. Corprew, D. Dazlich, F. G. Hall, Y. Kerr, R. Koster, S. Los, K. Mitchell, J. McManus, D. Myers, K.-J. Sun, and P. Try, 1996: The ISLSCP Initiative I global datasets: Surface boundary conditions and atmospheric forcings for land-atmosphere studies. *Bull. Amer. Meteorol. Soc.,* **77,** 1987-2005.

Walker, J., 1999: *Estimating Soil Moisture Profile Dynamics From Near-Surface Soil Moisture Measurements and Standard Meteorological Data.* Dissertation Thesis, Dept. of Civil, Surveying and Environmental Engineering, The University of Newcastle.

Walker, J. P., and P. R. Houser, 2001: A methodology for initializing soil moisture in a global climate model: Assimilation of near-surface soil moisture observations. *J. Geophys. Res.,* **106,** 11761-11774.

LAND DATA ASSIMILATION SYSTEMS

PAUL R. HOUSER
NASA Goddard Space Flight Center
Greenbelt, Maryland 20771 USA

1. Introduction

Land surface temperature and wetness conditions affect and are affected by numerous climatological, meteorological, ecological, and geophysical phenomena. Therefore, accurate, high resolution estimates of terrestrial water and energy storages are valuable for predicting climate change, weather, biological and agricultural productivity, and flooding, and for performing a wide array of studies in the broader biogeosciences. In particular, terrestrial stores of energy and water modulate fluxes between the land and atmosphere and exhibit persistence on diurnal, seasonal, and interannual timescales. Furthermore, because soil moisture, temperature, and snow are integrated states, errors in land surface forcing and parametrization accumulate in the representations of these variables in operational numerical weather forecast models, which lead to incorrect surface water and energy partitioning. Therefore, accurate re-initialization of water and energy state variables in these models is crucial.

Large-scale land surface fields for weather forecast model initialization are scarce and often unreliable. However, innovative new ground- and space-based observation systems are in various stages of development, which will provide data to constrain modelled land surface states. Constraints can be applied in two ways. First, by forcing the land surface primarily by observations, biases in atmospheric model-based forcing can be avoided. Second, by employing data assimilation techniques, observations of land surface states can be used to reduce the likelihood of unrealistic model states.

Through innovation and an ever-improving conceptualization of the physics underlying Earth system processes, land surface models have continued to evolve and to display improved ability to simulate complex phenomena. Concurrently, increases in computing power and affordability are allowing global simulations to be run more routinely and with less processing time, at spatial resolutions that were intractable as little as five years ago.

Recognizing the need for global, high resolution, near-real time land surface state fields and the opportunity created by the availability of new observations, sophisticated land surface models, and powerful yet affordable computers, scientists at the National Aeronautics and Space Administration's Goddard Space Flight Center (NASA-GSFC) and National Oceanic and Atmospheric Administration's National Centers for Environmental Prediction (NOAA-NCEP) set out to develop a powerful, "state-of-the-art", Global Land Data Assimilation System to produce accurate global maps of land surface energy and moisture stores and fluxes. Towards this goal four component objectives were identified: (1) modelling, (2) observation, (3) data assimilation, and (4) validation.

R. Swinbank et al. (eds.), Data Assimilation for the Earth System, 345–360.

Many land surface models have been developed and enhanced during the past decade with varying core themes, such as subgrid variability, community-wide input, advanced physical representations, and compatibility with atmospheric models. Several of these models have come to the forefront and are included in the Global Land Data Assimilation System (GLDAS). The Global Land Data Assimilation System programming code has been developed in a modular fashion to facilitate the incorporation of multiple land surface models.

Global atmospheric model predictions provide baseline forcing for the Global Land Data Assimilation System, but, whenever and wherever possible, the modelled forcing fields are replaced or corrected by observation-based fields. In particular, the Global Land Data Assimilation System program code has the capability to accept satellite-derived precipitation and solar insolation data and repopulate the modelled forcing fields with those data as they are available. In this way, biases in the atmospheric models are impeded from distorting the land surface model predictions without sacrificing spatial and temporal constancy of forcing.

Data assimilation techniques merge observed data fields, which may include spurious errors, with a model prediction. In this way land surface models can be constrained to a best estimate of the state of the natural environment, thereby improving subsequent predictions. Soil moisture and temperature assimilation strategies based on the Kalman filter approach (Kalman, 1960) are being implemented in the Global Land Data Assimilation System (see chapter *Assimilation of land surface data*). In addition, simpler correction techniques are being developed including a snow correction which relies on snow cover derived from Moderate Resolution Imaging Spectroradiometer (MODIS) observations from the Earth Observing System Terra satellite.

Assimilation of observation-based surface fields will continually steer the modelled states towards our understanding of reality. However, only a few variables can be observed directly, such as skin temperature. Other variables, such as sub-surface soil temperature, will be affected by the assimilation, but their representation is dependent on imperfect model physical assumptions. Therefore, output surface states and fluxes will be compared with independent observations. Through the validation process, the quality of the resulting land surface fields will be evaluated, and, if necessary, changes to the model structure, forcing, or parameters will be explored.

2. North American land surface data assimilation system

The Global Land Data Assimilation System has its basis in the North American Land Data Assimilation System project (Mitchell *et al.* 1999). The North American Land Data Assimilation System was initiated in 1998 with the goal of modelling land surface states and fluxes, while relying as much as possible on observation-based parameter and forcing fields in order to avoid biases that are known to exist in forcing fields produced by atmospheric models. The study region for the North American Land Data Assimilation System encompasses the conterminous United States and parts of Mexico and Canada. The land surface models implemented in the North American Land Data Assimilation System are run at 1/8th degree latitude by 1/8th degree longitude resolution. Separate versions of the system have been developed at the National Aeronautics and Space Administration's Goddard Space Flight Center, the National Oceanic and Atmospheric Administration's National Centers for Environmental Prediction, Princeton University, the University of Washington, and National Oceanic and Atmospheric Administration's Office of Hydrology. Each group runs their land surface models both in real time, retrospectively using the same high quality parameter and forcing fields, thus enabling unambiguous intercomparison of the land surface model simulations. Results are being validated by researchers at Rutgers University, using time series of observed variables, including soil moisture and temperature, to validate the strengths and weaknesses of each model. Much of the Global Land Data Assimilation System program code was derived from Goddard Space Flight Center's North American Land Data Assimilation System program code, and many of the project specifications are identical.

3. Methods and specifications

One of the primary objectives of the Global Land Data Assimilation System was to develop a system that would allow users to run multiple land surface models without specific knowledge of the models' architectures or physics. Currently, program code for three land surface models has been installed. Designing a Global Land Data Assimilation System simulation only requires modification of a single, simple interface file, which includes switches and variables for many run time options (summarized in Table 1). The Global Land Data Assimilation System program code interprets the forcing data to the individual input requirements of each respective land surface model, so that the same data can be used to force multiple land surface models. Thus, the influence of discrepancies in forcing data can be eliminated when comparing land surface fields simulated by different land surface models.

Table 1. Options available in the Global Land Data Assimilation System user interface.

Spatial Resolution Options				
Degrees Longitude	0.25	0.5	1.0	2.5
Degrees Latitude	0.25	0.5	1.0	2.0
Land Surface Model	Mosaic; Community Land Model; Noah			
Forcing	various model and satellite-derived products			
Initialization	None (constant value); restart file; forcing data			
Subgrid Variability	1 to13 tiles per grid cell (constant or fractional cutoff)			
Elevation Adjustment	temperature; pressure; humidity; longwave radiation			
Data Assimilation	surface temperature; snow cover			
Soil Classification	lookup table; Reynolds et al. [1999]			
Leaf Area Index	lookup table; satellite-derived			
Inland Water Tiles	Community Land Model lakes option			

As a standard, all Global Land Data Assimilation System models run on a common 0.25 degree longitude by 0.25 degree latitude grid which is nearly global, covering all of the land north of latitude 60 degrees South[1]. The Global Land Data Assimilation System also is able to run on 0.5 degree longitude by 0.5 degree latitude, a 1 degree longitude by 1 degree latitude, and a 2.5 degree longitude by 2.5 degree latitude global grid. Subgrid variability is simulated using a vegetation-based tiling approach, as described in the next section. The model time step is user-defined (15 minutes is standard). Forcing data is typically available on 0.25 degree longitude by 0.25 degree latitude to 1 degree longitude by 1 degree latitude grids with three or six hourly resolution. The Global Land

---

[1] Given increases in production-line computer random access memory which have become available only very recently, Global Land Data Assimilation System scientists hope to increase the core spatial resolution to 1/8th degree longitude by 1/8th degree latitude in the near future.

Data Assimilation System includes spatial and temporal interpolation routines based on commonly accepted algorithms. The output interval is also user-defined (3 hours is standard).

A high quality vegetation classification map is critical to the Global Land Data Assimilation System for three reasons. First, the Mosaic land model, the Community Land Model, and Noah land model incorporate soil-vegetation-atmosphere transfer schemes, so that the fluxes and storages of energy and water at the land surface are strongly dependent on the properties of the vegetation. Second, the vegetation type dictates other parameters such as albedo and roughness height. Third, the Global Land Data Assimilation System simulates subgrid scale variability by dividing each grid cell into a set of tiles based on the distribution of vegetation classes within the cell. Recent observations from the Advanced Very High Resolution Radiometer (AVHRR) aboard the National Oceanic and Atmospheric Administration 15 (NOAA-15) satellite have allowed scientists to classify vegetation globally at very fine 1 kilometre scales. The Global Land Data Assimilation System uses a static, 1 kilometre resolution, global vegetation classification dataset produced by the University of Maryland (Hansen *et al.*, 2000) from the Advanced Very High Resolution Radiometer data. The University of Maryland dataset includes thirteen vegetation classes.

The Global Land Data Assimilation System also employs a satellite observation-based, 1 kilometre resolution climatology and, when available, a time series of leaf area index. These were generated using three information sources: (1) an 8 kilometre resolution time series of leaf area index (currently spanning July 1981 through July 2001), which was derived by scientists at Boston University (Myneni *et al.*, 1997) from Advanced Very High Resolution Radiometer measurements of normalized difference vegetation index and other satellite observations, (2) a climatology based on the 8 kilometre time series, in which leaf area index is indexed by 10 degree latitude zone, month of year, and vegetation type, and (3) the 1 kilometre University of Maryland vegetation type classification. The information is blended so that the resulting 1 kilometre pixel values vary by vegetation type while the Boston University 8 kilometre average leaf area indexes are maintained. The Global Land Data Assimilation System scales the 1 kilometre data to the selected model resolution and adjusts for fractional vegetation cover. Preliminary tests indicate that output land surface states have been improved by the inclusion of the satellite-derived leaf area index data.

The soil parameter maps used in the Global Land Data Assimilation System were derived from the global soils dataset of Reynolds *et al.* (1999). That dataset includes 5 minute resolution global maps of porosity and the percentages of sand, silt, and clay, which are based on the United Nations Food and Agriculture Organization Soil Map of the World (FAO 1990) linked to a global database of over 1300 soil pedons. A soil pedon is defined as the smallest volume of soil that is large enough to permit study of all its distinct horizontal layers. This soil information was spatially resampled to the 0.25 degree longitude by 0.25 degree latitude Global Land Data Assimilation System spatial grid and vertically interpolated to 0-2 centimetre, 2-150 centimetre, and 150-350 centimetre depths from the original 0-30 centimetre and 30-100 centimetre depths. Those depths were chosen mainly to match the set of depths most commonly assigned in the original version of the Mosaic land model. Furthermore, the thickness of the surface layer will facilitate assimilation into the Global Land Data Assimilation System of future soil moisture fields derived from Advanced Microwave Scanning Radiometer (AMSR) satellite observations. Certain parameters employed by the land surface models are indexed based on the United States Department of Agriculture soil texture class, including the saturated soil potential and the *b* parameter. Therefore the Global Land Data Assimilation System includes a routine to classify the texture based on the percentages of sand, silt, and clay in a given grid cell. The Community Land Model also employs soil color as a parameter. The Global Land Data Assimilation System soil color map was interpolated from a 2.0 degrees latitude by 2.5 degrees longitude global map produced at the National Center for Atmospheric Research (NCAR).

The Global Land Data Assimilation System uses the Global 30 Arc-Second Elevation Data Set (Verdin and Greenlee, 1996) as its standard. The Global 30 Arc-Second Elevation Data Set was averaged onto the 0.25 degree longitude by 0.25 degree latitude Global Land Data Assimilation System grid. By default, the Global Land Data Assimilation System corrects the modelled temperature, pressure, humidity, and longwave radiation forcing fields based on the difference between the Global Land Data Assimilation System elevation definition and the elevation definition of the model that created the forcing data. Because some land surface models, including the Mosaic land model, ingest surface or bedrock slope as a parameter, geographic information systems software was used to assess the slope at each Global 30 Arc-Second Elevation Data Set pixel, and from those values the mean slope within each Global Land Data Assimilation System grid cell was computed.

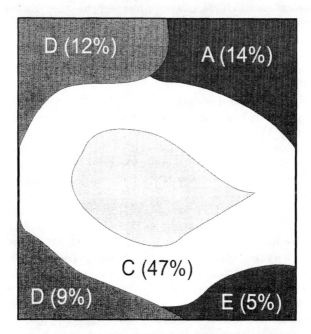

Figure 1: The spatial coverage of five vegetation types (A-E) within an illustrative grid cell is shown. Given this information along with user-defined cutoffs of 3 = maximum tiles per grid and 10% = minimum grid coverage to define a tile, Global Land Data Assimilation System would (1) eliminate vegetation type E (less than 10%), (2) eliminate vegetation type B (not among 3 predominant vegetation types), (3) normalise the weights of the three remaining tiles to 100%, resulting in three tiles: A (17%), C (57%), D (26%).

The Global Land Data Assimilation System land surface models run on a series of independent soil columns, or tiles, where each tile represents one vegetation class within a grid cell. Gridded output is returned by weighting each tile by its fractional coverage within the cell. Users select a maximum whole number of tiles to be defined per grid cell. This can be as many as 13, which is the number of vegetation types in the University of Maryland vegetation classification. In addition, users select the smallest percentage of a cell for which to create a tile. For instance, given settings of 8 tiles and a 5% cutoff, a tile would be created for any vegetation type that covers at least 5% of a grid cell, for up to 8 tiles per cell. Any vegetation type that covers less than 5% of the cell or is not among the 8 most common types would be ignored, and the resulting tile areas would be normalized to 100%. Figure 1 illustrates this example. Tests have shown that constraining the number of tiles per grid using the percentage cutoff is more efficient (*i.e.*, the global sum of the all the subgrid areas containing a vegetation type that was ignored is smaller given the same global total number of tiles) than setting the maximum number of tiles per grid to some arbitrary number less than thirteen. Accordingly, a standard of 13 for the maximum number of tiles (i.e., no constraint) and 10% for the minimum tile area percentage has been established for the Global Land Data Assimilation project.

4. Global land data assimilation

A technique has been developed to allow the Global Land Data Assimilation System to assimilate remotely sensed, 6-hourly skin temperature observations (Ottle and Vidalmadjar, 1992) from the Television Infrared Observation Satellites (TIROS) Operational Vertical Sounder (TOVS) instrument. This capability is currently included for use with the Mosaic land model and the Community Land Model land surface models. The technique relies on an optimal interpolation routine to calculate an analysis increment, which is used to perform an incremental, semi-daily, or daily skin temperature bias correction. The bias correction continually steers the modelled state towards the observation. The observed-minus-forecast value for the skin temperature is first calculated by the Global Land Data Assimilation System and passed into the optimal interpolator to retrieve the analysis increment. This is then relayed, along with a bias correction term, to the land surface model code for proper energy budget considerations. Testing has shown that the technique effectively constrains the modeled surface temperature (Radakovich *et al.*, 2001), though the results are limited by the temporal sparseness of the Television Infrared Observation Satellites Operational Vertical Sounder data. However, the same technique can be applied using 3-hourly skin temperature data from the International Satellite Cloud Climatology Project.

Techniques for assimilating surface soil moisture data have been designed and tested by members of the Global Land Data Assimilation System science team (*e.g.* Walker and Houser, 2001), but the capability has not yet been implemented in Global Land Data Assimilation System due to a lack of global observations of soil moisture. This deficiency is expected to be remedied when Advanced Microwave Scanning Radiometer instruments, aboard National Aeronautics and Space Administration's Earth Observation System Aqua satellite and the joint National Aeronautics and Space Administration (NASA) and National Space Development Agency of Japan (NASDA) Advanced Earth Observing Satellite II (ADEOS-II) satellite, begin to deliver global C-band microwave observations (both satellites were successfully launched in 2002). It has been demonstrated that remotely sensed C-band microwave fields can be used, given certain land cover conditions, to derive near surface (0-2 centimeters) soil moisture (Owe *et al.*, 2001).

The Global Land Data Assimilation System is able to assimilate snow cover information derived from measurements made by the Moderate Resolution Imaging Spectroradiometer (MODIS) sensor aboard National Aeronautics and Space Administration's Earth Observation System Terra satellite. Among other products, the Moderate Resolution Imaging Spectroradiometer science team at Goddard Space Flight Center provides a daily, 0.25 degree longitude by 0.25 degree latitude global snow cover dataset. For each 0.25 degree longitude by 0.25 degree latitude grid cell over land, this product reports the percentages of snow-covered, non-snow-covered, and non-visible (*e.g.* obscured by cloud cover) subgrid pixels. The Global Land Data Assimilation System snow correction algorithm determines the snow cover state (snow-covered or bare) at each grid cell based on the ratio of snow-covered pixels to visible pixels. It also determines whether or not the snow cover state observed at a grid cell is reliable based on the number of visible pixels. If the grid cell is deemed unreliable or if the observation and model agree on the snow cover state in a grid cell, the modelled snow variables are not changed in that grid cell. Otherwise, if the Moderate Resolution Imaging Spectroradiometer shows that the grid cell is bare but the model shows snow, then the equivalent heights of water in all the model snow layers are set to zero. If the Moderate Resolution Imaging Spectroradiometer shows snow but the model shows no snow, then a thin cover of snow is added to the model grid cell and the albedo changes automatically. Testing is currently underway to determine how this correction affects the modelled energy and water balances and to improve the technique, possibly by adjusting near-surface temperatures, soil moisture, and/or the precipitation types (rain or snow) of recent events to reflect the observed snow cover state.

Other options are provided in the Global Land Data Assimilation System user interface file. One is the forcing data source, and another is the type of land surface model state variable initialization: (1) the user can declare a globally uniform value for each variable; (2) the values can be taken from a restart file produced by a prior run; (3) the Global Land Data Assimilation System can input the surface state variable fields produced by the land model coupled to the weather forecast model which produced the forcing data (forcing data initialization option). Although time to spin up a model to a stable state is still required with the third approach, this is greatly reduced. Therefore the forcing data initialization option is a valuable innovation, especially considering the high computational expense of a 0.25 degree longitude by 0.25 degree latitude Global Land Data Assimilation System run. A third additional option is the inclusion of a lake model in the Community Land Model, which enables the creation and modeling of inland water tiles.

Users choose one of the numerical weather prediction models as the baseline forcing source. Users then may choose observation-derived fields, including precipitation and longwave and shortwave radiation, to replace the corresponding model-based forcing fields. If the observation-derived fields are not available, the Global Land Data Assimilation System reverts to the model-based fields. This capability reduces reliance on modelled forcing fields, which may contain biases, while ensuring continuity of forcing. However, in choosing a forcing option, the observation-based fields' lack of bias must be weighed against their limitations (the methods used to derive these fields from raw satellite observations are far from perfect) and the strengths of the forecast modelling systems (including assimilation and quality assurance procedures). Variables required to force the land surface models are listed in Table 2.

Table 2. Global Land Data Assimilation System forcing and output fields.

Required Forcing Fields	Summary of Output Fields
total precipitation	soil moisture in each layer
convective precipitation	snow depth, fractional coverage, and water equivalent
downward shortwave radiation	plant canopy surface water storage
downward longwave radiation	soil temperature in each layer
near surface air temperature	average surface temperature
near surface specific humidity	surface and subsurface runoff
near surface wind speed	bare soil, snow, and canopy surface water evaporation
surface pressure	canopy transpiration
	latent, sensible, and ground heat flux
	snow phase change heat flux
	Snowmelt
	snowfall and rainfall
	net surface shortwave and longwave radiation
	aerodynamic conductance
	canopy conductance
	surface albedo
	vegetation greenness and leaf area index

The primary goal of the Goddard Earth Observing System Data Assimilation System (Pfaendtner *et al.*, 1995) is to support the National Aeronautics and Space Administration Earth Observing System (Atlas and Lucchesi, 2000) product retrievals. The Goddard Earth Observing System Data Assimilation System runs two production suites for each daily time period: a "first look" assimilation and, two weeks later, an improved, "late look" assimilation. The Global Land Data Assimilation System downloads the first look output for the operational simulations and replaces it in the forcing archive with the late look output when that becomes available. As the Goddard Earth Observing System Data Assimilation System begins a forecast cycle, it uses a restart file from the previous cycle with current boundary conditions (*e.g.* sea surface temperature) to produce a first guess, or three hour forecast. Then, using error statistics and various types of observations collected over the 6-hour window, it computes corrections to the first guess. Rather than applying the gridded corrections directly to the first guess to produce the analysis, as is the norm in meteorological data assimilation, the Goddard Earth Observing System Data Assimilation System performs an incremental analysis update by gradually inserting the corrections back into another integration of the model, resulting in 3-hourly assimilation data sets for the surface fields of interest to the Global Land Data Assimilation System. The Goddard Earth Observing System Data Assimilation System fields are produced on a 1 degree longitude by 1 degree latitude global grid. Based on comparisons with observations and other model output, the Goddard Earth Observing System Data Assimilation System has been chosen as the primary forcing source for Global Land Data Assimilation System.

The Global Data Assimilation System is the global, operational weather forecast model of the National Centers for Environmental Prediction (Derber *et al.*, 1991). The Global Data Assimilation System runs on a quadratic T170 Gaussian grid with 512 gridpoints in the zonal direction and 256 gridpoints in the meridional direction (about 0.7 degrees longitude by 0.7 degrees latitude resolution). There are 42 model atmospheric levels with the lowest pressure layer at 2 hPa, and 10 levels above 100 hPa. Due to computation limitations, 6-hour analyses are performed on a linear representation of the T170 grid that is essentially equivalent to a 1 degree longitude by 1 degree latitude global grid. The analysis utilizes ground-based, radiosonde, airborne, and satellite-derived observations. The Global Land Data Assimilation System makes use of Global Data Assimilation System 0 hour, 3 hour, and, as needed, 6 hour forecasts, which are produced at six-hour intervals.

The European Centre for Medium-Range Weather Forecasts (ECMWF) produces operational, global analyses for four synoptic hours: 00, 06, 12, and 18 Coordinated Universal Time. The model is run on a T_L511 triangular truncation, linear reduced Gaussian grid, which has 553/384 (zonal/meridional) surface grid points (approximately 39 kilometre resolution). It includes 60 model atmospheric levels with the lowest pressure layer at 0.1 hPa, and increased resolution in the boundary layer (see chapter *Atmospheric modelling*). The analysis incorporates both in situ conventional and satellite-derived data using a four-dimensional multivariate assimilation approach.

The Global Land Data Assimilation System estimates global, downward shortwave and longwave radiation fluxes using a procedure from the Air Force Weather Agency's Agricultural Meteorology modelling system. It utilizes the Air Force Weather Agency's Real Time Nephanalysis 3-hourly, 48 kilometre resolution cloud maps (Hamill *et al.*, 1992), and the Air Force Weather Agency's daily, 48 kilometre resolution snow depth maps (Kopp and Kiess, 1996) to calculate surface down-welling shortwave radiation based on the algorithms of Shapiro (1987). These cloud and snow products are derived primarily from observations made by Defense Meteorological Satellite Program (DMSP) and National Oceanic and Atmospheric Administration (NOAA) satellites. The Air Force Weather Agency's Real Time Nephanalysis cloud analysis is also used to calculate surface down-welling longwave radiation following Idso (1981). Both the shortwave and longwave schemes are implemented within the four "floating" Air Force Weather Agency's Real Time Nephanalysis cloud layers. For each layer, atmospheric transmissivity and reflectivity, in respect to shortwave radiation, and longwave radiation emitted by clouds are calculated as functions of cloud type and amount.

Near-real time satellite-derived precipitation data is obtained from two research groups: one at Goddard Space Flight Center and one at the United States Naval Research Laboratory (NRL). United States Naval Research Laboratory produces precipitation fields based on both geostationary satellite infrared cloud top temperature measurements and microwave observation techniques (Turk *et al.*, 2000). The microwave product merges data from the Special Sensor Microwave/Imager (SSM/I), the National Aeronautics and Space Administration and the National Space Development Agency of Japan's Tropical Rainfall Measuring Mission (TRMM), and the Advanced Microwave Sounding Unit (AMSU) instruments. Both United States Naval Research Laboratory products have a spatial resolution of 0.25 degree longitude by 0.25 degree latitude and a temporal resolution of 6 hours and both cover an area from 60 degrees South to 60 degrees North. The Global Land Data Assimilation System began to archive those data in April 2001. When the United States Naval Research Laboratory precipitation option is chosen, the Global Land Data Assimilation System inputs a modelled precipitation field that has been interpolated to 0.25 degree longitude by 0.25 degree latitude (in order to ensure complete global coverage) and then overwrites it with infrared-based precipitation. The resulting precipitation field is then overwritten with any available microwave-based precipitation data, which tends to be more accurate but also more sparse than the infrared-based data. Consequently, the final forcing field is a combination of model-, infrared-, and microwave-based fields.

More recently, scientists at the Goddard Space Flight Center have started to provide a near-real time, satellite-based precipitation product (George Huffman, Goddard Space Flight Center, personal communication, 1992). It results from an advanced algorithm which optimally merges the more accurate microwave data with the more frequent infrared data. This product is more consistent in space than the field that results from the overlay procedure described in the previous paragraph. The Goddard Space Flight Center began production of this product in February 2002, and it is now included in the Global Land Data Assimilation System suite of forcing options. In the future the Global Land Data Assimilation System will also use precipitation fields based on the National Oceanic and Atmospheric Administration Climate Prediction Center's Merged Analysis of Precipitation pentad (5-daily) product, which blends satellite and gauge observations. This pentad, 2.5 degree longitude by 2.5 degree longitude product is interpolated to the Global Land Data Assimilation System time and space resolution using interpolation weights derived from numerical weather prediction fields.

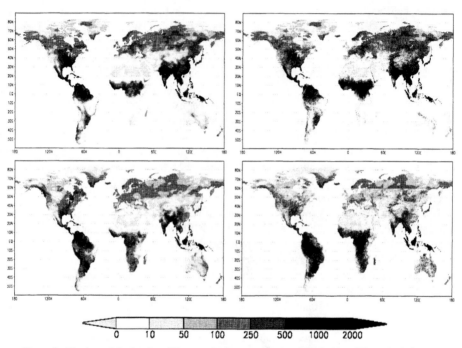

Figure 2. Total precipitation in millimetres from two operational Global Land Data Assimilation System simulations. Left: Control Run: Global Earth Observing System modeled precipitation. Right: Derived Forcing Run: United States Naval Research Laboratory satellite-derived precipitation. From top to bottom: March-April-May 2001, June-July-August 2001, September-October-November 2001, December-January-February 2001-02.

5. Results

The Global Land Data Assimilation System runs daily in an operational mode. The Mosaic land model is the current operational model, but parallel simulations with the Community Land Model and the Noah land model will be instituted in the near future. The Goddard Earth Observing System Data Assimilation System is currently the baseline forcing source. The Goddard Earth Observing System Data Assimilation System (GEOS-DAS) precipitation and radiation fields are overwritten with the observation-based fields, when and where these are available. The model spatial resolution is 0.25 degrees longitude by 0.25 degrees latitude, and 10% is the minimum tile area allowed. The model time step is 15 minutes, and output is 3-hourly. Typically the daily near real time runs are complete within 36 to 48 hours of real time. The time delay is mainly attributed to delays in observation availability; the actual Global Land Data Assimilation System computation completes in less than one hour.

Results are presented from two simulations: a Control Run and a Derived Forcing Run. Each run started on 1 January 2001. The Mosaic land model was used for both runs with the operational settings defined in the previous paragraph, except that the combination of forcing fields varied. The forcing data initialization option was used so that the Goddard Earth Observing System Data Assimilation System provided the initial surface energy and water storage states. Evidence suggests that this allowed the model to spin up and achieve reasonable stability in about three months. The Control Run relied on the Goddard Earth Observing System Data Assimilation System forcing exclusively. The Derived Forcing Run used the United States Naval Research Laboratory observation-based precipitation fields and the observation-based downward shortwave and longwave radiation fields.

The greater fine scale variability of the observation-based precipitation (Figure 2) is reflected in the fine scale patterns of soil moisture in the Derived Forcing Run. Because rainfall tends to be spatially heterogeneous at local to regional scales and soil moisture shows a high degree of variability at all scales (e.g. Famiglietti et al., 1999), the fine scale soil moisture variability evident in the Derived Forcing Run results may be preferable to the Control Run results. However, the exact locations of the fine scale features are unlikely to be reliable due to the imprecision of precipitation maps derived from satellite infrared observations of cloud top temperatures.

If global, reliable observations of root zone soil moisture existed they could be used to validate the spatial distribution of wet and dry areas, which is highly dependent on the precipitation forcing, the statistical distribution of soil water content on global, continental, and regional biases, which are apt to be a function of the soil properties and model physics, and the degree of fine scale heterogeneity. Of course, if these observations did exist, the development of the Global Land Data Assimilation System might not have been necessary.

Due to the high quality of the cloud observations that contribute to the Air Force Weather Agency's Agricultural Meteorology modelling system radiation fields, their fine scale patterns are more likely to be more accurate than those of the observation-based precipitation fields. Consequently, on a local to regional scale the derived radiation forcing may improve estimates of surface temperature and evapotranspiration. Figure 3 focuses on North America and Southeast Asia so that the small scale patterns of shortwave radiation and evapotranspiration can be easily distinguishable. It is apparent that moisture availability limited the evapotranspiration rate in the southwestern United States on that particular day, while input energy was the limiting factor along the western border with Canada.

One of the most ambitious activities of the Global Land Data Assimilation System project has been the assemblage of an archive of global, operational weather forecast model output and observation-based data fields for parameterizing and forcing land surface models (Table 3). Most of the time series begin around January 2001 and continue up to present. The most recent fields are downloaded daily from forecast centers and groups that process satellite data. Many of these fields become inaccessible from their original sources not long after initial release, being archived to tape or not saved at all. However, the Global Land Data Assimilation System active archive is public to the extent allowed by the data providers, so that these data, once lost for all intents and purposes, are now retrievable through the Global Land Data Assimilation System website (http://ldas.gsfc.nasa.gov) or by request. The Global Land Data Assimilation System parameter information, including vegetation, soils, and elevation fields, is also available.

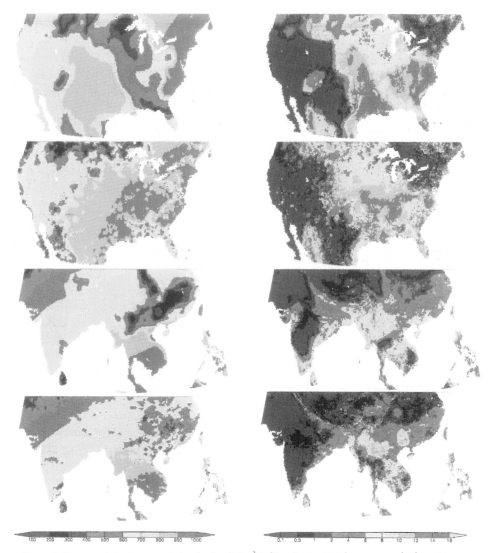

Figure 3: Downward shortwave radiation forcing (W/m²; left) and output total evapotranspiration rate (mm/day; right) from two operational Global Land Data Assimilation System simulations. From top to bottom: Control Run 18-21Z 31 July 2001; Derived Forcing Run 18-21Z 31 July 2001; Control Run 6-9Z 31 January 2002; Derived Forcing Run 6-9Z 31 January 2002. Top four: central North America; bottom four: Southeast Asia.

Output fields of land surface states and fluxes from the Global Land Data Assimilation System model simulations are also freely available to the public. The Global Land Data Assimilation System website includes a real time image generator which allows users to view the most recent output fields. Time series are available by request, subject to manpower limitations. It is the intention of the Global Land Data Assimilation System project to encourage broad use of Global

Land Data Assimilation System results: for education, policy making, and social, agricultural, and natural hazards planning, as well as scientific research.

6. Future directions

Several enhancements are scheduled for the Global Land Data Assimilation System in the near future. One is the inclusion of more land surface models. Additional forcing data options are currently being tested, and include new observation-based precipitation products. A 1 kilometre canopy greenness dataset will be estimated using an 8 kilometre fraction of photosynthetically active radiation dataset from Boston University along with the University of Maryland vegetation classification. In addition, when a Moderate Resolution Imaging Spectroradiometer derived climatology becomes available, its 1 kilometre leaf area index, photosynthetically active radiation, and albedo data will be incorporated into the Global Land Data Assimilation System. Building on the vegetation-based subgrid variability scheme, elevation and soils information will be linked to the tiles to increase their realism. A source-to-sink river routing scheme based on that of Olivera *et al.*, (2000) will be linked to the Global Land Data Assimilation System. Testing has begun on the coupling of a boundary layer model to the Global Land Data Assimilation System, which will allow near-surface meteorological variables to be modified by the Global Land Data Assimilation System land surface states, allowing more realistic development of the boundary layer and resulting fluxes.

The development phase of the Global Land Data Assimilation System project is nearly complete and the scientific research and applications phase soon will begin in earnest. The Global Land Data Assimilation System results will be compared and analysed from simulations that will use various combinations of land surface models, modelled and observation-derived forcing fields, data assimilation, and other options listed in Table 1. Output land surface fields will be compared with observations and with results from other modeling projects such as the North American Land Data Assimilation System. Data from ground-based monitoring stations and field experiments, including those affiliated with the upcoming Coordinated Enhanced Observing Period, will be compared with model location time series from the Global Land Data Assimilation System. The model location time series will be generated using the sub-grid tile results, so that the vegetation type can be chosen to match that of the monitoring station. Multi-year retrospective simulations will be staged using data from the International Satellite Land Surface Climatology Project, Initiative 2 (Hall *et al.* 2002) and other sources (*e.g.* Berg and Famiglietti, 2002). Lower resolution Global Land Data Assimilation System simulations may be implemented depending on the forcing data resolutions and processing time constraints.

Table 3. Summary of the Global Land Data Assimilation System data archive.

Type of Data	Source	Original Spatial Resolution	Time Period	Near Real-Time
Modelled Forcing	The National Aeronautics and Space Administration's Data Assimilation Office Goddard Earth Observing System	1 degrees longitude by 1 degrees latitude	12/2000 – present	✔
	National Centers for Environmental Prediction Global Data Assimilation System	~0.7 degrees longitude by ~0.7 degrees latitude	1/1999 – present	✔
	European Centre for Medium-Range Weather Forecasts forecasts and analyses	~39 kilometre square grid	10/2001 – present	✔
	Berg *et al.* (2002) bias corrected European Centre for Medium-Range Weather Forecasts reanalysis	0.5 degrees longitude by 0.5 degrees latitude	1/1979 – 12/1993	
Observation-Based shortwave and longwave radiation forcing	Derived at the National Aeronautics and Space Administration's Goddard Space Flight Center using United States Air Force Weather Agency cloud and snow analyses	0.25 degrees longitude by 0.25 degrees latitude	3/2001 – present	✔
Observation-based precipitation forcing	United States Naval Research Laboratory	0.25 degrees longitude by 0.25 degrees latitude	4/2001 – present	✔
	National Aeronautics and Space Administration's Goddard Space Flight Center Mesoscale Atmospheric Processes Branch	0.25 degrees longitude by 0.25 degrees latitude	3/2002 – present	✔
	National Oceanic and Atmospheric Administration Climate Prediction Center	2.5 degrees longitude by 2.5 degrees latitude	1/1979 – present	
Observation-based snow cover	Derived at National Aeronautics and Space Administration's Goddard Space Flight Center using Terra- Moderate Resolution Imaging Spectroradiometer observations	0.125 degrees longitude by 0.125 degrees latitude	11/2000 – present	
Observation-based leaf area index	Boston University Department of Geography	16 kilometre square grid	7/1982 – 5/2001	
Observation-based surface temperature	Television Infrared Observation Satellites Operational Vertical Sounder	~15 kilometre square grid	1/1998 – 12/1998	
Vegetation class	University of Maryland	1 kilometre square grid	static	
Soils	Reynolds *et al.* (1999)	5 minute longitude by 5 minute latitude	static	
Elevation	Global 30 Arc-Second Elevation Data Set	30 second longitude by 30 second latitude	static	

Fully coupled land-atmosphere-ocean models will be initialized with the Global Land Data Assimilation System land surface fields in order to test their influence on weather forecasts and climate predictions. The Global Land Data Assimilation System also will provide *a priori* knowledge of variations in the global distribution of terrestrial water mass, which will contribute to the optimization of global gravity field solutions based on observations from the National

Aeronautics and Space Administration's Gravity Recovery and Climate Experiment. Gravity Recovery and Climate Experiment (GRACE) observations will be used to estimate monthly changes in terrestrial water storage over regions larger than 200,000 km^2 (Rodell and Famiglietti, 1999, 2001). These estimates may prove to be useful for validating and constraining the Global Land Data Assimilation System simulations.

Scientists at the Goddard Space Flight Center have recently launched a new project whose primary goal is to use high performance, massively parallel computers to increase the horizontal resolution of Global Land Data Assimilation System towards the goal of 1 kilometre. This Land Information System will have a web-based user interface designed to enable data mining and visualization of modelled, remotely sensed, and ground-based fields.

References

Atlas, R. M., and R. Lucchesi, 2000: File Specification for GEOS-DAS Gridded Output. Available online: *http://dao.gsfc.nasa.gov/DAO_docs/File_Spec_v4.3.html.*

Berg, A. A., J.S. Famiglietti, J. P. Walker, and P.R. Houser, 2003: The Impact of Bias Correction to Reanalysis Products on Simulations of North American Land Surface States and Fluxes. *J. Geophys. Res.* In Preparation.

Derber, J. C., D.F. Parrish, and S. J. Lord, 1991: The new global operational analysis system at the National Meteorological Center. *Wea. and Forecasting,* 6, 538-547.

Famiglietti, J. S., J. A. Devereaux, C. A. Laymon, T. Tsegaye, P. R. Houser, T. J. Jackson, S. T. Graham, M. Rodell, and P. J. van Oevelen, 1999: Ground-based investigation of soil moisture variability within remote sensing footprints during the Southern Great Plains 1997 (SGP97) Hydrology Experiment. *Wat. Resour. Res.,* 35, 1839-1851.

Hamill, T.M., R.P. d'Entremont, and J.T. Bunting, 1992: A description of the Air Force real-time nephanalysis model. *Wea. Forecasting,* 7, 288-306.

Hansen, M.C., R.S. DeFries, J. R. G. Townshend, and R. Sohlberg, 2000: Global land cover classification at 1km spatial resolution using a classification tree approach. *International Journal of Remote Sensing,* 21, 1331-1364.

Idso, S.: 1981: A set of equations for the full spectrum and 8- and 14- micron and 10.5- to 12.5 thermal radiation from cloudless skies. *Wat. Resour. Res.,* 17, 295-304.

Kalman, R.E., 1960: A new approach to linear filtering and prediction problems. *Trans. ASME, Ser. D, J. Basic Eng,.* 82, 35-45.

Kopp, T.J., and R.B. Kiess, 1996: The Air Force Global Weather Central cloud analysis model. *AMS 15th Conf. on Weather Analysis and Forecasting,* Norfolk, VA, 220-222.

Mitchell, K., P. Houser, E. Wood, J. Schaake, D. Tarpley, D. Lettenmaier, W. Higgins, C. Marshall, D. Lohmann, M. Ek, B. Cosgrove, J. Entin, Q. Duan, R. Pinker, A. Robock, F. Habets, and K. Vinnikov, 1999: GCIP Land Data Assimilation System (LDAS) project now underway, *GEWEX News* 9(4), 3-6.

Myneni, R. B., C. D. Keeling, C. J. Tucker, G. Asar, and R. R. Nemani, 1997: Increased plant growth in the northern high latitudes from 1981 to 1991. *Nature,* 386, 698-702.

Olivera, F., J. S. Famiglietti, and K. Asante, 2000: Global-Scale Flow Routing Using a Source-to-Sink Algorithm. *Wat. Resour. Res.,* 36, 2197-2207.

Ottle, C., and D. Vidalmadjar, 1992: Estimation of land surface temperature with NOAA9 data. *Rem. Sens. Env.,* 40, 27-41.

Owe, M., R. de Jeu, and J.P. Walker, 2001: A Methodology for Surface Soil Moisture and Vegetation Optical Depth Retrieval Using the Microwave Polarization Difference Index. *IEEE Transactions on Geoscience and Remote Sensing,* 39, 1643-1654.

Pfaendtner, J., S. Bloom, D. Lamich, M. Seablom, M. Sienkiewicz, J. Stobie, and A. da Silva: 1995: Documentation of the Goddard Earth Observing System (GEOS) Data Assimilation System - Version 1, *NASA Technical Memorandum 104606* 4, 44 pp.

Radakovich, J. D., P. R. Houser, A. da Silva, and M. G. Bosilovich, 2001: Results from global land-surface data assimilation methods. *AMS 5th Symposium on Integrated Observing Systems,* Albuquerque, NM, 14-19 January, 132-134.

Reynolds, C. A., T. J. Jackson, and W. J. Rawls, 1999: Estimating available water content by linking the FAO Soil Map of the World with global soil profile databases and pedo-transfer functions. American Geophysical Union, Fall Meeting, *Eos Trans. AGU,* 80.

Rodell, M. and J. S. Famiglietti, 2001: Terrestrial Water Storage Variations over Illinois: Analysis of Observations and Implications for GRACE. *Wat. Resour. Res,.* 37, 1327-1340.

Rodell, M., and J. S. Famiglietti, 1999: Detectability of variations in continental water storage from satellite observations of the time dependent gravity field. *Wat. Resour. Res,.* 35, 2705-2723.

Turk, F. J., G. Rohaly, J. D. Hawkins, E. A. Smith, A. Grose, F. S. Marzano, A. Mugnai and V. Levizzani, 2000: Analysis and assimilation of rainfall from blended SSM/I, TRMM and geostationary satellite data. *AMS 10th Conf. On Sat. Meteor. and Ocean.*, Long Beach, CA, 9-14 January, 66-69.

Verdin, K. L., and S. K. Greenlee, 1996: Development of continental scale digital elevation models and extraction of hydrographic features. *Proceedings, Third International Conference/Workshop on Integrating GIS and Environmental Modeling,* Santa Fe, NM, January 21-26, National Center for Geographic Information and Analysis, Santa Barbara, CA.

Walker, J. P., and P. R. Houser, 2001: A methodology for initializing soil moisture in a global climate model: Assimilation of near-surface soil moisture observations. *J. Geophys. Res.,* **106**, 11761-11774.

REANALYSIS

RICHARD B. ROOD
NASA/Goddard Space Flight Center
Greenbelt, MD, USA

1. Introduction

Reanalysis is the assimilation of long time series of observations with an unvarying assimilation system to produce data sets for a variety of applications; for example, climate, chemistry-transport, and process studies. Because the longest, most complete set of observations exists for the atmosphere, reanalyses have focused on atmospheric assimilation. As the satellite records of chemical, land and oceanic parameters lengthen in time, reanalyses will be of interest to these fields. In this chapter discussion will be drawn from the experience of atmospheric reanalysis, but the issues raised are relevant to all types of reanalysis.

The provision of reanalyses was advocated by Bengtsson and Shukla (1988) and Trenberth and Olson (1988) in order to provide homogeneous data sets for climate applications and to encourage research in the use of satellite observations without the operational constraints of numerical weather prediction. Kalnay and Jenne (1991) proposed that a reanalysis be performed as a partnership between the National Meteorological Center (NMC, now part of the National Centers of Environmental Prediction, NCEP) and the National Center for Atmosphere Research (NCAR). This project required the preparation of the input data sets, the definition of the analysis system, and a data distribution plan. The analysis system was a version of the operational system used for weather prediction, but at lower resolution.

Three organisations performed a first generation of reanalyses in the spirit of Bengtsson and Shukla (1988) and Kalnay and Jenne (1991). Aside from the NCEP-NCAR reanalysis (Kalnay et al., 1996), the European Centre for Medium-range Weather Forecasts (ECMWF) executed the ERA-15 project (Gibson et al., 1997) and the Data Assimilation Office (DAO) at NASA's Goddard Space Flight Center provided a 17-year reanalysis (Schubert et al., 1993). These three reanalyses have been cited in many studies, which document successes as well as identifying a series of shortcomings that stand as the core of future research. New reanalyses are being produced by each of these organisations. NCEP and ECMWF are reanalyzing back to the 1950's, and NASA is performing reanalyses focused on specific observing systems.

The quality of the first-generation reanalyses has been documented in the proceedings from two workshops (WCRP, 1998; 2000; see also, Newson, 1998). Kistler *et al.* (2001) give an excellent overview of the NCEP/NCAR reanalysis project, and the discussions in that paper are relevant to all of the projects. Quantities that are

R. Swinbank et al. (eds.), Data Assimilation for the Earth System, 361–372.
© 2003 *Kluwer Academic Publishers. Printed in the Netherlands.*

directly impacted by the observations, *i.e.*, temperature, geopotential, and the rotational component of the wind, are consistent across the three reanalyses. At the other extreme, quantities that are only weakly constrained by the observations and are dependent upon the physical parametrizations of the assimilating models differ greatly. Further, these derived quantities, which include the divergent component of the wind, precipitation, evaporation, clouds, fresh-water runoff, and surface fluxes, have significant errors, as revealed either by independent validation or through applications in scientific studies.

The next section will discuss current and proposed reanalysis efforts. This will be followed by a discussion of the special challenges caused by the heterogeneity of the input data stream and biases between observations and model predictions. A summary is, then, presented.

2. Ongoing Reanalysis Activities

The lessons learned from the first-generation reanalyses provide the foundation for a second generation of reanalyses. These lessons can be summarised as general success in defining the major modes of variability on synoptic and planetary scales, as well as credible representation of the variability associated with longer-term, large-scale phenomena: e.g., monsoons, El Niño – La Niña, and the Madden-Julian oscillation. The deficiencies include fundamental problems in the hydrological cycle and the general circulation, and artifacts in the reanalysis data sets that are directly related to changes in the observing network. Reanalyses are not appropriate for trend studies, which is generally attributed to sensitivity of changes in the observing system.

There are a number of planned reanalysis activities. Some specific shortcomings in the first-generation reanalyses will be corrected, and there is also expected benefit from the progressive improvement of data assimilation systems. However, the most difficult issues associated with internal consistency of the model's dynamics and physics with the observations will require sustained, directed research activities that include development of more robust physical parametrizations and their interactions with observations. Those artifacts that are related to the heterogeneity of the observing network suggest problems that are at some level intractable, requiring study to characterize the observing system itself. Possible strategies for future reanalyses may include, therefore, optimisation of the observing system to focus on key problems and hierarchies of reanalyses that quantify the sensitivity to the observing system. An interesting discussion of the use of data assimilation to study the physical processes of the climate and the need to bridge weather and climate science is given in Morel (2001).

An important product from the first-generation reanalyses is the quality-controlled input data record (see, for example, Onogi, 2000). This examination of the input data record results from comparing the input data stream with model estimates of expected values as well as with neighbouring observations, and provides information on both global and local observing systems. For instance, it is possible to establish jumps in mean quantities as satellite instrumentation changes as well as to quantify changes in instrument performance. For the radiosonde network, measurement differences between the instruments used by different countries and provided by different manufacturers are quantified. For other types of observations, for example shipboard observations, it is possible to identify systematic errors that establish that the observing sensor is not at the reported altitude above the sea's surface. As the different organisations push forward with new reanalyses, they are committed to sharing these quality-controlled input datasets. This will improve the robustness of future conclusions drawn from reanalysis datasets as one source of non-geophysical variability will be reduced.

There are some other unique lessons learned from the reanalysis activities. One is that modern assimilation systems applied to the historical observations improve forecasts. A number of notable forecast failures in the pre-satellite era have been studied and forecast quality is greatly improved. This validates that research investments in model development and the evolution of assimilation methodology have beneficial impact. Another result of note is that methods of data treatment that have been applied in weather prediction might have to be reconsidered in climate applications. For instance, direct consideration of aerosol radiative effects on infrared observations might be important during periods of volcanic activity to assure the accurate use of radiances. Finally, the reanalyses help to focus attention of those observations needed to address the key uncertainties in moisture and tracer budgets, providing guidance for future observing systems.

Reanalysis activities continue in both the U.S. and Europe. The following activities are noted.

2.1 NCEP/DOE

NCEP and the Department of Energy (DOE) produced an update of the NCEP/NCAR reanalysis called the NCEP/DOE AMIP-II reanalysis. The reanalysis is to be used in the Atmospheric Model Intercomparison-II (AMIP-II) project. The primary goal was to correct known errors in the first-generation reanalysis as well as to include some improvements in the physics package that might address some of the common deficiencies of all of the reanalyses. The producers emphasise that it is not a next-generation reanalysis, as it does not include advanced treatment of the observations and additional observations that have been incorporated into the NCEP operational assimilation. A summary of the project and results is given in Kanamitsu et al. (2002).

2.2 ECMWF

The purpose of the ERA-40 Project is to produce a comprehensive set of global analyses describing the state of the atmosphere, land, and ocean waves from mid-1957 onwards. In addition, a primary product will be the quality controlled observational archive. ERA-40 will use a 3-D variational analysis scheme. Compared to the ERA-15 reanalysis the stratosphere will be resolved and from 1979 onwards there will be an ozone assimilation. The radiances from satellite temperature sounders will be used instead of retrieved products. In addition, when available, scatterometer and altimeter observations will be used as will satellite information on sea-ice. The reanalysis will be performed in segments that are broadly synchronised with different epochs of data coverage. Overlap across these segments will help quantify the impact the observing system has on the analysis. A number of external validation teams have been set up to focus on specific topics, and workshops are used to bring together the results of the validation and to report on the quality of the analyses. More information can be found in Simmons and Gibson (2000).

2.3 NASA

The current strategy for reanalysis at NASA's Data Assimilation Office (DAO) is to focus on particular periods associated with major satellite missions. The decision to perform a reanalysis is based on the expectation that the addition of a particular observation type will have an important impact or that a particular development in the analysis system will have an impact on an important customer application. Based on these criteria there are two ongoing reanalyses. The first is the TRMM reanalysis, using precipitation observations from the Tropical Rainfall Measurement Mission (TRMM), which was launched in November 1997. The second is the Reanalysis for Stratospheric Trace Gas Studies (ReSTS). The baseline period for ReSTS is May 1991 through April 1995, which corresponds to the part of the Upper Atmospheric Research Satellite (UARS) mission that has the greatest data density. The baseline assimilation will be extended to the present. In addition the DAO plans to perform a consistent reanalysis using conventional operational observations from the time of the Earth Observing System (EOS) Terra Launch (December 18, 1999).

The primary point of contact for the TRMM reanalysis is Arthur Hou (Arthur.Y.Hou@nasa.gov). The primary point of contact for ReSTS is Steven Pawson (Steven.Pawson.1@gsfc.nasa.gov). Additional information can be found at (http://userpages.umbc.edu/~pawson/rests.html).

2.4 U.S. NATIONAL REANALYSIS PROGRAM

A proposal entitled "National Program for Analysis of the Climate System (NPACS)" has been submitted to both NASA and NOAA to develop both a science and an implementation plan. The point of contact for information is Siegfried Schubert (Siegfried.D.Schubert@nasa.gov). The National Reanalysis Program calls for a commitment to sustained research and the production of consistent multi-year assimilated data sets. Long-term data sets extending back into the pre-satellite era would be generated on the time scale of 4-5 years. The decision to generate such a data set would follow scientific investigation that suggests a new reanalysis might have significantly improved quality relative to previous products. Collaboration and partnership with activities external to the U.S. is desired, possibly with the staggering of the generation of data sets. The proposed U.S. program also calls for the generation of shorter length data sets to investigate specific geophysical events and to investigate the impact of specific observing systems. Within the U.S., the activity is expected to be multi-agency as the different agencies hold key resources that can contribute to the success of the program. The project is expected to include a core team and a set of validation teams with commitments to timely validation of the reanalysis data sets.

Continued activity in data archaeology would be important for improving the quality of the reanalysis, especially in the pre-satellite era. This might allow pushing the starting point for the reanalyses to earlier times. Continual improvement of a quality controlled observational data set will be an important product of the program. It is anticipated that this will build from the current collaborations between ECMWF and the U.S. agencies. Other activities will include the investigation of assimilation methodology for climate assimilation and whether or not this is sensitive to the comprehensiveness of the observing system.

2.5 SPECIAL ASPECTS OF THE REANALYSIS PROBLEM

Two related aspects that provide special challenges to reanalysis are heterogeneity of the input data stream and bias. These will be discussed more fully in this section.

2.5.1 Heterogeneity of the input data stream

The input observations used in reanalysis come from many sources. Historically, the bulk of the assimilated measurements are extracted from those collected, operationally, for weather forecasting. These measurements include observations of the surface conditions over land and ocean, observations from weather balloons and airplanes, and remotely sensed observations from satellites. The instruments used to make these observations were not designed with calibration standards to establish long-term, climate-quality data sets. Further, observing systems deployed by different countries and different agencies within countries were not (and are not) procured and deployed in a way to assure consistent accuracy.

In addition to those observations collected for operations, there are those observations collected for research. A present and growing practice is to use research observations in operational applications. Reanalysis projects, however, are ideally suited to include research-data streams that were not appropriate for real-time applications when they were originally collected. Some of these research observations were collected in campaigns of limited temporal span and spatial extent. Others have been collected during multiyear satellite missions. Data archaeology, pioneered by Roy Jenne at the National Center for Atmospheric Research in the United States, recovers some of these research data so they can be brought to bear on reanalysis problems. These recovered data sets are especially important for the quality of the reanalyses during the 1950's.

The original proposed goal for reanalyses efforts, to remove the artifacts from assimilated data sets associated with changes in the data assimilation system, has been realised. With that success, the significant impact of the changing observing system was revealed. Mean quantities of observed parameters, e.g. temperature, change in, perhaps expected, ways. Derived quantities, e.g., precipitation, which are already known to have large errors, change in unexpected ways. It is likely that the most valuable and lasting scientific information to come from the reanalyses is the examination of the observing system through data assimilation.

Two examples of the impact of heterogeneity of the input data stream will be shown. In Figure 1, time series of the temperature anomaly from both the NCEP/NCAR reanalysis and the ERA-15 reanalysis are shown (Figure provided by M. Fiorino, Lawrence Livermore National Laboratory). The anomaly is calculated at each altitude by removal of the time mean. There are two features of note. The first is the jump in 1979 at all altitudes in the NCEP/NCAR time series. Below 100 hPa the transition is from cold anomalies to warm anomalies. This coincides with the introduction of global satellite temperature observations. Prior to this time, the upper air observing system was dominated by order 105 radiosonde observations per day. The radiosonde observations were (and are) concentrated in the northern hemisphere. Besides differences in spatial and temporal coverage, the jump in 1979 is related to specific characteristics of the profile-by-profile observations. For example, the vertical resolution of the radiosondes is much higher than that of the satellite observations. One result of this is that near the tropical tropopause the poorer resolution of the satellite observations manifests itself as a positive temperature bias. There are numerous sources of bias between radiosonde and satellite temperature observations, and these vary with space and time.

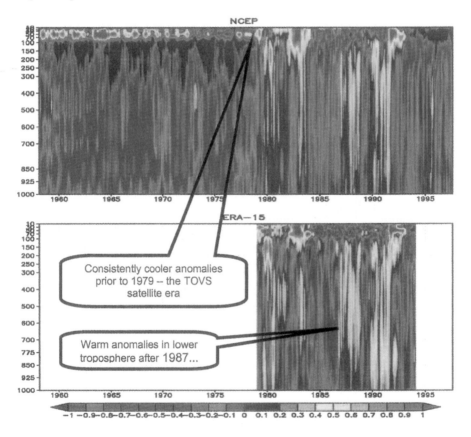

Figure 1. This figure provided by M. Fiorino (Lawrence Livermore National Laboratory). Temperature anomaly calculated by removal of the time mean as a function of altitude. Top: NCEP/NCAR reanalysis Bottom: ERA-15 reanalysis.

The second feature of note in Figure 1 is the warm anomaly evident in the ERA-15 time series after 1987. After this time, the tropical temperatures in ERA-15 are persistently higher than those from the NCEP/NCAR reanalysis for the next 7 years. This problem is ultimately traceable to a change in the performance of a single detector channel on the Microwave Sounding Unit (MSU). The ERA-15 Reanalysis uses a bias correction routine based on the radiosonde observations and historical instrument characteristics. When the detector behaviour on the MSU changed, this bias correction was no longer accurate, and a deleterious impact is realised. Changes in instrument performance over the course of multiyear satellite missions are an important source of variability in the data record. While such changes leave artifacts in the reanalysis data sets, the identification and quantification of these changes is often first revealed through assimilation. This results, ultimately, in the improvement of the fidelity of primary data sources.

The second example of the impact of input data heterogeneity is from the radiosonde network itself. Radiosondes provide what many consider to be the single most important class of observations of the upper air. This might be arguable in the current era of high quality satellite observations, but there is no argument that the radiosonde network is of paramount importance prior to the satellite era. The radiosonde measurements have benefited from much scrutiny, and strategies to develop climate quality data sets have been exercised. Different countries use different types of radiosondes, and within a country, several manufacturers of radiosondes are used. There is not consistent calibration of radiosondes.

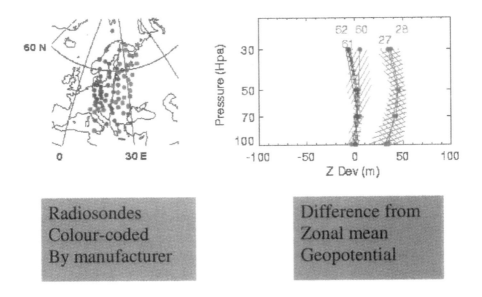

Figure 2. From Lait (2002). The left panel shows the distribution of radiosondes observations over Eastern Europe colour-coded by manufacturer. The right panel shows the difference of the radiosonde heights from the zonal mean analysis. The different types of radiosondes group together, and a spurious circulation separates the different types of radiosondes.

Lait (2002) examines the impact of the heterogeneity of the radiosonde network on the quality of the assimilation analysis. Lait subtracts the zonal mean geopotential height from that of the radiosonde observation. This reveals persistent anomalies clustered by radiosonde type. A regional aspect of this impact is shown in Figure 2. The left panel shows the radiosondes over Eastern Europe, colour coded by manufacturer. The right panel shows the difference of the geopotential height from the zonal mean, still, colour coded by manufacturer. The eastward lying observations are between 30 and 40 geopotential meters higher than the westward lying observations. This height gradient is persistent with altitude. A wind error of order 5 m/sec is consistent with this height gradient in a part of the atmosphere where the expected wind speed is order 10 m/sec. Lait (2002) identifies persistent wind patterns, seemingly spurious rivers of air, surrounding regions of differing radiosonde instrumentation. Again, this is directly related to biases in the observations of fundamental geophysical parameters.

The two examples discussed above bracket the nature of the data heterogeneity problem. In the first example, when a new global observation type was added to the observing system, large changes in the assimilation occurred. In the second example, subtle biases between different types of radiosondes were shown to have large enough impact on the analysis of wind to impact the quantification of atmospheric transport. The granularity of the heterogeneity in the radiosonde network is small, and will be difficult to eliminate. Between these two extremes are a whole set of impacts that might be expected when new data types are introduced. For example, the introduction of scatterometry data to define the ocean surface winds or precipitation observations to define the hydrological cycle will, no doubt, improve the quality of the assimilated data product. However, these improvements will be accompanied by changes in mean quantities such as surface pressure, precipitation, and outgoing longwave radiation; hence, leaving a signal in the climate record that is not of geophysical origin.

2.5.2 Impact of bias

Data assimilation theory has been developed, primarily, under the assumption that the information from the observations is unbiased relative to the information from the model. That is, given a parameter such as temperature, the mean of the observations subtracted from the mean of the model prediction is zero. However, as the previous discussion on heterogeneity in the observing system shows, the observations themselves are biased relative to each other even within the same nominal instrument type, e.g., the radiosondes and the succession of operational satellites. Different observing systems measuring the same geophysical parameter are expected to have bias between each other. Given that there are systematic errors in the models, the assimilation quality is impacted by the bias between model prediction and observations as well as the bias between different pieces of the observing system.

The impact of bias is large and limits the range of applications of reanalysis data sets. The conclusion from the cited studies (WCRP, 1998, 2000: Newson, 1998; Kistler et al., 2001) is that the atmospheric data sets from the reanalyses are inappropriate for trend determination. This is attributed to the heterogeneity of the input data stream and is, ultimately, a statement that the bias between the different types of observations leaves signals in the determination of the mean state of the atmosphere that are of similar or greater magnitude than any geophysical signal. The inability to determine trends is manifested in a well-observed quantity such as temperature, which is also a primary predicted variable in the model. It has been shown that the quality of these primary variables is of much higher quality than that of derived parameters such as precipitation. Precipitation is determined to first order by the estimation of temperature and humidity and the use of these estimates by the physical parametrizations of the model. Often it is the case that when the physical parametrizations utilise the corrected temperature and humidity that comes from the assimilation, precipitation far in excess of that which is observed is diagnosed. This biased estimate of precipitation suggests that fundamental processes in the model are not well represented on the scale of the observations; *i.e.*, there is substantial model error. Imbalances such as those documented with precipitation are often called "spinup" problems and are characteristic of biases that exist between modelled and observed physical parameters at the time of analysis.

From the point of view of short-term prediction, directly assimilating information that corrects the physical parametrizations can have a large positive impact. Hou *et al.* (2001, 2002) have shown that assimilating satellite precipitation observations improves both forecast skill and the estimate of important metrics of the climate system, for example, outgoing longwave radiation. Still, however, the physical processes in the model are always tending towards their biased state, and the correction by the insertion of observations is not without consequence. The general circulation, the time averaged, spatially averaged dynamics of the atmosphere, is where the consequence is usually realised. One way to quantify this impact is to examine long lived trace species, which are useful for determining, for instance, the exchange of mass across the equator or between the stratosphere and troposphere. Two recent papers by Douglass *et al.* (2003) and Schoeberl *et al.* (2003) present convincing evidence that the large-scale transport in the atmosphere is corrupted by data insertion. These authors conclude that despite the ability to use winds and temperatures derived from assimilation analyses to predict trace gas distributions on weekly to seasonal time scales, multiyear assessments are better performed with model simulations that do not assimilate observations because the physical integrity of the general circulation is better represented. Again, this is ultimately a problem of bias between model prediction and observations.

In the end, the quality of assimilation analyses will be dependent on eliminating the bias between the model and the observations. Assuming that the observations can be corrected in some way to eliminate the bias between different instrument types, the elimination of bias between the model and observations relies on improved model quality. Much of this improvement will come from better physical parametrizations and will require reformulation of physical parametrizations. Such development will be based on improved, more complete observations and modelling algorithms that can utilise the observed information. In the meantime, however, there is potential benefit derived from bias correction.

Figure 3 demonstrates a prescribed, idealized system and an estimate of that system by model-data assimilation. The smooth line shows the known mean state, *i.e.*, climate. The segmented line shows a series of model forecasts corrected intermittently by a set of observations that, over time, are randomly distributed around the known mean state. In the top plot the model forecast is unbiased; in the bottom plot the forecast is biased. In both plots the observations are unbiased. At a given time, 1979, the observing system is changed so that more observations are taken. This is symbolic of the increase in temporal and spatial resolution that occurred when satellite observations became operational (see Figure 1). In the top plot when the model predictions are unbiased, the mean error in the analysis remains essentially the same before and after the change in the observing system. In the bottom plot, where the model prediction is biased, the increase in density of the observations reduced the mean error in the analysis by half, leaving a jump in the estimate of the mean state. Therefore, even if the mean state of the observations is homogenised prior to assimilation through some calibration procedure, as long as there is model error, reanalyses will be subject to errors based simply on improved data coverage.

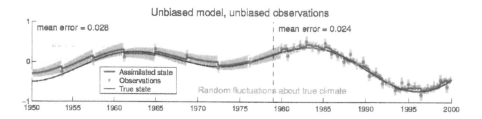

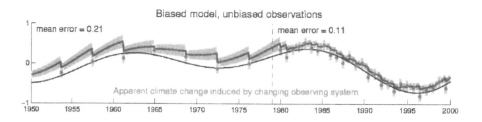

Figure 3. This figure provided by Dick Dee (Data Assimilation Office). The solid line represents a known true state of an idealized climate system. The red dots are observations of system. The blue lines are model forecasts of the mean state following assimilation of the observations into the model. In the top frame the model is not biased. In the bottom frame the model is biased.

In some cases it is not difficult or expensive to estimate bias and apply a correction algorithm (Dee and da Silva, 1998). This can prove the quantitative integrity of the assimilated data set, and have positive impact, especially on prediction of parameters that are being assimilated. However, the bias correction is ultimately compensating for shortcomings in the model. This implies that the model physics (or chemistry) are not correct, and this will ultimately manifest itself somewhere in the assimilated data set. Often this manifestation is in the time and spatially averaged circulation that is of exquisite importance in climate change but of little importance in prediction problems of less than seasonal scale.

3. Summary

Assimilation analysis of long time series of observations into a non-varying assimilation system is called reanalysis. A number of reanalyses have been performed, and there are active reanalysis efforts. The reanalyses have found productive application in a broad range of geophysical applications. These studies have also revealed a series of problems that need to be addressed. Two classes of problems, however, offer special challenges. The first is the heterogeneity of the input data stream. This comes from changes in the observing system and leads to realisation that the addition of new observations that might formally improve the quality of the analysis also causes spurious jumps in key climate parameters. Because of this, reanalysis data sets are not suitable for trend detection. The second special challenge comes from the assimilation in the presence of bias. Even if all of the observations are unbiased relative to each other, any remaining bias between the model and the observations will lead to spurious representation of the mean state, *i.e.*, the climate. Future reanalysis activities are focused on generating data sets further back in time and in support of special observing systems.

Acknowledgments. I thank Mike Fiorino and Leslie Lait for the figures used in the paper and their discussion of their experiences with using reanalyses. I thank Siegfried Schubert for information about the U.S. National Reanalysis Program. Finally, I thank Dick Dee for providing the figures from his studies of bias and bias correction as well as for his contributions to the discussion.

References

Bengtsson L., and J. Shukla, 1988: Integration of space and in situ observations to study global climate change. *Bull. Amer. Meteorol. Soc.*, **69**, 1130- 1143.

Dee, D. P., and A. da Silva, 1998: Data assimilation in the presence of forecast bias. *Q. J. R. Meteorol. Soc.*, **124**, 269-295.

Douglass A. R., M. R. Schoeberl, R. B. Rood, and S. Pawson, 2003: Evaluation of transport in the lower tropical stratosphere in a global chemistry and transport model. *J. Geophys. Res.*, **108**, to appear.

Gibson, J. K., P. Kållberg, S. Uppala, A. Nomura, A. Hernández, E. Serrano,1997: ERA Description, ECMWF Re-analysis Final Report Series, 1.

Hou, A. Y., S. Q. Zhang, A. M. da Silva, W. S. Olson, C. D. Kummerow, and J. Simpson, 2001: Improving global analysis and short-range forecast using rainfall and moisture observations derived from TRMM and SSM/I passive microwave sensors. *Bull. Amer. Meteorol. Soc.*, **81**, 659-679.

Hou, A. Y., S. Q. Zhang, O. Reale, 2002: Variational continuous assimilation of TMI and SSM/I rain rates: Impact on GEOS3 analysis and forecast. *Mon. Wea. Rev.* (Manuscript in preparation, contact Arthur.Y.Hou.1@gsfc.nasa.gov). Kalnay, E., and R. Jenne, 1991: Summary of the NMC/NCAR Reanalysis Workshop of April 1991. *Bull. Amer. Meteorol. Soc.*, **72**, 1897-1904.

Kalnay E., M. Kanamitsu, R. Kistler, *et al.*, 1996: The NCEP/NCAR 40-year reanalysis project. *Bull. Amer. Meteorol. Soc.*, **77**, 437-471.

Kanamitsu, M., W. Ebisuzaki, J. Woolen, S.-K. Yang, J. J. Hnilo, M. Fiorino, and G. L. Potter, 2002: NCEP-DOE AMIP-II reanalysis (R-2). *Bull. Amer. Meteorol. Soc.*, **83**, 1631-1643.

Kistler R., E. Kalnay, W. Collins, *et al.*, 2001: The NCEP-NCAR 50-year Reanalysis: Monthly means CD-ROM and documentation. *Bull. Amer. Meteorol. Soc.*, **82**, 247-267.

Lait, L. R. 2002: Systematic differences between radiosonde measurements. *Geophys. Res. Lett.*, **29**, 10.1029/2001GL014337.

Morel, P., 2001: Why GEWEX? The agenda for a global energy and water cycle research program. *GEWEX News*, **11**, 1-11.

Newson, R., 1998: Results of the WCRP First International Conference on Reanalysis. *GEWEX News*, **8**, 3-4.

Onogi, K., 2000: ERA-40 Project Report Series 2. The long-term performance of the radiosonde observing system to be used in ERA-40, European Centre for Medium-range Weather Forecasts August 2000, 77 pp.

Schoeberl, M. R., A. R. Douglass, Z. Zhu, and S. Pawson, 2003: A comparison of the lower stratospheric age-spectra derived from a general circulation model and two data assimilation systems. *J. Geophys. Res.*, **108**, 10.1029/2002JD002652.

Schubert S. D., R. B. Rood, and J. Pfaendtner, 1993: An assimilated dataset for earth-science applications. *Bull. Amer. Meteorol. Soc.*, **74**, 2331-2342.

Simmons, A. J. and J. K. Gibson, 2000: ERA-40 Project Report Series 1. The ERA-40 Project Plan, European Centre for Medium-range Weather Forecasts, March 2000, 62 pp.

Trenberth, K. E., and J. G. Olson, 1988: An evaluation and intercomparison of global analyses from NMC and ECMWF. *Bull. Amer. Meteorol. Soc.*, **69**, 1047-1057.

WCRP, 1998: Proceedings of the First WCRP International Conference on Reanalyses (Silver Spring, MD, USA, 27-31 October, 1997), WMO/TD-N 876.

WCRP, 2000: Proceedings of the Second WCRP International Conference on Reanalyses (Wokefield Park, nr. Reading, UK, 23-27 August 1999), WCRP-109, WMO/TD-N 985.

INDEX

R. Swinbank et al. (eds.), Data Assimilation for the Earth System, 373–377.
© 2003 *Kluwer Academic Publishers. Printed in the Netherlands.*